MAPPING THE SUBJECT

The human subject is difficult to map for numerous reasons. How do you map something that does not have precise boundaries, that is a set of different, intersecting and sometimes conflicting positions, that is always on the move and only partially locatable in space and time? The essays collected in this book untangle these difficulties in new and exciting ways through revealing case study material and sophisticated theoretical expositions.

Mapping the Subject contains a wide-ranging review of the literature on subjectivity across the social and human sciences. Essays are subdivided under four main headings: constructing the subject, sexuality and subjectivity, the limits of identity and the politics of the subject. Part I establishes the idea that the subject is constructed and makes this clear through detailed histories of the subject. In Part II, in their research on the place of sexuality in subjectivity and subjectivity in sexuality, the authors show that sexuality cannot be assumed to be natural. Authors continually come up against the limits to subjectivity. Part III, therefore, takes issue with the idea of a singular, self-contained identity, and asks how is it possible to make sense of ourselves when the boundaries which seemingly tell us who 'we' really are appear incoherent, or fragmented, or fuzzy, or somehow unreal, or fluid or on the move. Power relations and the effects of power are consistent themes which run throughout this book, so in the fourth and final part, authors make space for a politicised subject, dealing explicitly with relations of power, whether organised around 'gender', 'race', 'class' or other kinds of difference.

The authors gathered in this collection take up the challenge to consider the place of the subject anew. There is a commitment to mapping the subject – a subject which is in some ways detachable, reversible and changeable; in other ways fixed, solid and dependable; located in, with and by power, knowledge and social relationships. This book is, moreover, about new maps for the subject: it seeks new spaces, new politics, new possibilities.

Steve Pile is Lecturer in the Faculty of Social Sciences at the Open University. **Nigel Thrift** is Professor of Geography at the University of Bristol.

MAPPING THE SUBJECT

geographies of cultural transformation

**Edited by
STEVE PILE and
NIGEL THRIFT**

LONDON AND NEW YORK

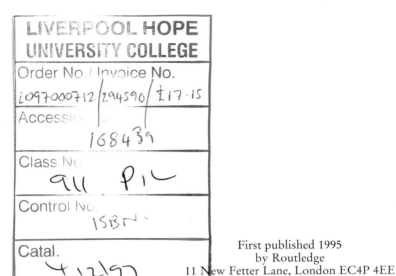
First published 1995
by Routledge
11 New Fetter Lane, London EC4P 4EE

Simultaneously published in the USA and Canada
by Routledge
29 West 35th Street, New York, NY 10001

Reprinted in 1996

Typeset in Garamond by
Solidus (Bristol) Limited
Printed and bound in Great Britain by
Biddles Limited, Guildford and King's Lynn

British Library Cataloguing in Publication Data
A catalogue record for this book is available from the British Library

Library of Congress Cataloguing in Publication Data
Mapping the subject: geographies of cultural transformation/edited
by Steve Pile and Nigel Thrift.
p. cm.
Includes bibliographical references (p.   ) and index.
1. Subject (Philosophy) 2. Power (Philosophy) 3. Identity.
4. Sex. 5. Political culture. I. Pile, Steve.
II. Thrift, N. J.
BD223.M37 1995
302'.1–dc20   94–23747

ISBN 0–415–10225–1 (hbk)
ISBN 0–415–10226–X (pbk)

CONTENTS

Part II Sexuality and subjectivity

Introduction

Part III The limits of identity

Introduction

Part IV The politics of the subject

Introduction

LIST OF FIGURES

LIST OF CONTRIBUTORS

David Bell is a part-time lecturer and part-time researcher at Staffordshire University. His research interests are in the space of sex and the sexes of space, and he is co-editing a forthcoming volume on sexual geographies.

Julia Cream is a research student in the Department of Geography, University College, London. She is interested in constructions of the body and her most recent work focuses on women on the pill in the 1950s/1960s. She has published papers on various aspects of feminism and sexuality.

Marcus Doel is a lecturer in Human Geography at the School of Social Science, Liverpool John Moores University. He has written widely on new theoretical directions in contemporary human geography and is presently working on the social, cultural and political geographies of extreme events. He is presently completing a book entitled *Deconstruction/Geography/ Postmodernism*.

Stephen Frosh is a senior lecturer in Psychology at Birkbeck College, University of London, and a consultant clinical psychologist at the Tavistock Clinic, London. He is author of several books, amongst which are *Identity Crisis: Modernity, Psychoanalysis and Self* (1991) and *Sexual Difference: Masculinity and Psychoanalysis* (1994).

Michael Keith is Principal Research Officer in the Centre for Urban and Community Research and the Department of Sociology at Goldsmiths College, University of London. He is the author of *Race, Riots and Policing: Lore and Disorder in a Multi-racist Society* (1993) and co-editor of *Hollow Promises: Rhetoric and Reality in the Inner City* (1991), *Racism, the City and the State* (1993) and *Place and the Politics of Identity* (1994). His current interests focus on racism and forms of contemporary urbanism and the politics of urban renewal. He is currently the Assistant Director of the Deptford City Challenge Evaluation Project which is carrying out an evaluation of the work of Deptford City Challenge.

David Matless is a lecturer in Geography at the University of Nottingham. He has published widely on cultures of landscape and on the history and philosophy of geography. He is currently preparing a book entitled

Landscape and Englishness, 1918 to the Present and is co-author of *Writing the Rural* (1994).

Miles Ogborn is a lecturer in Geography at Queen Mary and Westfield College, University of London. He has written on the interconnections between space, state formation and identity and is currently preparing a book on eighteenth-century London under the title *Spaces of Modernity*.

Hester Parr is a postgraduate student completing a Ph.D. thesis at the Department of Geography, University of Wales, Lampeter. Her research pivots around the deinstitutionalisation of mental health care and the creation of alternative 'care in the community' arrangements. Her work seeks to centralise the voices of people labelled as 'mentally ill', and also to theorise contemporary spaces of mental health provision, 'insanity' and identity through ethnographic encounters.

Chris Philo is a lecturer in Human Geography at the Department of Geography, University of Wales, Lampeter. His research includes inquiries into the historical geographies of asylums, madhouses and other institutional responses to the phenomenon of 'madness'. As well as authoring a number of papers, he has co-authored *Approaching Human Geography: An Introduction to Contemporary Theoretical Debates* (1991) and has co-edited *Selling Places: The City as Cultural Capital, Past and Present* (1993).

Steve Pile is a lecturer in the Faculty of Social Sciences at the Open University. He has published work relating to identity, politics and space. He authored *The Private Farmer* (1990) and co-edited *Place and the Politics of Identity* (1993). He is writing a book about psychoanalysis and geography under the provisional title *The Body and the City*.

Nigel Rapport is a lecturer in Social Anthropology at the University of St Andrews. He has conducted participant-observation research in England, Newfoundland and Israel, and is author of *Talking Violence: An Anthropological Interpretation of Conversation in the City* (1987), *Diverse World-Views in an English Village* (1993) and *The Prose and the Passion: Anthropology, Literature and the Writing of E. M. Forster* (1994).

Paul Rodaway is currently a lecturer in Human Geography at Edge Hill College of Higher Education. His research focuses upon everyday experiences of space and place identity in contemporary urban and rural settings. He has recently published *Sensuous Geographies: Body, Sense and Place* (1994).

Gillian Rose teaches Social, Cultural and Feminist Geographies at the University of Edinburgh. Her research interests focus on the politics of the production of geographical knowledges. She is the author of *Feminism and Geography: The Limits of Geographical Knowledge* (1993) and co-editor of *Writing Women and Space: Colonial and Post-colonial Geographies* (1994).

Victor Jeleniewski Seidler is Reader in Social Theory in the Department of Sociology at Goldsmiths College, University of London. He has written widely in the fields of moral theory and sexual politics. He is the editor of the Male Orders series for Routledge. His most recent work includes *The Moral Limits of Modernity* (1991), *Unreasonable Men* (1993) and *Recovering the Self* (1994). He also edited *Men, Sex and Relationships* (1992).

David Sibley teaches Human Geography at Hull University. His interest in cultural and spatial boundary questions began when writing about relations between Gypsies and the state. He has subsequently developed these ideas in relation to disparate topics, including the boundaries of knowledge and boundaries in the home.

Carolyn Steedman is Reader in the Centre for the Study of Social History, University of Warwick. Her books include *The Tidy House* (1982), *Landscape for a Good Woman* (1986), *Childhood, Culture and Class in Britain* (1990) and *Past Tenses* (1992).

Nigel Thrift is Professor of Geography at the University of Bristol. His interest in the space of the subject goes back to the 1970s. His books include *Times, Spaces and Places* (1979), *Writing the Rural* (1994) and *Spatial Formations* (1995).

Gill Valentine is a lecturer in Geography at Sheffield University. She has researched widely around gender, crime and insecurity. Her current work concerns lesbian geographies. She is co-editor of a forthcoming book on geographies of sexualities.

Valerie Walkerdine is Professor of the Psychology of Communication in the Department of Media and Communications at Goldsmiths College, University of London. She has written widely on children's, and especially girls', cognitive development, within differing social, familial and schooling contexts. Her previous works include *The Mastery of Reason* (1988), *Schoolgirl Fictions* (1991) and, with Helen Lucey, *Democracy in the Kitchen* (1989).

PREFACE

Since the 1960s, geographers have been aware of the importance of people's subjectivity in directing their spatial behaviour. By the early 1970s, geographers had begun to experiment with different models of (what was then called) 'man'. These models were primarily drawn from the disciplines of cognitive psychology, political economy and philosophy. In the early 1980s, time-geography and structuration theory seemed to have provided a solution to many of the questions being raised at the time. But the recent deluge of writing which draws on, or takes issue with, notions of postmodernism or postmodernity has rekindled the question: 'Who or what is the subject?' And it has added a further, more nihilistic one: 'Who or what comes after the subject?'

To their great surprise, geographers have suddenly found that spatial metaphors are in vogue as ways into these questions. Somewhat to their astonishment they find that cultural geography is now being read outside the discipline. This book stands at the gateway between these spatial metaphors and the interdisciplinary question of the constitution of the subject: hence the title, *Mapping the Subject*. We hope that this book will prove timely in relation to developing debates concerning space and the subject of subject formation.

In some ways, producing an edited collection cannot be divorced from thinking as a reader or reviewer about the contributions it contains. Though it does no good to second-guess what other people might say, we have both thoroughly enjoyed and learnt a great deal from each and every chapter gathered here. We would, therefore, like to thank whole-heartedly the contributors to this book, for the work they have put in, for their enthusiasm for the project and for providing a collection which we hope is both coherent and stimulating.

Steve Pile and Nigel Thrift
August 1994

1

INTRODUCTION

Steve Pile and Nigel Thrift

The human subject is difficult to map for numerous reasons. There is the difficulty of mapping something that does not have precise boundaries. There is the difficulty of mapping something that cannot be counted as singular but only as a mass of different and sometimes conflicting subject positions. There is the difficulty of mapping something that is always on the move, culturally, and in fact. There is the difficulty of mapping something that is only partially locatable in time-space. Then, finally, there is the difficulty of deploying the representational metaphor of mapping with its history of subordination to an Enlightenment logic in which everything can be surveyed and pinned down.

There is, however, another way of thinking of mapping, as wayfinding. This is the process of 'visiting in turn all, or most, of the positions one takes to constitute the field ... [covering] descriptively as much of the terrain as possible, exploring it on foot rather than looking down at it from an airplane' (Mathy 1993: 15) and it is this meaning that is deployed in this introductory chapter. In spirit, wayfinding is probably closest to Deleuze and Guattari's distinction between mapping and mere tracing;

> What distinguishes the map from the tracing is that it is entirely oriented towards an experimentation in contact with the real. The map does not reproduce an unconscious closed in upon itself; it constructs the unconscious ... The map is open and connectable in all of its dimensions; it is detachable, reversible, susceptible to constant modification. It can be torn, reversed, adapted to any kind of mounting, reworked by an individual group or social formation ... A map has multiple entryways, as opposed to the tracing which always comes back 'to the same'. The map has to do with performance, whereas the tracing always involves an alleged 'competence'.
>
> (1988: 12)

This introductory chapter is therefore an attempt to find a way through the forests of literature on the subject. The first section of this chapter discusses the subject which figures in geographical discourse, arguing that subjectivity has been examined only rarely and in very specific ways. In the latter section, we move on to consider the matter of terms; even wayfinding requires some landmarks. This path is not meant to be definitive, but to raise questions about commonly assumed notions – the body, the self, the person, identity and the subject – which this collection of essays refuses to take for granted. Inevitably, we fail to cover all the terrain but, hopefully, we will have provided the reader with the beginnings of a map or, more accurately perhaps, a map of beginnings.

GEOGRAPHIES OF THE SUBJECT

In everyday life, certain words are bandied around with thoughtless abandon, such as body, self, person, identity and subject. When these words are used in this way, they become solid triangulation points with which it is possible to map 'the subject' into the social landscape. Phrases including these words appear obvious – the words themselves seem to need no further elaboration. For example, the line in the old song – 'If I said you had a beautiful body, would you hold it against me?' – plays on the ambiguity of the phrase 'hold against', while the 'it' of 'a beautiful body' is cheerfully assumed. Other expressions, however, speak of greater anxiety about the everyday experience and understanding of these co-ordinates of the subject: what does it mean to say 'She is beside herself with anger'? She may be angry because someone has tried to chat her up using that line from the old song, but how can she be 'beside' her own self – has she cracked in two or was she already split? Or has she lost control, but control over what? This is not just an expression – it speaks of an experience of self, body, identity and subjectivity that cannot be quite so easily contained within dictionary definitions or commonplace understandings of what these words mean. The question becomes how terms such as body, self, subjectivity and so on, are to be mapped; crudely, positions have been taken up in relation to a particular dualism, namely structure/agency. This dualism expresses the problem of subject formation in relation to, on the one side, social rules, sanctions and prohibitions and, on the other side, the individual's feelings, thoughts and actions.

In geography, in the first article to directly address the problem of subject formation, Thrift outlines a research agenda based on the possibility of a theory of social action which recognises both the determinations of structure on the actions of individuals and the determination of individuals to do things, sometimes differently (Thrift 1983a). What is at stake is most famously summarised by Marx's 'unobjectionable' aphorism: 'people make history, but not in circumstances of their own choosing' (Marx 1852). The problem lies in the precise relationship between 'people', 'history', 'circumstances' and 'choosing'. Thrift shows that, in trying to understand the position of the individual about the social world, social theory has usually decided to resolve the issue either on the side of structure or on the side of agency. To simplify greatly: on the side of structure, it is argued that circumstances by and large determine what people choose to do – from this position, it is a short step to believe that circumstances determine what people do and that people are unwitting dupes to the dominant logic of the social structure (whether this is named as capitalism or patriarchy or ... and so on); on the side of agency, it is argued that people make history, though bound by certain constraints – from here, it is a short step to believe that people are completely free to choose what to do, without constraint on their actions.

These positions can be caricatured still further.

From the perspective of structure, our triangulation points of the subject (body, self, person, identity, subjectivity) have no meaning outside their relationship within a system of social relations: thus, the body only has meaning as, for example, 'labour-power' or 'male', or the self only has meaning in relationship to 'class consciousness' or 'masculine', for example. Whatever the theory of the dominant system of meaning and power, it is this that fills the empty containers of body, self, person, identity. Outside the dominant system – whether it be capitalism, patriarchy or something else – these components are assumed to do nothing. The challenge, then, is to change the system.

From the perspective of agency, the co-ordinates of body, self, person, identity and subjectivity have their own internal meaning, though this is commonly taken to be hidden under a great depth of received ideas, which are usually understood to disguise their true meaning. The individual's experiences of body, self, person, identity and subjectivity are seen as central to understanding their (true) meaning. The body (for example) can mean anything that an individual takes it to mean, taking on different qualities at different times: thus, it could be experienced as male, weak, white, old and so on depending on the way the body is coded in a social setting and the way that setting is decoded and recoded. Though cloaked in the meaning of culture, the task is to strip the body (for example) to its bare essentials. Because the subject's body, self, person, identity and subjectivity are assumed to derive their deep or true meaning either from their own inherent qualities or from the intersubjective experiences of the individual, they are open to contest through changing their meaning.

The problem, for Thrift (1983a), was to conceptualise body, self, person, identity, subjectivity in terms of both structures and agency: after all, social structures could not exist without human subjectivity; on the other hand, social structures at least set the parameters within which humans behave and at most set the rules for 'allowed', 'prohibited' and 'enabled' thoughts and actions. For Thrift, 'human agency must be seen for what it is, a continuous flow of conduct through time and space constantly interpellating social structure' (1983a: 31). The individual acts in time and space – located, moving, encountering, interpreting, feeling, being and doing.

Through the processes of socialization, the extant physical environment, and so on, individuals draw upon social structure. But at each moment they do this they must also reconstitute that structure through the production or the reproduction of the conditions of production and reproduction. They therefore have the possibility, as, in some sense, capable and knowing agents, of reconstituting or even transforming that structure.

(Thrift 1983a: 29)

Through this dualism – structure/agency – it is possible to locate co-ordinates such as the body and the self not only in relation to structural determinations and the meanings they give to the lives of individuals but also in terms of the relationship between the meanings people give to their lives and the choices they subsequently make. It is clear that history must be viewed from both sides of the coin – and the problem Marx's aphorism raises appears to have been solved.

However, the structure/agency dualism has not exhausted the mapping of the subject: more recently, the debate has been recast onto the terrain of language, or more properly 'discourse'. One outstanding problem with the way the structure/agency dualism operated was that it still seemed unable to interrogate 'everyday life' as simultaneously real, imaginary and symbolic. The assumption that terms such as the body or the self had identifiable meanings informed the structure/agency debate. This assumption was challenged in poststructuralist thought.

By concentrating simultaneously on 'discourse' as an identifiable practice or institution and the interanimations between different discursive practices, it was possible to argue that the co-ordinates of subjectivity were constituted by the practices that they seemingly described. Words such as body and self seem to describe things, but in fact disguise their constitution by those very words. Institutional practices such as the madhouse, prisons, schools and universities, rather than containing particular subjects, actually and actively create them: thus, prisons create prisoners, universities create students. Prisoners and students are inconceivable outside of the institutions that give them meaning. The structure/agency debate had been twisted: 'discourse' was neither structure nor agency and both structure and agency. From this perspective, the body or the self becomes a location within various power-riddled discursive positions, but where the body or the self is not a passive medium on which cultural meanings are merely inscribed; they are neither a thing nor a free-floating set of attributes. Aware of the discursive production of subjectivity and the facts of life, Elspeth Probyn proposes that the self:

is a doubled entity: it is involved in the ways in which we go about our everyday lives, and it puts into motion a mode of theory that problematizes the material conditions of those practices. Unlike the chickens which are presumably sexed one way or the other, once and for all, a gendered self is constantly reproduced within the changing mutations of difference. While its sex is known, the ways in which it is constantly re-gendered are never fixed or stable. One way of imaging this self is to think of it as a combination of acetate transparencies: layers and layers of lines and directions that are figured together and in depth, only then to be rearranged.

(1993: 1)

Thus, the focus of an analysis of the self or body has changed from identifying their location on the continuum between structural and personal determination to looking at the ways in which subjectivity is reproduced in time and space: for example, the truth of sex is 'performatively produced and compelled by the regulatory practices of gender coherence' (Butler 1990: 24). The co-ordinates of subjectivity are, thereby, reproduced both through discursive practices and through power-laden regulatory practices. 'Gender is the repeated stylization of the body, a set of repeated acts within a highly rigid frame that congeal over time to produce the appearance of substance, of a natural sort of being' (Butler 1990: 33).

Questions still remain – and they relate to the map of the subject: co-ordinates – such as body, self, identity, subjectivity – appear to tell us who the individual is, what they are like, whether we like them, but this is dependent on the kind of map on which we place these co-ordinates. The 'mapping' metaphor, which appears to tell us a great deal, actually hides other relationships. Thus, Catherine Nash shows how a flat, two-dimensional map articulated masculinist and colonial desire to control the land and place its subjects within places which it controlled (1993). She suggests that post-colonial discourses gave the map 'volume and height', where the map subverts its own authority by disclaiming its ability to re-present the true, real world (1993: 52). Thus, the representation of topography 'becomes a shifting ground, a spatial metaphor which frees conceptions of identity and landscape from repressive fixity and solidity' (Nash 1993: 52). Under this conception, the map must suffer continual renaming and remapping in order to prevent its closure around one dominant cartography of meaning and power.

This book seeks to take apart the cotton-woolled security surrounding maps of the subject, to release the co-ordinates of subjectivity from static, uniform, transparent notions of place and being, which seemingly inform the way the subject is thought of. From this perspective, it is inappropriate to think the co-ordinates of subjectivity as being like lines and directions on layers and layers of transparent acetate sheets. Problems lie in the seeming stability, transparency and autonomy of each layer of the self – and a self seems to stand outside the layers choosing the arrangement of the layers. At the very least, the self would appear to be constituted through these layers and unable to 'shuffle' identity in this way. Other metaphors will need to be found, other maps need to be drawn, which are more capable of elucidating the fixity and fluidity, the ambivalence and ambiguity, the transparency and opacity and the surface and depth of the mapped subject.

TERRITORIES OF THE SUBJECT

Mapping the subject usually begins as a journey away from the forbidden territory of Enlightenment thought, from the Cartesian division of mind and body (or reason and nature) and the tenets of humanism (especially the privileging of the human, the individual, consciousness, agency, self-knowledge and experience). But making this journey means negotiating a whole series of interconnected terms – the body, the self, identity – the person, the subject – which are both the main terrains of inquiry and the chief cartographic tools that we have to hand. It is no surprise, then, that these terms are usually equivocal, often ambiguous, sometimes evasive and always contested. What follows is not, therefore, a set of definitional terms. Rather, it is an attempt to get a preliminary feel for the lie of the land.

Whilst we do not believe that there is a requirement for any absolute exactitude, it is, nevertheless, dangerous to avoid any attempt to define these tools. Therefore in this section, we attempt to give at least a minimum of form to these ideas which will constitute the contested territories of this book.

The first term is *the body*. Harré has highlighted the enormous number of ways in which the body can be used in societies:

> we use our bodies for grounding personal identity in ourselves and recognising it in others. We use other bodies as points of reference in relating to other material things. We use our bodies for the assignment of all sorts of roles, tasks, duties and strategies. We use our bodies for practical action. We use our bodies for the expression of moral judgements. We use the condition of our bodies for legitimating a withdrawal from the demands of everyday life. We use our bodies for reproducing the human species. We use our bodies for artwork, as surfaces for decoration, and as new material for sculpture. We use human bodies for reproducing the human species. We use human bodies for the management of the people so embodied. We use our own bodies and those of others to command the cosmos. We use our bodies as message boards, and their parts as succinct codes. We use our bodies for fun, for amusement and for pastimes.
>
> (1991: 257)

Given this bewildering variety, it is still possible to identify at least five related but distinct approaches to the study of the body. The first of these sees the body as a part of a general temporal and spatial logic, an 'order of connection': 'this is the order found in nature's logic which perpetuates the living, a logic of multitudinous paths that interect, which works through living things rather than imposes itself upon them from outside and above' (Brennan 1993: 86). This is the kind of order found in time-geography and similar attempts to map the logic of corporeality. The second approach identifies the body as part of a prediscursive realm through an emphasis on bodily movement. As Merleau-Ponty (1962a: 140) puts it:

> our bodily experience of movement is not a particular case of [theoretical knowledge]; it provides us with a way of access to the world and the object, with a 'praktognosia' which has to be recognised as original and perhaps as primary.

A third approach considers the body as an origin. This is a notion often found in psychological theory, whether as identification with the father, or, latterly, to make up for the 'originary absence' in Freudian theory, the Mother (Irigaray 1985a). A fourth approach to the body sees it as a site of cultural consumption, a surface to be written on, 'an externality that presents itself to others and to culture as a writing or inscriptive surface' (Grosz 1989: 10). Thus, for example, women's clothes may inscribe 'maternity' on their bodies. In this approach the body becomes significant 'only insofar as it is deemed to be by factors external to the body, be they social systems (Turner), discourse (Foucault) or shared vocabularies of body (Goffman)' (Shilling 1993: 99).

What is clear is that the body, understood as a biological entity, has undergone significant spatial augmentation. On one level, there is the physical extension of capacities made possible through the various media of telecommunications. The body is able, as a result, to act at a distance. At another level the body now has much greater capacities for peripatetic movement through the development of transportation. Fifth and finally, the body can be physically constructed in ways that were not available before. 'Medical' developments like plastic surgery mean that the body can be continually re-presented.

We have identified logical, prediscursive, psychological, cultural and social approaches to the body. So far as the next term, *the self* is concerned, the range of approaches that can be adopted are not just complex but bewildering in their variety. In the literature, the meaning of the self constantly slides from the simply ego of 'folk psychology' to complex, even heroic, projects of self-creation of the Nietzschean and, more recently, Foucaldian kind: '*one thing is needful* – to "give style" to one's character – a great and rare act!' (Nietzsche, *The Gay Science*, cited in Glover 1988: 131). Yet, at heart, much work on the self can be placed on a continuum between the Lockean–Humean understanding of personal identity, in which 'I' cannot be considered as an *a priori* unity of experiences, but refers simply to a series of experiences, and a Kantian–Cartesian understanding of personal identity, in which self-awareness is part of a continuing biography that tags each experience as belonging to a distinct self. These two traditions have become mixed together, most particularly in Freud.

> Freud followed Kant in seeing that I must be aware of a frontier between myself and other things. But he avoided the Cartesian side of Kant, and accepted the bodily frontier. He followed Hume in accepting that actions are caused by desires, and that decisions to act are not taken by a 'will' that escapes the causal process. But, unlike Hume, he saw that I can subject even my strongest present desire to mutual criticism. This criticism is not free-floating: it is based on other devices. These devices have to do with the conception I have of the life I want and the sort of person I want to be. Seeing the importance of these devices enables Freud, while accepting that what we do is causally determined, to stress that we can be active in taking charge of our lives.
>
> (Glover 1988: 130)

Modern examples of the two traditions are not easy to find because of the degree of mixing signified by the example of Freud and because of the intervention of new offshoots, for example the work of Husserl, Heidegger, Merleau-Ponty and Wittgenstein. However, it is possible to point to the work of Derek Parfit, whose magisterial *Reasons and Persons* (1984) eloquently expresses the Lockean view and Charles Taylor whose equally magisterial *Sources of the Self* (1989) defines the modern self as a turn 'inward' prompted by our efforts to define and reach the good based on moral ideals of self-mastery and self-exploration. In doing so, he continues the Kantian–Cartesian tradition, though in a radically augmented fashion. In particular, Taylor identifies the central paradox of the modern 'punctual' self as stemming from the reflexive turn made by Romanticism (and philosophies of vitalism and expressivism) and Modernism (and philosophies of flux and freedom) in the face of increasingly systematised societies:

> The modern ideal of disengagement requires a reflexive stance. We have to turn inward and become aware of our own activity and of the processes which form us ... We had to be trained (and bullied) into making it, not only of course through imbibing doctrines, but much more through all the disciplines which have been inseparable from our modern way of life, the disciplines of self-control, in the economic, moral and sexual fields. This vision is the child of a peculiar reflexive stance, and that is why we also have been forced to understand and judge ourselves in its terms and naturally describe ourselves with the reflexive expressions which belong to this stance: the 'self', the 'I', the 'ego'.
>
> That, at least, is part of the story. Another is that the set which emerges from the objectification of and separation from our given nature cannot be identified with anything in this given ... this ungrounded extra-worldly status of the objectifying subject accentuates the existing motivation to describe it as a 'self'. All other appellations seem to place it somewhere in the roster as one among others. The practical argument seems nothing else but a 'self', an 'I'.
>
> Here we see the origin of one of the great paradoxes of modern philosophy. The philosophy of disengagement and objectification has helped to create a picture of the human being, at its most extreme in certain forms of materialism, from which the last vestiges of subjectivity seem to have been expelled. It is a picture from a completely third person perspective. The paradox is that this severe outlook is connected with, indeed based on, according a central place to the first person stance. Radical objectivity is only intelligible and accessible through radical subjectivity.
>
> For us the subject is a set in a way he or she couldn't be for the ancients. Ancient moralists frequently formulated the injunction 'Take care of yourself', as Foucault has frequently reminded us ... They can sometimes *sound* like our contemporaries. But, in reality, there is a gulf

between us and them. The reason is that the reflexivity that is essential to us is radical ... Disengagement requires the first person stance.

... The turn to oneself is now also and inescapably a turn to oneself in the first person perspective – a turn to the self as self. That is what I mean by radical reflexivity. Because we are so deeply embedded in it, we cannot but search for reflexive language.

(Taylor 1989: 175–6)

The person can now be understood as a description of the cultural framework of the self. In all the writing on persons, perhaps the classic work is Mauss's (1985) last essay, first published in 1938, on the nature of the person. Much influenced by Durkheim, Mauss was concerned to show that societies can hold very different notions of the person and he documented a number of these variations. Since the essay was written this work of cultural translation has become both more pressing and, at the same time, more suspect, on two related counts. First, we still cannot be sure if the presupposition of a fundamental category of the person is absolutely necessary (Lukes 1985). It may be useful as a 'cross-cultural background' but it does not follow that the category of the person is a necessary part of all societies. In some societies this may not be the case. In Japan, for example, some authors argue that the boundaries between self and social are sufficiently different (because the self is considered as multiple, moving and changing) that the very notion of personhood is challenged (Rosenberger 1992). Second, notions of the personal have become explicitly political. For example, some have argued that the nature of a person is a conception so bound up with western presuppositions that became embedded in the colonial project that it may be irreversibly tainted. New ideas of the enunciation of the intersubjective may be required (Bhabha 1994).

Certainly it is clear that any discussion of the politics of the personal also requires discussion of *identity*. There has, of course, been much discussion in recent years of identity politics. Such a politics is usually thought of as arising from historical change. Agents, so the story goes, once had a defined and recognisable location in all-encompassing and rigid social structures like class and the family. The 'problem' of their location in these structures therefore did not arise in any fundamental way. But these 'traditional' social structures have now begun to 'detraditionalise' (Beck 1992; Giddens 1991), as a consequence of a number of changes including new social movements. As a result it has become a normal part of life to question identities, to construct them reflexively rather than simply recognise them. Thus, the univocal becomes the polyvocal. Consequently, social conflicts are no longer seen as just the epic clash of antagonistic social blocs but as a distributed deconstruction and reconstruction of social identities.

The processes by which identification occurs are seen in a number of ways in the literature, but two stand out. One is essentially psychoanalytic and centres around a lack which is seen as 'at the root of any identity: one needs to identify with something because there is an originary and insurmountable lack of identity' (Laclau 1994: 3). The other is essentially dynamic. Building on a theory of qualitative multiplicity which can never be reduced to one principle, the self and identity are seen as an affirmative, active flux, an image set in direct opposition to a monolithic and sedentary image of self and identity which is seen as clearly deriving from a phallologocentric system. This is the kind of stance now associated with writers like Butler, Castoriadis, Deleuze and Irigaray.

Nowadays, identity is often hedged about with spatial metaphors, with what Gilroy (1993: 195) calls the 'spatial focus'. Most particularly, of late, metaphors of mobility, transculturation and diaspora have become current. These metaphors are intended to capture the possibilities of hybrid identities which are not essentialist but which can still empower people and communities by producing in them new capacities for action. The ethnic absolutism of 'root' metaphors, fixed in place, is replaced by mobile 'route' metaphors which can lay down a challenge to the fixed identities of 'cultural insiderism', metaphors like diaspora:

> Diaspora refers to the scattering and dispersal of people who will *never* literally be able to return to the places from which they came; who have to make some kind of difficult settlement with the new, often oppressive cultures with which they were forced into contact, and who have succeeded in remaking themselves and fashioning new kinds of cultural identity by, consciously or unconsciously, drawing on more than one cultural repertoire. These are people who as Salman Rushdie writes in his essay on *Imaginary Homelands*, 'having been borne across the world ... are translated men and women'. They are people who belong to more than one world, speak more than one language (literally and metaphorically); inhabit more than one identity, have more than one home; who have learned to 'negotiate and translate' between cultures and who, because they are 'irrevocably the product of several interlocking histories and cultures' have learned to live with, and indeed to speak from difference. They speak from the in-between of different cultures, always unsettling the assumptions of one culture from the perspective of another, and thus finding ways of being both the same as and different from the others amongst which they live. Of course, such people bear the marks of the particular cultures, languages, histories and traditions which 'formed' them, but they do not occupy these as if they were pure, untouched by other influences, or provide a source of fixed identities to which they could ever fully 'return'.
>
> They represent new kinds of identities – new ways of 'being someone' in the late modern world. Although they are characteristic of the cultural strategies adopted by marginalised people in the latest phase of global-isation, more and more people in general – not only ex-colonised or marginalised people – are beginning to think of themselves, of their identities and their relationship to culture and to place in these more open ways.

(Hall 1995: 47–8)

Other metaphors of mobility abound. Thus Gilroy's chronotope of *The Black Atlantic* is an intermediate space of cultural criss-cross which spans the boundaries of nation states, which can stand for blackness without being reduced to it, and which immediately offers a notion of a new kind of cultural and political condition:

> I have settled on the image of ships in motion across the spaces between Europe, America, Africa and the Caribbean as a central organising principle for this enterprise and as my starting point. The image of the ship – a living micro-cultural, micro-political system in motion – is especially important for historical and theoretical reasons ... Ships immediately focus attention on the middle passage, on the various projects for redemptive return to an African homeland, on the circulation of ideas and activists as well as the movement of key cultural and political artefacts; tracts, books, gramophone records, and choirs.
>
> (Gilroy 1993: 4)

Finally, then, what about *the subject*? About the exactitudes of this term there is remarkably little agreement, except that the subject is a primary element of being and that the Cartesian notion of the subject as a unitary being made up of disparate parts, mind and body, which is universal, neutral and gender-free, is in error. Nowadays, the subject and subjectivity are more likely to be conceived of as rooted in the spatial home of the body, and therefore situated, as composed of and by a 'federation' of different discourses/persona, united and orchestrated to a greater or lesser extent by narrative, and as registered through a whole series of senses, not just what Descartes conceived of as the 'noblest of senses', sight, with its implicit Cartesian perspectivalism (which, in turn, produced an orientation 'to be a spectator rather than an actor' (Descartes, cited in Jay 1993: 101)).

What is quite clear is that, in recent work, what counts as the subject and subjectivity is being extended. Most particularly, the field of subjectivity increasingly encompasses 'the object world', as evidenced by actor-network theory, or the work of Haraway (1991) and Strathern (1992). As Latour (1993) argues, we need a new 'anthropological matrix' in which the object world has its place, and in which old ideas of the subject and agency are replaced by 'variable geometry entities' which translate between and across categories rather than purify within them. This expanded idea of subjectivity is nowhere more valuable than in discussions of an 'ecological self' (Mathews 1991; Plumwood 1993) which posits connections of mutuality between 'human' and 'non-human' and cleaves to an ethic of care. Thus, for example, 'wilderness' is redefined as 'not a place where there is no interaction between self and other, but one where self does not impose itself' (Plumwood 1993: 164). Of course, the concept of an ecological self is not without its problems: it can be interpreted as simply an expression of a western egoism. But it does at least force us to consider the question of the boundaries to, and categories of, the subject anew.

The authors gathered in this collection take up this challenge to consider the subject anew. Not all of the exploring on foot leads in the same direction, and some would rather fly by plane, but there is a commitment here to mapping the subject; a subject which is in some ways detachable, reversible and changeable; in other ways fixed, solid and dependable; located in, with and by power, knowledge and social relationships. This map, this subject, this book are not the same: they seek new paths, new performances, new politics.

2

MAPPING THE SUBJECT

Steve Pile and Nigel Thrift

If ways are to be found to map the subject, then it is clear that many things have to be kept in mind: possibly, too many things. In order to find our path through the forests, we have started with 'the map'. In this sense, the map is clearly a metaphor for 'the subject', but it cannot be naively presumed and so we have highlighted some ideas which are implied by the use of 'mapping', such as being somewhere, going somewhere, maybe meeting people – face-to-face – in specific ways, for particular reasons. Behind this story lie the ghosts of other stories – and these relate to power. We have decided not to isolate particular regimes of power, however, because these always seem to be there (more or less): power – whether organised through knowledge, class, 'race', gender, sexuality and so on – is (at least partly) about mapping the subject; where particular sites – for example, the body, the self and so on – become 'points of capture' for power. In this way, we aim to be ever mindful of the so many things which subjects must find their way through – 'relations of power'/'sites of power' – in order to find out where they are.

In this chapter, we provide six different pathways over the terrain of the subject, based on six different motifs: position, movement, practices, encounters, visuality and aesthetics/ethics. These mappings hardly exhaust the field. There are significant areas of literature on the subject which we have not been able to find space for, all the way from Althusser through Bakhtin and Vygotsky to Žižek. Here we only want to point to three omissions. The first of these is the ethogenic theory of the self, founded in the work of G. H. Mead, and developed most particularly in the dramaturgical interaction of Goffman and the ethno-methodological approach of Garfinkel (see Burkitt 1991; Heritage 1984), although it is worth pointing out that this ethogenic theory is one of the determinants of both social constructionism and actor-network theory which we map out below. The second omission is the object relations theory of Winnicott and others (Winnicott 1974, 1975). Winnicott's reworking of the process of self and othering is clearly very significant because of its emphasis on the other. But again, it has echoes in social constructionism, especially in the equation of Winnicott's 'space of play' and the constructionist 'third' space of joint action. Third, we have not paid any attention to the general historical dynamics of the subject, and especially the general history of 'containment' of the subject found in Elias, Foucault and some readings of Lacan. Partly this is because such an emphasis would have made this chapter unbearably long and partly it is because, for all the complexity of the theoretical mode of appropriation (whether in Elias's notion of figuration or Foucault's notion of genealogy), the underlying historical template – a history of progressive constraint – is quite simple while the necessary apparatus for grasping the symbolic construction of self through history is in both cases incomplete (see Burkitt 1991, 1994; McNay 1992).

What seems certain about the six pathways that we travel is that they all start from much the same place, and that is a critique and rejection of the Cartesian model of the subject. This model has now become deeply embedded in western thought. The idea of a disengaged first-person-singular self calls on each of us to become a responsible thinking mind, self-reliant for her or his judgements on life, the universe and everything. In turn, this means that:

> we easily tend to see the human agent as primarily a subject of representations: representations about the world outside and depictions of ends desired or feared. This subject is a monological one. She or he is in contact with an 'outside' world, including other agents, the objects she or he and they deal with, her or his own and other bodies, but this contact is through the representations she or he has 'written'. The subject is first of all an 'inner' space, a mind to use the old terminology, or a mechanism capable of processing representations if we follow the more fashionable computer-inspired models of today. The body, other people or objects may form the content of my representations. They may also be causally responsible for some of these representations. But what I am is definable independently of body or other. It is a centre of monological consciousness.
>
> (Taylor 1993: 49)

This 'stripped-down' conception of the subject is, by now, backed up by an enormous vocabulary of the psychological interior which has been reinforced by Kantian and Husserlian notions of the subject. Thus,

> we speak with ease and confidence of our thoughts, beliefs, memories, emotions and the like. We also possess an extended discourse through which we render accounts of the relationships among aspects of the mental world. We speak of ideas, for example, as they are shaped by sense data, bent by our motives, dropped into memory, recruited for the process of planning and so on. And we describe how our emotions are fixed by our ideas, suppressed by our conscience, modified by our memories and seek expression in our dreams. In effect, we have at our disposal a full and extended ontology of the inner region. When asked for accounts of self, participants in Western culture unfailingly agree that emotions, ideas, plans, memories and the like are all significant. Such accounts of the mind are critical to who we are, what we stand for and how we conduct ourselves in the world.
>
> (Gergen 1989: 70)

Yet, with the benefit of hindsight and much anthropological research, this monological idea of the subject now seems an extraordinarily ethnocentric one. As Geertz puts it,

> the Western conception of a person as a bounded, unique, more or less integrated motivational and cognitive universe, a dynamic centre of awareness, emotions, judgement and action, organised into a distinctive whole and set contrastively against a social and natural background is, however incorrigible it may seem to us, a rather peculiar idea within the context of the World's cultures.
>
> (1989: 229)

Increasingly, then, the monological conception of the subject bars the way to a richer and more adequate understanding of what the human self can be like. In turn, it also debars us from a fuller appreciation of the variety of differences between human cultures.

What pathways might we therefore turn to, which might give us a richer and more adequate understanding?

POSITION AND THE POLITICS OF LOCATION

Being an 'intellectual' is a difficult and fraught task nowadays. It is rather like playing a game of snakes and ladders. Having climbed to the top of one ladder, a snake beckons and produces a precipitate Fall. Thus, the intellectual climbs the ladder of taking 'the people's' side. But there's the snake. Writing about people too often involves an heroic assumption of getting closer to the people when the intellectual may just be slipping further away as a result of an intellectualist bias which construes the world as a set of significations to be interpreted rather than as a concrete set of problems that people have to solve practically (Bourdieu and Wacquant 1992). Meaghan Morris provides a recent example of this slippage when she examines the tendency in cultural studies to characterise people as possessing 'an indomitable capacity to "negotiate" readings, generate new interpretations, and remake the materiality of culture' (Morris 1988: 17). More likely, she says, people are there to:

> represent the most creative energies and functions of critical reading. In the end they are not simply the cultural student's object of study, and his native informants. The people are also the textually delegated, allegorical emblem of the critic's own activity. Their *ethos* may be constructed as other, but it is used as the ethnographer's mask.
>
> (Morris 1988: 17)

Another ladder. It is clear that an intellectual has to be critical. But the snake is that the intellectual can very easily become just another voyeur. This is a charge made most often nowadays about 'feminist' men (although it is a charge that is also levelled at intellectuals, and in relation to white middle-class women's relation to working-class women and women of colour). What we see are men intent on 'getting a bit of the other' (Moore 1988), men who are into 'critical cross-dressing' (Showalter 1987), and men who are rampant careerists:

> Boone, by seizing the right, 'oppressed' side of the well-known series of patriarchal binary oppositions, and by placing them without his own professional context, is trying to pass every unknown male critic off as silent, 'invisible', 'powerless' – in short, as feminine, and therefore also as feminist. The implicit equation between femininity and feminist is more than obvious in itself. But even more important is the fact that this manoeuvre is possible because Boone attributes all the characteristics on the left, 'oppressive' side of the oppositions to older famous males. But how would these young 'silent' men fare if they were cast opposite young *women*? And what would happen to Boone's rhetorical structure if these unknown male intellectuals actually *became* visible, known, and so on? Would that make them less feminist? ... 'Visible' female feminist critics are let off the hook on the grounds that their expertise makes up for their 'visibility'. A 'visible' man, however, can have nothing interesting to say.
>
> (Moi 1989a: 187–8)

One more ladder. The intellectual needs to be reflexive and, in particular, reflexivity is a crucial tradition of modern work on the subject. These maxims are crucial to the practice of modern ethnography, up to and including the vogue for auto-ethnography (e.g. Steedman 1986; Fiske 1990). The snake is that, too often, the results of a writer's attempts to use reflexivity to interrogate the self/other relationship come perilously close to narcissism and solipsism. Every early childhood slip, every parental flaw, every departmental tiff, every conference slight, becomes grist to a 'falsely radical' mill (Bourdieu and Wacquant 1992: 72). We end up with something remarkably like the confessional, romantic hero of yore that the writer has just spent blocks of print criticising, but now reconsecrated by the act of self-criticism (Probyn 1993). The result is that the writer's subject becomes the writer's object and the writer's object slides gently away. As Probyn puts it, concerning Dumont's study of the Panare, 'as Dumont-the-author emerges within the text, the Panare seem to disappear' (1993: 75).

Siding with the people becomes constructing a people that exist only in the writer's imaginary, 'polemologic' (de Certeau 1984) becomes patronising; epistemic reflexivity is turned into a narcissistic textual reflexivity ... The list of snakes and ladders can no doubt be extended indefinitely. The point is that these are all perils of a world where *positioning* has become a crucial element of everyday intellectual practice. In the past, intellectual positioning was achieved by an appeal to a series of 'great thinker' ancestors. Now it is as often achieved by accounts which (inevitably it seems) appeal to parts of a person's history which allow them to count themselves as an 'outsider', even in the cases where this can be difficult to see. The politics of kinship has been succeeded by a politics of position.

How, then, to go beyond this game of snakes and ladders which is largely (but not only) a crisis of the white, western, heterosexual male self? How to help to kindle rather than hinder the construction of an orientation to community and solidarity which is not, at the same time, a takeover bid by this self?

The strategy that is usually adopted is one of injecting the equivocal and the personal into work through, to use Rich's (1986b) well-known phrase, a politics of location. This is a politics that makes no claims to second-guessing others' experience but still allows people to speak for themselves. Some of the ingredients of this kind of politics are clear; the body, the self (as a practice and as a speaking position), experience, the validation of 'minor' knowledges (and especially the 'thinking' of the body).[1] What is less clear is what they all mean when put into practice. For example, Probyn writes that,

> the critic's experience may be tuned into an articulated position which allows him or her to speak as an embodied individual within the process of cultural interpretation. This does not mean that critical activity becomes focused on a reflexive account of one's experiences of oneself; this is not a proposal for an endless deconstruction of the subject/text relation. As an enunciative position within cultural theory, the self can be used to produce a radical rearticulation of the relationship of critic, experience, text and the conjunctural moments that we construct as we speak of that within which we live.
>
> (1993: 31)

But such a statement is, at best, programmatic. It clearly requires the construction of a discursive image of the self which: is not located in the traditional discourses of individualism; is located in an historical analysis of what self and experience can consist of at particular conjunctures; is relational; is embodied; insists on difference as a qualitative multiplicity; and can provide new, empowered speaking positions. In other words, it requires the construction of:

> a multiple, shifting and often self-contradicting identity, a subject that is not divided in, but rather at odds with, language; an identity made up of heterogeneous and heteronomous representations of gender, race, and class and often indeed across languages and cultures; an identity that one decides to reclaim from a history of multiple assimilations and that one insists upon as strategy.
>
> (de Lauretis 1986a: 9)

What novel kinds of discursive images of the self and experience, what different kinds of identity, what fresh image-concepts, what new maps of subjectivity, which new *figurations* are available? Many have been proposed. In what follows we shall note just a few of these available figurations drawn from feminist and postcolonial writings. One such is found in the work of Irigaray who proposes the figuration of 'woman' as marking a specific form of transcendence, a female humanity with its own discursive presence based on the sexed body. In other words, Irigaray is searching for an alternative female symbolic, accompanied by an alternative female genealogy. Another figuration can be found in the work of Wittig (1977). In contrast to Irigaray, she rejects any idea of femininity because it is founded on a concept of sexual difference. She suggests the elimination of woman as a category, and its replacement by a figuration of the lesbian as the 'third sex', a position beyond gender because it is 'subtracted' from identities based on the phallus. Finally, there is Haraway's (1990) figuration of the 'cyborg'. Haraway's intent is very wide. Like Irigaray and Wittig, she wants to produce a new materialism based around rethinking the subject's bodily roots, where the body stands for the radical materiality of the subject. But she differs in the imaginary that she uses to accomplish the task – science – and in the conclusions she arrives at. Haraway wants us to think about what new kinds of bodies are being constructed in the modern scientific world, that is, what new kinds of gender systems are being produced. Thus, Haraway is sceptical about Foucault, in that she believes that Foucault's account of the disciplining of docile bodies is both out-of-date and androcentric. To it, she wants to oppose a new figuration which both more clearly captures current systems of domination and also the possibilities of escape from them. This figuration is the cyborg, an overconnected and hybrid body-machine which can replace old ontologies by redefining the old dualisms like mind–body and which can help to produce new forms of 'literacy' with which to decode today's world.

Much the same impulses to find new places to speak from as are found in feminist writings we also find in writings on postcolonialism. The articulation of postcolonial subjectivities are similarly attempts to produce new figurations, positive identities from which to speak which are neither faintly pencilled in nor permanently etched in stone. It is the search for a 'partial identification' (Bhabha 1994), which can nonetheless be a source of personal and political agency. This new hybrid postcolonial subject (Hall 1991a, 1991b; Spivak 1991) is in part a recognition that subjects are found 'in-between' domains of difference like race, class and gender, in the interstices where these domains intersect. It is also in part a political ambition, an attempt to find a 'third space' for the exchange of values, meanings and memories between communities which may never be collaborative and dialogical and which indeed may well be antagonistic to one another, or even incommensurable. In other words, the postcolonial subject is a way of representing difference as not just a set of pregiven and calcified ethnic or cultural traits but also as a process of negotiation, in which self and experience are never totalised and always on-going. From such an 'interstitial perspective', new notions of solidarity and community can be posed and new subjectivities forged. In turn, the liminal spaces in which this interstitial subject can thrive are not just literary allegories. They are quite clearly related to borders and frontiers, to migrants and diasporas, to the colonised, to political refugees and to the consequent refiguring of notions of 'home' and 'nation'. In other words, as the 'unhomely' becomes the norm, replacing the sovereignty of national cultures, or the universalism of a human culture, so new subjectivities are needed.

As the latter part of this discussion already makes clear, what is conspicuous about all the writings on the politics of position is the degree to which space figures. It figures in the metaphors that are now almost ritually incanted – position, location, centre, margin, local, global, border, boundary, interstice ... It figures in the degree to which actual interstitial spaces and subjects tend to become the examples that are drawn on. And it figures in the degree to which space is used both to ground discussions and to take them on, all the way from Irigaray's discussion of place as a crucial element in her conception of woman, set up in opposition to the theft by man of 'the tissue or texture of her spatiality' (Irigaray 1985b: 123), to Haraway's discussion of 'situated knowledges' which allows her to argue for a multifaceted foundational theory and an anti-relativistic acceptance of differences (1988). We now want to look more closely at how space figures, by considering metaphors of movement.

<h2 style="text-align:center">MOVEMENT</h2>

In the preceding section we pointed to the current emphasis on spatial metaphors as a way of comprehending the subject, metaphors that can reanimate body, self, identity, person and subject. Of these different metaphors, some of the most fertile have tended to cluster around ideas of movement and mobility, journeying and travelling, what Wallace (1993) calls the 'peripatetic' mode of signification. Such metaphors can be used, or so it is hoped, to construct new, more open figurations of the subject. Politically, that might mean the creation of new sites of action and subject constitution, sometimes called 'third spaces' (see Pile 1994). For gender, that might mean the constitution of new hybrid sexualities, which are less concerned with domination. In work on ethnicity, that might mean the construction of a new, more open notion of self and society, based on the cosmopolitan flux of the modern metropolis. These kind of attempts at the remetaphorisation of theory and practice have now become a common theme in the social sciences and humanities alike. Yet, to an extent, they are as confusing as they are revealing because they mask a whole series of concerns, which jostle with each other but are not necessarily coincident. What are these concerns? We will consider just some of them here.

First of all, there is a concern for capturing being as a process of provisional and open-ended movement. This kind of work on lived time clearly has two main modern inspirations. The first of these is Bergson's theory of time which depends on a notion of body/image in action.

All objects have a bodily form, and contrary to the usual privileging of consciousness, bodies – the human body included – are sites of action, influencing each other in movement. Perception is not qualitatively different from image-body; it is these images, referred. Action rather than consciousness characterises bodies: 'my body is a centre of action, it cannot give birth to representation' (Bergson 1950: 5). The subject lives the material world; it is of that world and produced by it. We are not the source of meaning or representation, but in the movement of relations between bodies, change is always possible (Game 1991: 11).

The second, later inspiration, is Heidegger's view of time. Heidegger rebelled against a view of time in which the present is the dominant dimension in favour of a view of time which is oriented to the future: 'Heidegger's time is lived time, organised by a sense of the past as the source of a given situation, and the future as what my action must co-determine' (Taylor 1989: 463).

A second concern is difference, understood as a non-hierarchical, qualitative multiplicity, which can realise continuity without assimilation. In a sense, or so it is argued, the metaphor of movement can convey this more open sense of difference. Such claims usually depend, directly or indirectly and knowingly or unknowingly, on a Bergsonian model of difference: 'Bergsonian difference defines, above all, the principle of the positive movement of being, that is the temporal principle of ontological articulation and differentiation. Bergson does not ask what being is, but how it moves' (Hardt 1993: 113).

A third concern follows on from this notion of difference. That is a rethinking of 'experience'. Experience is, of course, a notoriously slippery concept. But in recent writing, especially by feminist theorists, the concept has undergone something of a revival as a means of engaging with the felt facticity of material social being as being something more than a theory of signification. For Probyn (1993), this means retelling experience as a specific enunciative practice, as a mode of speaking which galvanises self and context in a new way, just as can the experience of travel.

> As an active articulation of ontological and epistemological levels, the experience may enable an enunciative position which puts forward a level of being as the conditions of that being are problematised. In this mode of speaking, the self is put forward not to guarantee a true referent but to create a *mise-en-abyme* effect in discourse. In distinguishing these two levels at which the experiential may be made to work, I want to enable a use of the self which neither guarantees itself as an authentic ground nor necessarily rejects the possibility of a ground.
>
> (Probyn 1993: 29–30)

Such an emphasis on experience as a repositioning of self and context has recently been most prevalent in two areas. The first of these is so-called 'postmodern' ethnography (Marcus and Fisher 1986; Clifford 1988, 1992), where metaphors of movement have become commonplace as both a description of ethnographic practice and a symbolic quest:

> the travel metaphor seems quite appropriate to ethnography. To put it simply, ethnography is always about traversing the difference between the familiar and the strange. The ethnographer leaves her home (the familiar) and then travels to the other home (the strange), and then returns home to make sense of it in her writing.
>
> (Grossberg 1988: 23)

The second area, one which tries to do without what has been called the 'ontological egotism' (Probyn 1993: 80) of postmodern ethnography, concentrates on travel writing, most especially as a history of 'imperial meaning-making' (Pratt 1992: 4). This strand of postcolonial work pays particular attention to the ways in which land and space are represented in texts as both the subject and object of expansionist energies and imaginings.[2] A particular, divergent variant of this literature concerns the case of women travellers in imperial times, showing how the spatiality and gendering of travel are intertwined (Mills 1991; Blunt 1994).

A fourth concern with metaphors of movement follows on again. It is with the possibilities of mutable sharing. Whether we are talking of new, more open forms of politics and an expansion of what is regarded as 'the political' (Laclau and Mouffe 1985), the meeting of people in global cities (Hannerz 1992), the intermixing of bodies and fluids in sexual activity (Mort 1988) or the varied outcomes of imperial and postimperial contact zones (Pratt 1992) the outcome tends to be an appeal to ideas of hybridity, as a description of new cultures and subjects formed by the juxtaposition and co-presence of different cultural forces and discourses and their effects. This emphasis on the body, like travel, as an active mode of making new connections (drawn in part from previous notions of articulation) has been most marked in the work of writers like Paul Gilroy and Stuart Hall who have argued that in a world of movement, 'it ain't where you're from, it's where you're at' (Gilroy 1991).

> The notion that identity . . . [can] be told as two histories, one over here, one over there, never having spoken to one another, never having anything to do with one another, when translated from the psychoanalytic to the historical terrain, is simply not tenable any more in an increasingly globalised world. It is just not tenable any longer.
>
> (Hall 1991b: 48)

Both Gilroy and Hall are arguing for a 'politics of transfiguration' (Said 1992), founded on an ethics of difference, which can express and encourage an openness of outlook based upon a freedom to move across border and boundaries in pursuit of new senses of self and other.

A fourth concern is with alterity. Ideas of movement and travel are bound up with a sense of something other around the corner, a new image-concept that will produce a new subject position or a new subjectivity. In particular, forging such an image-concept requires the recognition of new spatialities and, here, exploration and explorers have proved particularly useful as figures for conceptualising such a process:

> the explorers were not simply travel writers; for, unlike the purveyors of picturesque places, they travelled without records. What they described, then, was not a succession of places, but the plotting of a travel along which historical time might later flood in on a tide of names. Hillis Miller has pointed out that one etymology for diegesis, the narration of events, gives as the world's original meaning the

redrawing of a line already drawn. Plotting in this context is the desire to bring out the meaning of that line, to endow it with form, to bend it perhaps into the ring of eternal return. But the activities of the explorers preclude this: their task is to draw the we for the first time, to give space a narrative form and hence the possibility of a future history, a history that will subdivide, and even efface their own narratives in the interest of a thousand domestic plots.

(Carter 1992: 23)

More particularly, the forging of new subject positions and subjectivities out of ideas of movement and travel has been bound up with the history of photography, and especially the cinema. The study of the prehistory of photography and the cinema has shown the way in which new subject positions and subjectivities had to be formed (often from visual regimes associated with travelling and journeying) before these technologies could operate, and indeed the subject positions were a part of the process of their invention.[3] These new subject positions and subjectivities, resulting from a realignment of urban space, the body and desire are perhaps best understood through the work of Deleuze[4] and Virilio[5] on the cinema, and, in particular, Deleuze's reworking of Bergson through the concept of the movement-image.

Cinematic pleasure belongs to the range of erotic pleasures of the nomadic gaze first known to the traveller and the flâneur and then embodied, by way of panoramic spatio-visuality, in the modes of inhabiting space of transitional architectures. Suggesting these historic aspects of the fascination of the apparatus, and highlighting its fantasmatic connection to travel and landscaping, one looks back not only to early cinema but also 'back to the future'. This view offers numerous avenues for future studies. One can place the art of 'unconscious optics' in the context of contemporary forms of inter-cultural travelling and sites of spatio-temporal tourism, of which airplane cinema is the ultimate metonymy. For if the unconscious is 'housed', it is also 'moving'.

Embodying the dynamics of journey, cinema maps a heterotopic photography. Its heterotopic fascination is to be understood as the attraction to, and habitation of, a site without geography, a space capable of juxtaposing in a single real place several possibly incompat-ible sites as well as times, a site whose system of opening and closing both isolates it and makes it penetrable, as it forms a type of elsewhere/nowhere, where 'we calmly and adventurously go travel-ling'. Thus, we female spectators, in the midst of our old enclosed prison world, may go travelling. As we move through filming architecture, as in street walking through the mater-polis, our mother-city, we reclaim forbidden places – wandering through erotic geographies.

(Bruno 1993: 57)

Such thoughts also prefigure much of the current work on the new 'visual' regions of cyberspace in which the links between vision and the sexual geopolitics that Bruno points to are made even clearer by linking the crisis of representation that the advent of cyberspace seems to prefigure with the contemporary crisis of masculinity. Thus, on the one hand, the practices of cyberspace have often been inflected with fantasies of power and control, while, on the other hand, they also seem to threaten the centred nature of the masculine subject .[6]

Metaphors of movement are also, sixth, concerned with agency. There is now a general swing back in the social sciences and humanities from extreme forms of poststructuralist thought, in which the subject is only an effect of discourses, to a consideration of forms of subjectivity which, although limited and contingent, can still exert a degree of agency. The reasons for this state of affairs are twofold. The first is political (Bordo and Moussa 1993). As Hartsock's now famous question puts it:

why is it that just at the moment when so many of us who have been silenced begin to demand the right to name ourselves, to act as subjects rather than as objects of history, that just then the concept of subjecthood becomes problematic?

(1990: 163)

The second is theoretical. This is that what Smith (1988) calls the subject's self-narrative only makes sense as a source of action. Thus, as Laclau and Mouffe argue, 'the analysis (of the subject) cannot simply remain at the level of dispersion, given that "human identity" involves not merely an ensemble of dispersed positions but also the forms of overdetermination existing among them' (1985: 117). Further, 'the "subject's" self-interest is in part what has to be articulated in (this) inevitably complex way for someone to be able to act at all' (Smith 1988: 158). Limited, contingent but still potent forms of subjectivity are clearly difficult to grasp and theorise, but the task is not impossible. For example, Deleuze's notion of the fold is simultaneously an attempt to theorise a pleat or crease in the space-time continuum within which embodied presence can exert force in some way and, at the same time, an attempt to describe what it means to live in the modern 'compressed' world (Deleuze 1993b). Chambers ably summarises these kinds of political and theoretical ambitions:

we imagine ourselves to be whole, to be complete, to have a full identity and certainly not to be open or fragmented; we imagine ourselves to be the author, rather than the object, of the narratives that constitute our lives. It is this imaginary closure that permits us to act. Still, I would suggest, we are now beginning to learn to act in the subjunctive mode, as if we had a full identity, while recognising that such a fullness is a fiction, an inevitable failure. It is this recognition that permits us to acknowledge the limits of our selves and with it the possibility of dialoguing across the subsequent differences – the boundary, or horizon, from which, as Heidegger points out, things unfold; both towards and away from us.

(1994: 25–6)

Finally, there is a concern with the process of thinking itself, as a mobile, fluid and vertiginous activity. This refashioning of what it is to think reaches its climax in the work of Deleuze, who constantly uses spatial metaphors to describe thought. For Deleuze, theories are not 'objects' but living territories of contemplation, constantly on the move. It follows that:

> thought is made of sense and value, it is the force or level of intensity, that fixes the value of an idea, not its adequation to a pre-established normative model. Philosophy as critique of negative, reactive values is also the critique of the dogmatic image of thought; it expresses the force, the activity of the thinking process in terms of a typology of forces (Nietzsche) or an ethology of passions (Spinoza). In other words, Deleuze's rhizomatic style brings to the fore the affective foundations of the thinking process. It is as if beyond/behind the positional content of an idea there lay another category – the affective force, level of intensity, desire, or affirmation – that conveys the idea and ultimately governs its truth value. Thinking, in other words, is to a very large extent unconscious, in that it expresses the desire to know, and this desire is that which cannot be adequately expressed in language, simply because it is that which sustains language.
>
> (Braidotti 1994: 165)

Hopefully, it is now clear that the motifs of movement, journeying, travelling ... figure large in contemporary writings on the subject – and for many reasons. But such motifs are not without their problems.

Three of these problems seem particularly pressing. First, and most familiarly, there is the constant danger in travelling theory of a form of gender or ethnic tourism in which the white male subject 'gets a bit of the other' (cf. Woolf 1993). What is still a landscape of constraint for most people is redefined as a landscape of movement and mobility by those for whom movement and mobility are unproblematic. Second, much of the writing on mobility and movement comes perilously close to reinventing the kind of modernism that celebrates speed, flow and vibration for their own sakes (Gergen 1991; Taylor 1989). Under modernism, 'the epiphanic centre of gravity begins to be displaced from the self to the flow of experience, to new forms of unity, to language concerned in a variety of ways ... an age starts of decentring subjectivity' (Taylor 1989: 465). Not only is the language of movement and mobility therefore nowhere near as radical as is often imagined but it can often simply displace rather than reformulate questions of subjectivity. As Taylor goes on to write, 'for all the genuine discoveries which we have made in this mode, the impetus to enter it is in large part the same as that which turned us inward. Decentring is not the alternative to inwardness; it is its complement.' Finally, there is a problem of empirical accuracy. Descriptions of the contemporary world are often casually thrown around that are predicated upon a few simple master(sic)-concepts – time-space compression, globalisation, postmodernism, hyper-reality – which are, in fact, highly contested (Thrift 1994a). Concepts like these four certainly include substantial elements of exaggeration, a dubious modernist ancestry, dashes of Eurocentrism and the like. Part of the reason for their acceptance is no doubt the prevailing academic division of labour which, for example, too often assumes that subjectivity is 'cultural' rather than 'economic' (and which, in turn, means that many writers in cultural studies uncritically appropriate 'economic' concepts). Another part of the reason is that, too often, commentators seem unwilling to make counter-propositions: 'What if it wasn't like this?'

These are serious objections to the ideas of movement and mobility but, that said, their usage can be effective if it is moderated. As Miller has written, 'what I think we need ... is a less utopian, less arrogant, and less messianic theorisation of movement, a positive cosmopolitanism that remains meticulously aware of localities and differences, a more convincing ethic of flow' (1993: 33). That this is a possibility can be seen from a number of literatures. One example is the turn to an examination of science fiction as a way of thinking alternate futures/subjectivities in an age which often 'sees itself *as* science fiction' (Bukatman 1993a: 6). First of all, it is possible to use the prospect of new technologies found in science fiction as analogies for the crisis in current thinking. Thus, the practices of photography could be interrogated as a 'representation' of 'reality'. Now, the new information technologies, in which a field of digital data replaces the old perspectivalist point of view, can act as an analogue for new, hybrid subjects constantly on the move. Second, science fiction can denaturalise language, producing both a heightened reflexivity and an emphasis on the interconnection between textuality and thought (Kuhn 1990). Third, science fiction can provide ideas for a new politics of the subject, as in Haraway's (1991, 1992) reading of feminist science fiction for 'cyborgs', or Bukatman's (1993a, 1993b) search for 'terminal identities' (see Piercy 1992). Fourth, science fiction offers some tentative solutions to some of the aesthetic problems posed by new technologies.

Not surprisingly, the appeal to the subject through science fiction is often made via the deployment of particular notions of spatiality. Most commonly, this is the conceit of 'cyberspace',

> now the inertial shell of the personal computer replaces the thirsty power of the Saturn V as the emblem of technological culture. Invisible spaces now dominate, as the city of the modernist era is replaced by the non-place urban realm and outer space is replaced by cyberspace.
>
> (Bukatman 1993a: 5–6)

The search is on in the pages of science fiction novels for models of new 'in-between' subjects that can move in these new cyberspaces, new human architectures that can mobilise new electronic architectures. The urban figures prominently in this search, as both the prime object of the new subject's attention and as an analogue for the new subject's make-up. But it is an urban scene which cannot be understood from afar, or on high (de Certeau 1984). This is, in other words, an underground urban which mirrors de Certeau's 'tactics'; 'the street finds its own uses for things – uses the manufacturers never imagined' (Gibson 1991: 29).

Of course, the literature on the subject in science fiction can be awful – a few casual opinions backed up by a few casual references is not an abnormal model in the field – but it can also produce new insights, especially, when there is, as there is more and more often, an historical imagination at work.[7]

PRACTICES

The body is in constant motion. Even at rest, the body is never still. As bodies move they trace out a path from one location to another. These paths constantly intersect with those of others in a complex web of biographies. These others are not just human bodies but also all other objects that can be described as trajectories in time-space: animals, machines, trees, dwellings and so on.

In embryo, this is a description of the time-space demography (or time-geography, as it is more commonly known) of Torsten Hägerstrand, the Swedish geographer. Yet, as a written description, it precisely misses Hägerstrand's main aim, which was to find a geographical vocabulary that could describe these prelinguistic movements prelinguistically. That was the purpose of his now famous time-space diagrams. He often compared these diagrams to a musical score which is a similar set of marks of movement, producing similarly complex existential effects. More than this, Hägerstrand took pains to point out that these diagrams, like a musical score, could stand for a different kind of (non-intellectual) intelligibility.

One more point tends to be made concerning Hägerstrand's work. That is that it is inherently dialogical. In opposition to a number of critics in geography (e.g. Rose 1993a) who have seen it as a robustly individualistic approach, Hägerstrand clearly saw time-geography in precisely the opposite terms. His stress was constantly on the congruences and disparities of *meeting*, that is on the *situated interdependence* of life. His intent was, in other words, to capture the pragmatic sense of possibility inherent in practical situations of 'going-on'. In consequence,

> rather than implying an idealist framework of intentional action shaping the resulting totality, it should be evident that Hägerstrand pointed to competitive allocation and displacement effects which made the total outcome anything but the sum total of intentions at the level of actors (be they organisms, human individuals, groups, organisations, or even states).
>
> (Carlstein 1982: 61)

Hägerstrand's maps of everyday coping can best be placed, therefore, in a line of thinking which stretches from Heidegger and Wittgenstein, through Merleau-Ponty, to, most recently, Bourdieu, de Certeau and Shotter, who have tried to conjure up the situated, prelinguistic, embodied states that give intelligibility (but not necessarily meaning) to human action. What Heidegger called the primordial or preontological understanding of the common world, our ability to make sense of things, what Wittgenstein knew as the background, what Merleau-Ponty conceived of as the space of the lived body and what Bourdieu means by the habitus. Each of these authors is concerned, in other words, to get away from Cartesian intellectualism, with its understanding of being as a belief system implicit in the minds of individual subjects, and return to an understanding of being as 'the social with which we are in contact by the mere fact of existing and which we carry with us inseparably before any objectifications' (Merleau-Ponty 1962a: 362). Thus, in this view being is not an entity but a way of being which constitutes a shared agreement in our practices about what entities can show up.

In each case, what these authors have in common is that they see the subject as primarily derived *in practice*.

> In the mainstream epistemological view, what distinguishes the agent from the inanimate entities which can also effect their surroundings is the former's capacity for inner representation, whether these are placed in the 'mind' or in the brain understood as a computer. What we have which inanimate beings don't have – representations – is identified with representations and the operations we effect on them. To situate our understandings in practices is to see it as implicit in our activity, and hence as going well beyond what we manage to frame representations of. We do frame representations: we explicitly formulate what our world is like, what we aim at, what we are doing. But much of our intelligent action, sensitive as it usually is to our situation and goals, is usually carried on unformulated. It flows from an understanding which is largely inarticulate.
>
> (Taylor 1993: 49–50)

Thus understanding of the subject in practice is fundamental in two ways. First, this kind of subjectivity is always present. Sometimes we frame representations. Sometimes we do not. But the practical intelligence is always there. More to the point, and second, the kind of representations we make are only comprehensible against the background provided by this inarticulate understanding. 'Rather than representations being the primary focus of understanding, they are islands in the sea of our unformulated practical grasp of the world' (Taylor 1993: 50).

This kind of thinking about the subject's understanding of the world (and note straight away how what constitutes the subject has already been problematised) has four main characteristics.

First of all, the subject's understanding comes from the ceaseless flow of conduct, conduct which is always future-oriented. In terms of practice, understanding does not come from individual subjects moving deliberately and intentionally through spaces in a serial time. That would be to revive the subject–object relation. Rather, subjects display absorbed coping or, to use a Heideggerian term, 'comportment'. Comportment differs in at least five ways from an action-directed view of understanding (Dreyfus 1991). First, it is an open mode of awareness which 'is not mental, inner, first person, private, subjective experience … separate from and directed towards non mental objects' (Dreyfus 1991: 68). Second, it is adaptable. Comportment manifests dispositions shaped by a vast array of previous dealings but does so in a flexible way. Third, comportment is understanding as 'aspect-dawning' (Wittgenstein 1953). That is, it depends upon the orientation to a particular activity, what Heidegger calls the 'towards-which', and is typified by instant recognition/description. Fourth, if something goes wrong with comportment, it produces a startled response because future-directed activity is being interrupted. Fifth, and related, if something goes awry, conduct becomes deliberate and acquires a sense of effort.

A second characteristic of the subject's understanding of the world is that it is intrinsically corporeal. Following Merleau-Ponty, the socialised body is not an object but the repository of a generative, creative capacity to understand. How can this be?

> Adapting a phrase of Proust's one might say that arms and legs are full of dumb imperatives. One could endlessly enumerate the values given the body, *made* body, by the hidden persuasion of an implicit pedagogy which can instil a whole cosmology, through injunctions as insignificant as 'sit up straight' or 'don't hold your knife in your left hand', and inscribe the most fundamental principles of the arbitrary content of a culture in seemingly innocuous details of bearing and physical manner, so putting them beyond the reach of consciousness and explicit statement.
>
> (Bourdieu 1990b: 63)

Further, embodiment also produces temporality (and spatiality). As Merleau-Ponty wrote,

> in every focusing movement, my body invites present, past and future … My body takes possession of time; it brings into existence a past and a future for a present, it is not a thing but creates time instead of submitting to it.
>
> (1962a: 239–40)

A third characteristic of the subject's understanding is that it is worked out in *joint action*. Many actions require co-operation to complete. Many actions assume the presence of others. All actions are bound together by mutual dispositions and shared understandings which they both take from and contribute to. In other words, dialogical action is a fundamental determinant of the intelligibility of social life; understanding comes from 'we', not 'I'. 'My embodied understanding doesn't exist only in me as an individual agent; it also exists in me as the co-agent of common actions' (Taylor 1993: 53). Often language's function is simply to set up the intersubjective spaces for these common actions, rather than to represent them. Further, dialogical action presupposes moral judgements:

> we can see that in the ordinary two-way flow of activity between them, people create, without a conscious realisation of the fact, a clinging sea of moral enablements and constraints, of privileges and entitlements, and dysfunctions and sanctions – in short, an ethos. And the changing settings created are practical-moral settings because the different places or positions they make available have to do, not so much with people's 'rights' or 'duties' (for we might formulate its ethical nature in different ways, at different times) as with the nurturance to the basic being of a person. For individual members of a people can have a sense of 'belonging' in that people's reality only if the others around them are prepared to respond to 'reality', only if the others around them are prepared to respond to what they do and say seriously.
>
> (Shotter 1993b: 31)

A fourth characteristic of the subjects' understanding of the world is its situatedness. The subject can only 'know from'. Therefore abstracting subjectivity from time and space becomes an impossibility because practices are always open and uncertain, dependent to some degree upon the immediate resources available at the moment they show up in time and space. Thus, each action is lived in time and space, and part of what each action is is a judgement on its appropriateness in time and space. Further, following any kind of social 'rule' about practice always involves some measure of openness and uncertainty associated with each movement:

> a rule doesn't apply itself; it has to be applied, and this may involve difficult, finely tuned judgements. This was the point made by Aristotle and under his understanding of the virtue of phronesis. However situations arise in infinite varieties. Determining what a norm actually amounts to in any situation can take a high degree of insightful understanding. Just being able to formulate rules will not be enough. The person of real practical wisdom is marked out less by the ability to formulate rules than by knowing how to act in each particular situation ... In its operation, the rule exists in the practice it 'guides'. But we have seen that the practice not only fulfils the rule, but it also gives it concrete shape in particular situations.
>
> (Taylor 1993: 57)

It follows that there is a major emphasis in theories of practice on the specificities of place. Particular contexts are crucial elements of the practical sense because dispositions have to be constantly tuned to the indeterminacy of each context, often in creative ways, so the 'rule' never stays quite the same. In other words place is constitutive of the subject's understanding of the world:

instead of denigrating Aristotle for his limited appraisal of the role of body in place, it would be more profitable to involve his idea of place's inherent power and say that a considerable portion of this power is taken on loan, as it were from the body that lives and moves in it. For a lived body energises a place by its own idiosyncratic dynamism, intersecting that place's ideological character. If we were to begin to think in this direction, our understanding of place itself – place as lived and imagined and remembered – would gain by deepening.

Just as there is no place without body – without the physical or psychical traces of body – so there is no body without place. This is so whether we are thinking of body in relation to its own proto-place, its immediately surrounding zonal places, its oppositional counter-places, its congenial commonplaces, or in relation to landscaped regions as configurated by such things as landmarks and lakes, towns and trees. For the lived body is not only locatory ... it is always already implaced.

(Casey 1993: 103–4)

It is quite clear that this kind of thinking about the subject's understanding of the world, with its emphasis on the flow of practice, embodiment, joint action and situatedness produces its own epistemological stance. Most particularly, theories are seen as highly provisional 'tool-kits', temporary constructs providing different images of the world. This is consonant with the general attempt in theories of practice to get away from the intellectual bias of so much social theory, which tends towards the objectifying gaze (Game 1991), associated with seeing the world as a set of significations to be interpreted, towards theory which grasps the world as a set of situated concrete problems to be solved practically (and which, as a number of commentators have pointed out, is not so different in many of its features from certain kinds of North American pragmatism or from Haraway's (1991) idea of situated knowledges). Nowadays, these theories of practice have taken on a wide variety of forms, not all of which it is possible to cover in a brief introduction.

However, five main schools are currently particularly well represented in the literature. Each of these we will consider briefly, namely the work of Bourdieu, the writings of de Certeau, the work by 'discursive' social psychologists like Harré and Shotter, the programme known as actor-network theory made popular by authors like Callon and Latour, and, perhaps surprisingly, the work of Deleuze. Bourdieu's work is in the tradition of Heidegger, Wittgenstein and Merleau-Ponty, all of whom he cites as intellectual mentors: 'Merleau-Ponty, and also Heidegger, opened the way for a non-intellectualist, non-mechanistic analysis of the relationship between the agent and the world' (Bourdieu 1990b: 10). In Heidegger, it is clear that 'everyday coping (primordial understanding as projecting) is taken over by each individual by socialisation into the public norm (the one) and this forms the clearing that governs people by determining what possibilities show up as making sense' (Dreyfus and Rabinow 1993: 37). But his description of social being is highly abstract. This is where Wittgenstein and Merleau-Ponty can be brought in.

> Heidegger is not interested in how the clearing – the understanding of being – is instituted and how it is picked up by individuals and passed along from one generation to the next. Wittgenstein, with his emphasis on forms of life, and Merleau-Ponty, with his description of the lived body, help us to see that Heidegger's ontology can be extended to the ontic realm – that is, the domain of social and historical analysis. To fill in being in the world one must see that what Heidegger is talking about are *social practices* (Wittgenstein) and that these practices are *embodied skills* that have a common style and are transposed to various domains (Merleau-Ponty). This makes possible an account of how durable and transposable bodily dispositions are appropriated and 'projected' back into the situation without appeal to conscious or unconscious representations. Such is Merleau-Ponty's account of embodiment, relating action and the perceptual field by way of an intentional arc. 'The life of consciousness – cognitive life, the life of desire or perceptual life – is subtended by an "intellectual one" which projects round about us our past, or future, or human setting, or physical, ideological and moral situation, or rather which results in our being situated in all these respects' (1962: 136).
>
> (Dreyfus and Rabinow 1993: 38)

In effect Bourdieu's notions of *field* and *habitus* ground these ideas. Thus a social field is a domain consisting of a set of objective relational configurations between positions based in certain forms of power. Each field prescribes its own particular values and possesses its own regulative principles which agents struggle to change or to preserve. The habitus 'anchors' the social field. The formal definition of the habitus is:

> the strategy generating principle enabling agents to cope with unforeseen and ever-changing situations ... a system of lasting and transposable dispositions which, integrating past experiences, functions at every moment as a matrix of perception, appreciations and actions and makes possible the achievement of infinitely diversified tasks.
>
> (Bourdieu 1977: 72, 95)

Less formally, the habitus is a kind of embodied unconscious:

> habitus reacts to the solicitations of the field in a highly coherent and systematic manner. As the collective individual through embodiment or the biological individual 'collectivised' by socialisation, habitus is akin to the intention in action of Searle or to the 'deep structure' of Chomsky except that, instead of being an anthropological invariant, this description is a historically constituted, institutionally grounded, and thus socially variable generative matrix. It is an operator of

rationality, but of a practical rationality inherent in a historical system of social relations and therefore transcends the individual. The strategies it 'manages' are systemic, yet ad hoc because they are 'triggered' by the encounter with a particular field. Habitus is creative, inventive, but within the limits of its structures, which are the embodied sedimentation of the social structures which produced it.

(Bourdieu and Wacquant 1992)

In other words, there is an 'ontological complicity between habitus and the social field' (Bourdieu 1990b: 194). Or as Dreyfus and Rabinow (1993: 38) put it even more succinctly, 'our socially inculcated dispositions to act make the world solicit action, and our actions are a response to this solicitation'.

Michel de Certeau (1984, 1986) has written on Bourdieu. He praises Bourdieu's 'ethnological' work on the everyday practices of the Kabyle and the Béarnais but is unable to find the same kind of subtlety in Bourdieu's 'sociological' work on the French education system, where the subtle energies of habitus are absorbed in a complex but still recognisable reproduction model. Most particularly, he points to the way that Bourdieu throws a blanket 'over tactics as if to put out their fire by certifying their amenability to socio-economic rationality or as if to mourn their death by declaring then unconscious' (de Certeau 1984: 59). Perhaps this is because of Bourdieu's need for an

> other (Kabylian or Béarnian) which furnishes the element that the theory needs to work and 'to explain everything'. This remote foreign element has all the characteristics that define the habitus: coherence, stability, unconsciousness, territoriality ... It is represented by the habitus, an invisible place where, as in the Kabylian dwelling, the structures are inverted as they are interiorised, and where the writing flips over again in exteriorising itself in the form of practices that have the deceptive appearance of being free improvisations.
>
> (de Certeau 1984: 58)

De Certeau's answer to this dilemma is interesting. It is to concentrate on the importance of tactics by emphasising the importance of *space*. De Certeau tries to surmount the problem of Bourdieu's implicit rejection of the tactical properties of practices by emphasising how space intervenes both in constituting tactics and in forming the other. Thus, 'a tactic insinuates itself into the other's place, fragmentarily, without taking it over in its entirety, without being able to keep it at a distance' (de Certeau 1984: xix). For de Certeau practices are always spatial-symbolic practices which can be discerned via spatial-symbolic metaphors like walking, pathways and the city. Through the movements of the body and the powers of speech which jointly provide the possibility of converting one spatial signifier into another the subject (now a walker) is able to call up transformative tactical resources. New places and meanings, 'acts and footsteps', 'meanings and directions' are produced and they produce liberated spaces that can be occupied. A rich indetermination gives them ... the function of articulating a second poetic geography on top of the geography of the literal, forbidden or permitted meaning. They insinuate other routes into the functionalist and historical order of movement.

(de Certeau 1984: 105)

Space intervenes in another way too, in the production of narratives. For de Certeau,

narrative structures have the status of spatial syntaxes. By means of a whole panoply of codes, ordered ways of proceeding and constraints, they regulate changes in space (or moves from one place to another) made by stories in the form of places put in linear or interlaced series ... More than that, when they are represented in descriptions or acted out by actors (a foreigner, a city dweller, a ghost), these places are linked together more or less tightly or easily by 'modalities' that specify the kind of passage leading from the one to the other.

Every story is a travel story – a spatial practice. For this reason, spatial practices concern everyday tactics, are part of them, from the alphabet of spatial indication ('It's to the right', 'Take a left'), the beginning of a story the rest of which is written by footsteps, to the daily 'news' ('Guess who I met at the bakery'), television news reports ('Teheran: Khomeini is becoming increasingly isolated ...'), legends (Cinderellas living in hovels), and stories that are told (memories and fiction of foreign lands or more or less distant times in the past). These narrated adventures simultaneously producing geographies of actions and drifting into the common places of an order, do not merely constitute a 'supplement' to pedestrian enunciations and rhetorics. They are not satisfied with displacing the latter and transposing them into the field of language. In reality, they organize walks. They make the journey, before or during the time the feet perform it.

(de Certeau 1984: 115–16)

In the latter parts of his career, de Certeau emphasised these spatial stories as a vital constituent of the other, specifically through consideration of practices of Empire and Colonisation.

Many of de Certeau's objections to Bourdieu have been answered in other ways. Most specifically, there are responses from the fields of social psychology and sociology. In social psychology, there is the social constructionist tradition which is, in effect, an attempt to foreground Wittgenstein's background, and to provide an account of its motive forms of life, seen as clusters of material and symbolic practices:

a contrast is drawn between cognitivist approaches to language, where texts, sentences and descriptions are taken as depictions of an externally given world, or as realisations of underlying cognitive descriptions of that world; and the discursive approach where versions of events, things, people and so on are studied and theorised primarily in terms of how those versions are constructed in an occasioned manner to accomplish social actions.

(Edwards and Potter 1992: 8)

Modern social constructionism claims a number of forebears. There are the Russian psychologists like Vygotsky, Luria and Volosinov. There are those who have taken up the pragmatist tradition of Dewey and Mead. There are philosophers like Wittgenstein and, more recently, Foucault and Taylor. In social psychology, the chief proponents of social constructionism have been Harré,[8] Gergen[9] and Shotter[10] who have all propounded what has come to be known as a discursive or dialogical psychology (see, for example, Parker 1992, Edwards and Potter 1992). This newer form of social constructionism depends on four important principles. Most importantly of all, it concentrates on the third space 'between' the individual psyche and the abstract systems of principles which supposedly characterise the external world. This is the space of everyday social life, a flow of responsive and relational activities that are joint, practical-moral and situated in character. This is the space of 'joint action' in which 'all the other socially significant dimensions of interpersonal interaction with their associated modes of subjective or objective being, originate and are formed' (Shotter 1993b: 7).

Second, social constructionists assign a crucial role to the use of language, not as a communicative device for transmitting messages from the psyche or social structures, but as a rhetorical-responsive means of moving people or changing their perceptions. Thus, in social constructionism the account of language that is offered is 'sensuous' – language is a communicational, conversational, dialogical means of responding to others – and

all of what we might call the person-world, referential-representational, dimensions of interaction at the moment available to us as individuals – all the familiar ways we have of talking about ourselves, about our world(s), and about their possible relationships which in the past we have taken as in some way primary – we now claim must be seen as secondary and derived, as emerging out of the everyday, conversational background to our lives.

(Shotter 1993b: 8)

Thus, and third, social constructionism is clearly a highly situated view of human life. Moreover, situations 'exist as third entities, between us and the others around us' (Shotter 1993b: 9). Yet, remarkably, geographers have (with the exception of Thrift 1986) never really drawn on the approach.

Fourth and finally, in the social constructionist account, cognitive abilities are constantly being formed in joint action. These abilities hail certain kinds of persons and not others and thereby produce a 'political economy of developmental opportunities' (Shotter 1984). Thus,

in *Personal Being*, I developed an account of the nature of persons in which they are seen as the products of the imposition of the structures of language on the natural endowments of the 'general animate being'. Among the most salient of these endowments are consciousness awareness, agentive powers and recollection. I simply assume that these features of the infant are capacities it has by virtue of a developing nervous system. But to become a person the infant's natural endowments must be synthesized into a coherent and unified structure. It is the great achievement of Vygotsky to have realised that this system … comes about by the acquisition of both symbolic and practical skills in symbiosis with more competent members of the infant's immediate circle. In particular, conscious awareness becomes self-consciousness, agency becomes moral responsibility, and recollection becomes the ordered memories of an autobiography through the acquisition, above all, of ways of making indexical reference to self and others, in short the pronoun system and its equivalents.

(Harré 1993: 6)

In sociology, some of these same kinds of ideas about language as practice and practice as language have gained currency, but in a different and more expansive form: actor-network theory.[11] Actor-network theory uses the metaphor of the network to consider how the social agency is constituted. The provenance of actor-network theory is poststructuralism by symbolic interactionism out of recent philosophers of science. As Law has it,

the provenance of actor-network theory lies in poststructuralism: the vision is of many semiotic systems, many orderings, jostling together to generate the social. On the other hand, actor-network theory is more concerned with changing recursive *processes* than is usual in writing influenced by structuralism. It tends to tell *stories*, stories that have to do with the processes of ordering that generate effects such as technologies, stories about how actor-networks elaborate themselves, and stories which erode the analytical status of the distinction between the macro and micro-social.

(1994: 18)

Actor-network theory has three main characteristics. First, agents – which can vary in size from individual subjects to the largest organisations – are treated as relational effects. Second, however, agents are not unified effects. They are contingent achievements. Many of the stories of actor-network theorists recount 'how it is that agents more or less, and for a period only, manage to constitute themselves. Agency, if it is anything, is a precarious achievement' (Law 1994: 101). Third, the social world is fragmenting. It is a set of more or less related bits and pieces which are the result of endless attempts at ordering, some of which are currently relatively successful, some of which are currently the equivalent of the faded silk flowers in the attic. The 'social' is the outcome of this 'recursive but incomplete performance of an unknowable number of intertwined orderings' (Law 1994: 101).

Clearly, achieving agency requires the mobilisation of all manner of things and this is probably where actor-network theory makes its most original contribution (Thrift 1994b). In actor-network theory things other than human agents – like tools and texts – are given their due, with two main results. First, and as a matter of principle, actor-network theory recognises networks as collectivities of all manner of 'actors' which all contribute in their way to the achievement (and attribution) of agency. In other words, actor-network theorists argue for a 'symmetrical anthropology' which is more likely to recognise (and value) the contribution of the non-human by shifting our cultural classification of entities. Latour (1993) goes so far as to argue for the necessity of a new constitution which will complete 'the impossible project undertaken by Heidegger' (Latour 1993: 67), both by correcting Heidegger's archaic bias, and also by restoring the share of the 'anthropological matrix' of actors other than human agents which has been lost. Thus, says Latour,

> all collectives are different from one another in the way they divide up beings, in the properties they attribute to them, in the mobilisation they consider acceptable. These differences constitute countless small divides, and there is no longer a great divide to take one apart from all the others. Among these small divides, there is one that we are now capable of recognising as such, one that has distinguished the official version of certain segments of certain collectives for three centuries. This is our constitution, which attributes the role of nonhuman to one set of entities, the role of citizens to another, the function of an arbitrary and powerless God to a third, and cuts off the work of mediation from that of purification.
>
> (1993: 107)

It is this constitution that Latour wants to say farewell to. He wants a new constitution that recognises hybrid or 'variable geometry entities', which restores 'the shape of things' and which redefines the human as 'mediator' or 'weaver'. Second, and following on from this latter point, because things are so intimately bound up in the production of networks that will last and spread, actor-network theory conjures up the idea of a world where 'the human' must be redefined as highly decentred (or as reaching further) and as unable to be placed in opposition to the non-human: 'the human is not a constitutional pole to be opposed to that of the nonhuman' (Latour 1993: 137). Thus, some of our most favoured dualities – like Nature and Culture or Nature and Society – fall away to be replaced by new hybrid representations and new ethical considerations:

> the human is in the delegation itself, in the pass, in the sending, in the continuous exchange of forms. Of course, it is not a thing, but things are not things either. Of course, it is not a machine, but anyone who has seen machines knows they are scarcely mechanical. Of course, it is not in God, but what relation is there between the God above and the God

below ... Human nature is the set of its delegates and its representatives, its figures and its messengers.

(Latour 1993: 138)

The kind of vivid, moving, contingent and open-ended cosmology that Latour and other actor-network theorists are trying to conjure up is perhaps most closely approximated by the work of Deleuze. Deleuze is not often thought of as a theorist of practice but we could claim that his work fills in important gaps in extant theories of practice: as he tries (with Guattari) to write a baroque theory of practice; one which, like Deleuze's notion of subjectivity, is full of swirls and whorls, pleats and folds: 'not ... an essence but rather ... an operative function' (Deleuze 1993b: 3).

Deleuze offers a number of insights for theories of practice. First, he produces a theory of practice out of an almost entirely different theoretical bloodline. His mentors include a recast Bergson (who enables Deleuze to displace consciousness with its function of casting light upon things by a new field of 'nomadic' singularities, intensive magnitudes which are preindividual and prepersonal), a reworked Spinoza (who provides an ethology of striving passions that can energise this field), a refitted Nietzsche and Foucault (who enable Deleuze to reflect on how subjectivity is constructed from the internalisation of 'outside' forces *without* reproducing a philosophy of interiority) and, latterly, a renovated Leibniz (who provides an account of the constitution of the 'individual').

Second, Deleuze concentrates, most especially via Spinoza and Nietzsche, on qualities of force and affect that have sometimes been neglected in other theories of practice that we might call, after Brennan (1993), the 'energetics' of 'activity, joy, affirmation and dynamic becoming' (Braidotti 1994: 164). Most particularly, that means that life is refigured as a slip-sliding flux of intersecting and impersonal forces. This allows Deleuze to rework ideas of the body, thinking and the self. Thus, the body becomes a 'complex interplay of highly constituted social and symbolic forces. The body is not an essence, let alone a biological substance. It is a play of forces, a surface of intensities; pure simulacra without originals' (Braidotti 1994: 163). Thinking also becomes an interplay of forces. Deleuze brings to the fore

the affective foundations of the thinking process. It is as if beyond/behind the propositional content of an idea there lay another category – the affective tone, level of intensity, desire or affirmation – that conveys the idea and ultimately gives it its value. Thinking, in other words, is to a very large extent unconscious, in that it expresses the desire to know, and this desire is that which cannot be adequately expressed in language, simply because it is that which sustains language.

(Braidotti 1994: 165)

Thus, the self becomes both disjunctive and nomadic, a highly variable speaking stance attuned to Deleuze's basic message, 'everything in the universe is encounters, happy or unhappy encounters' (Deleuze and Parnet 1988: 79).

Third, Deleuze produces a radically different idea of subjectivity, one which privileges intensity, multiplicity, productivity and discontinuity, one which is pitted against Lacan's negative vision of desire as lack, and one which hunts down all notions of interiority 'in search of an inside that lies deeper than any internal world' (Deleuze 1993b: 125). One might argue that what is left is simply the classical poststructuralist subject without much subject but this would be unfair. It would be more accurate to write that, like Latour, Deleuze wants to redefine 'human' around a new ethical constitution:

> in the wake of Spinoza's understanding of ethics, ethics is conceived of as the capacity of action and passion, activity and passivity; good and bad refer to the ability to increase or decrease one's capacities and strengths and abilities. Given the vast and necessary interrelation and mutual affectivity and effectivity of all beings on all others (a notion, incidentally, still very far opposed to the rampant moralism underlying ecological and environmental politics, which also stress interrelations, but do so in a necessarily prescriptive and judgemental fashion, presuming notions of unity, wholeness, integration and cooperation rather than, as do Deleuze and Guattari, simply describing inter-relations and connections without subordinating them to an over-arching order, system, or totality), the question of ethics is raised whenever the question of a being's, or an assemblage's capacities and abilities are raised. Unlike Levinasian ethics, which is still modelled on a subject-to-subject, self-to-other, relation, the relation of a being respected in its autonomy from the other, as a necessarily independent autonomous being – the culmination and final flowering of a phenom-enological notion of the subject – Deleuze and Guattari in no way privilege the human, autonomous, sovereign subject; the independent other; or the bonds of communication and representation between them. They are concerned more with what psychoanalysis calls 'partial objects', organs, processes, and flows, which show no respect for the autonomy of the subject. Ethics is the sphere of judgement regarding the possibilities and actuality of connections, arrangements, lineages, machines.
>
> (Grosz 1994: 196–7)

Let us take up Deleuze's sense of practice; where subjectivity is the folding of the outside into the inside, and the past into the present, for the sake of thinking the future; where the situated subject acts, and is acted upon, by numerous lines of force; where the self is a 'slow' inside space that is multiple, productive and continuous; where encounters are both exterior and interior.

ENCOUNTERS WITH OTHERS

The allegories of the map discussed so far – positionality, movement and practices – set out the modalities through which subjects come to place themselves into power-ridden, discursively-constituted, practically-limited, materially-bounded identities. The subject assumes, in both senses of the word, an identity on the basis of commonality with others and yet that subject, in both senses of the word, assumes that they are an individual: unique, sovereign. The formation of the subject also takes place, and fails, within the field of encounters with others – but this field is striated with simultaneous, different power relations. Some anecdotes will help illuminate these rather dense introductory remarks: each will be set within a context which sheds light on the question of mapping subjectivity in the spaces between the conflictual and incoherent self and the incommensurable and indissoluble other. There are five case stories.

(1) A man is sitting on a park bench, he is alone. Nothing stands in the way of the man's presumption that the park is there for him to look at. His eye can roam over the landscape without challenge, nothing disturbs his power to look at whatever pleases him. The man is at the centre of his world – he owns what he sees and, in this scene, he is also self-possessed because nothing upsets his thoughts. This 'megalomania' is shattered, however, by the intrusion of another into the park. The lord and master of all he surveys has suddenly become off-centred – for he has become the object of another's gaze. In this encounter, the lines of power have become reoriented: the man no longer controls the scene, lines of power converge on the intersubjectivity between the two people and between them and the scene of the encounter.

(2) Another man is sitting in a boat. He has decided to get away from it all for a day or two, to do something which does not require 'thinking': he has decided to go fishing. On this day, he is with some fishermen from a local village. The craft is frail and there is an element of danger, the man is enjoying sharing this danger with his fellow fishermen. The moment comes to pull in the nets, when one of the fishermen points to something floating in the sea. The object is sparkling as the sun mirrors off it. It is a sardine can, a can which once contained the kind of fish they were trying to catch. The fisherman cries to the man: 'You see that can? Do you see it? Well, it doesn't see you!' The fisherman found the incident highly amusing, the man in the boat however was disturbed. On thinking about it, he decided that the source of this anxiety was the fact that the fisherman was wrong: the can was, in fact, looking at him, but it was the fisherman who did not see him. In this encounter, the fisherman's joke highlights the fact that the man is out of place.

(3) A 4-year-old girl is sitting in a kitchen being fed by her mother when a window cleaner arrives at the back door. The mother leaves the child to talk to the man. The child is curious, she asks her mother about the man.

These questions reveal more than mere curiosity, however. The presence of this particular man has disturbed her, the girl is also confused, afraid and disgusted. While the man cleans the windows, the child whispers to her mother. A series of issues crop up in the conversation: why is the window cleaner cleaning windows? Why do people work for others? Why is he dirty, does he not wash? The girl is alarmed: the man is strange. To help understand this encounter, the mother provides the girl with a set of fictional narratives which describe the man's background in relation to their circumstances – involving paid labour, class relations and manhood – but the girl remains puzzled and her mother's answers never resolve her fear of this strange man looking through the window at her.

(4) A man is walking down the street, this is giving him some difficulty. A young child shouts out 'Look!' and points at him. He cannot suppress a quick smile as it flicks across his face. The child again exclaims 'Look!' The man is surprised, he is surprised that it amuses him again. Now the child cries 'Mama! I'm frightened!' The man is startled: frightened? He no longer feels amusement, laughter is impossible. The child was frightened of the difference between the two of them: the child was frightened of the man's body. The two of them were caught in a particular corporeal regime where one body is transparent, and the other visible; where one body is invisible, the other marked. In walking down the street, the man's body is surrounded by certainty (he certainly has the body he has) and uncertainty (how will people respond to the visible difference of that body?). In this encounter, the child and the man are placed in relation to one another, each responding to the same social map of power and meaning inscribed in the (fearful) body.

(5) A woman is trying to decide where to live: there are always difficulties in making this decision, but for her this problem is marked by the danger of the hostile encounter. Her perceptions of different places in the city are marked by the extent to which she feels she stands out, the extent to which she feels out of place. She has experienced violence in the past from neighbours – the tyres of her car had been slashed, rubbish thrown into her garden. She feels the need to choose a place to live where she thinks she can blend in, or a place which is tolerant, or where there are more women like her. The home gains significance in a dense network of social meanings, where her household is excluded, even reviled, by these values. In the everyday encounters with the other, this woman must be consciously aware that she is different, even where she looks the same. Even where encounters with strangers are rare, the danger of discovery carries the place of encounter beyond the front door; the home is not safe, the signifiers of difference have to be hidden from eyes that she has yet to meet.

Let us quickly dispense with the obvious: 'bodies are maps of meaning and power' (Haraway 1990: 222). These people have different bodies and they 'suffer' encounters with others in different ways. In each case, there is something insistent and excessive in the way the history of fleshly encountering appears to subject the subject. The subjection of the subject is instituted through the inscription of meaning and power through the never merely physical body: mastery, mind, skin, class, sexuality are systematically mapped onto the body of the same/other. The body becomes a point of capture, where the dense meanings of power are animated, where cultural codes gain their apparent coherence and where the boundaries between the same and the other are installed and naturalised (Douglas 1966; Butler 1990, 1993). It is now possible to specify these people's circumstances – in some cases, to name names. It is not that these individuals are paradigmatic, but that these stories reveal some of the ways in which the encounter maps the subject into discursively-constituted, embodied identities.

The first narrative is told by Jean-Paul Sartre (1943: 252–60), the second by Jacques Lacan (1973: 95–6), the third is related by Valerie Walkerdine and Helen Lucey (1989: 87–90; see also Walkerdine, this collection), the fourth by Frantz Fanon (1967: 110–15), and the last was revealed by the research of Gill Valentine (1993a: 397–400; see also Bell and Valentine, this collection). These stories reveal very different aspects of the encounter, but underlying them are a set of boundaries between the this-is-me and the that-is-not-like-me. The encounter provokes the subject into mapping subjectivity in a dual sense: the sovereign subject and the subjected subject. The bodies of these individuals become intensifying grids of meaning and power: the subject position of the one setting the frame for the meaning of the encounter with the other. It is now possible to work through these narratives to show how complex the vortices of meaning and power in the encounter can be.

The initial case tells of the existential crisis experienced by Jean-Paul Sartre. In the first moment Sartre feels himself free to dominate what he sees – at this point he does not fear the encounter, he is not threatened by the other. It is easy to speculate that his safety stems from his centrality to cultural norms: he is white, male, heterosexual and middle class. Alone, Sartre fantasises the control that his skin, gender, sexuality and class make invisible to him – this experience is then universalised. However, in the encounter, Sartre argues, the subject becomes an object for another and by this substitution vanishes as a subject: that is, the fantasy of mastery is revealed as an illusion in the encounter. Once Sartre was master of the world, now he is enslaved by the other. The subjection takes place through being mapped into a universal subject–object dichotomy: the encounter is equally annihilating for both people, as they look at one another.

Jacques Lacan's anecdote adds a further level of complexity to this story: his encounter involves another dimension – the exchange of meaning. Lacan is first disturbed by the blinking sardine can, second by the fisherman's laughter. For Lacan, the commonsense understanding of the situation is reversed: the sardine can looks at Lacan, while Lacan has become invisible to the fisherman. Installed at the heart of encounter is a two-headed primal terror: the somethingness (or agency) of the other and the dissolution of the self into nothingness. For Lacan, though, there are two kinds of 'other' in this encounter: first, the other as an object-for-the-self and, second, the Other as a moment in the exchange of meaning. The can acts in both senses: 'le can' as encountered defines Lacan and 'le can' is a moment of meaningful exchange between the fisherman and Lacan, where the fisherman identifies with the can in not seeing La-can. Where Sartre's encounter took place between two (presumed to be) universal subjects in an empty park, Lacan's encounter involves three figures (Lacan, the can and the fisherman) and takes place within the intersection between imaginary and symbolic exchanges between the three terms. For both Sartre and Lacan, however, the encounter is tragic: it annihilates, it terrorises.

The third story involves an encounter between two female members of a middle-class family and a working-class man. The story shows that the 4-year-old girl is already becoming a subject in relation to received maps of meaning. The girl cannot understand the material relations which bring the man to the outside of their home: she is fascinated – both captivated and terrified. Without being able to know the difference, she knows there is a difference: this man is paid by her mother, the man is dirty – she is not like this man, he is neither like her mother or her father. The terror of the encounter with the other within the field of power and meaning leaves the girl simultaneously fearing and desiring the difference, but not knowing the difference. The body is a surface of signification, where the boundaries between self-same and other-different become crucial in forestalling terror. The girl is mapped into (class) relations of power and meaning in a way that both leaves that privilege invisible and requires the softening of that terror through the (fictionalised) description of the other.

Another child is caught in the same trap: the trap of the other. This child sees Frantz Fanon walking down the street, this child proclaims that his skin is black: 'Mama, see the Negro!' The white child is fascinated – first captivated by the difference, then terrified by it, to the extent that the child needs to turn to its mother for support. Fanon is a marked man. This is not the same as the previous child's experience, however. We should note that there is a possibility of misreading these circumstances: this is not a straight reversal of these encounters. The girl and the black man are not interchangeable in these stories. While gender and race are fabrications which are inscribed on and through the surface of the body, they are situated differently in these encounters in relation to power and meaning.

The little middle-class girl identifies with her parents as she fears the other and desires not to be like the other; the black man – walking with difficulty down the street because he is trying to look white people in the eye – is continually forced to recognise himself as frightening and white people as better, civilised. The girl is asked to recognise the working-class man as different and to identify with her family; the black man is told to recognise himself as different but to identify with white power. The girl is allowed a place to be, but the black man is not permitted his place.

In the colonial situation Fanon dissects, the black man's visibility has a double effect: his skin allows him to be seen and marked as different (from whites), but it also separates him (from whites) in a way which makes him unknowable (to whites). As a strategy of colonial rule, the colonial master-subject separates and defines the colonial slave-subject, only to find that this makes the colonial slave-subject radically unknowable, because they have been differenced. In any case, the practice of authority which separates and defines the colonised suffers a double failure: separation fails both where the colonised identify with (supposed) civilisation which masters them and where the colonised define themselves as opposite to the coloniser, while the description of the colonised repeatedly stumbles over the fences of representation that the colonisers and colonised place between each other, in order that they should both know their place. The exchanges between coloniser and colonised involve the ambivalence of desire and fear, the failure of not only identification and anti-identification but also mutual misrecognition in the field of meaning, which amount to extraordinary efforts to police the boundaries between coloniser and colonised in and through the practice of power (see, for example, Spivak 1988).

So far we have stressed singular dimensions of power, each adding to the last: first, through an axis which assumes that the other is the same; second, through an axis which places the other within intersubjective exchanges; and, third and fourth, through axes which define the subject in terms of class and race, respectively. But the last example begins to unravel into a simultaneously fixed and dynamic situation: the ambivalence of power, the doubling of the effects of power, the ambivalence of the subordinate and the doubling of the effects of resistance are all implied in this situation, allegorically producing endless recombinations as other kinds of difference are brought into the picture. The body lies (in both senses of the word) at the centre of this allegory, a story implying other stories, of subjection and resistance.

The woman who is involved in the prosaic task of looking for somewhere to live is a lesbian. This woman is caught in two spaces at one and the same time: woman, lesbian. In the search for a home, she has a double life to lead. This woman is also middle class, a site of relative privilege. Such dimensions of subjectivity do not resolve themselves in one way: this woman may be 'out' as a lesbian, or she may be 'out' sometimes in some places, or she may not be 'out' at all (see Fuss 1991); indeed, she may not be 'a lesbian' all the time.

For example, her choosing to be among multiple sex partners engaging in a group bondage and whipping situation; though sometimes the scenario might include transsexuals (who might or might not be gay); or it might include gay men who might be 'clones' or 'queens' or whatever; sometimes the situation could emerge in terms of exchanging roles; or sometimes it could be a romp with two members of the same 'role', and so on, and so on, and so forth. Would this woman who had strayed from the path in such a manner, still be having a thing called lesbian-sex, if it included all these other referents? And what if she did most of these things half the time, but had a monogamous relationship (or none) during the other half? Should she be considered only half-a-lesbian? Which half?

(Golding 1993a: 215)

This home-hunter: a third woman, a third lesbian, a third middle-class professional? And what of age, skin, politics? Her subjectivity cannot be mapped onto a static, fixed, passive space, cartographies of the self cannot be plotted against socially-given bi-polar geometrics of power, such as male–female, straight–gay, bourgeois–proletarian, young–old, white–black, left–right and so on. Thus placed, the individual always escapes. This woman finds a home in a landscape which is replete with desire and danger, where she is mapped by others into complex positions of desire and disgust, and she maps herself into the world, into the fabric of the urban – in which she thinks, feels and acts – with desire and fear.

So, the subject is mapped, and maps, into interminable dimensions of power which subsist at all points, but the allegory of the map needs to be rethought, this map can no longer be thought of as simply a two-dimensional picture representing a specific interest in reality; instead the map becomes three-dimensional and fluid, on and through which bodies are the points of capture of multiple power relations, power relations which inhere in simultaneously real, imagined and symbolic encounters.

Encounters, then, appear to offer a tangled web of interactions between people as they are mapped into power-ridden discursively-constituted identities, where such interactions place individuals in complex positions in relation to power and meaning, where power and meaning are policed through bi-polar opposites, but where power and meaning cannot be contained by the violence of bisection. We might visualise this as an infinite number of spider's webs intersecting at infinite angles, each in dynamic relation within itself, where each change produces iterative changes else-where in the fable, fibril, febrile structure of the map of the subject. Such an interpretation overplays the fluidity of meaning and power and underplays the hard triangulations of mapping: the landscape of meaning and power is neither flat, nor static, nor isotropic.

REGIMES OF THE VISUAL

In the last section, five stories were told, each illuminating a different aspect of the encounter between the self and the other, but where the self and the other can no longer be understood in terms of that bi-polar opposite, they are always already mapped into other exchanges. Implicit in each of these stories was a kind of violence: implicit in these stories is another aspect of the allegory of the map: vision. In this section, the mapping of the subject will be seen to depend on power-ridden, discursively-constituted vision, where vision is far from neutral; concealed in cloaks of objectivity are unknown terrors.

First, we must dispense with the idea that vision is a cold, biological, universal fact – too many ambiguities exist in the discursive constitution of the field of vision for it not always already to imply power and meaning: (to) see and (the) see, (fore- and in-) sight (which sounds like site), the (mad) stare, the (casual) gaze, glancing (blow), vision (as spectacle, hallucination, manifestation or beauty). These words, which describe kinds of visual knowing, describe the way that the subject is naturalised and neutralised through practices of power which operate in, on and through the body. Without losing any of the sense of these ambiguities and ambivalences, or of their dynamism and incoherences, or of conflict between and within scopic regimes, visual practices fix the subject into the authorised map of power and meaning: it views the map and the subject, it is vigilant of boundary transgression and it is a vigilante wielding fear and terror. In and through dominant scopic regimes, sore-eyes are peeled which g(r)aze the self and the other.

We may now begin to disentangle the fable, fibril, febrile web of encounters by looking at their dependence on specific visual practices – using this insight, it is possible to specify particular ways in which the subject is mapped and where this mapping fails. This story will occasionally involve a play on the sounds-like of sight and site, saying that distance/depth is constitutive of authority and resistance, manifest and latent meaning. This analysis will examine two aspects of the scopic regime: first, the visual practice of seeing; and, second, the closure effects of that visual practice. It is now possible to look again at the situations described in the previous section.

Sartre's experience in the park speaks directly of the power of his sight to dominate the world that he sees. It is Sartre's presumption and fantasy that this visual practice is universal and universalisable that will fall foul of the encounter with the other. A particular visual practice structures this encounter, it structures what Sartre sees, what he feels about it and it is the contradictions within it that lead to his sense of annihilation: he cannot cope with the internal contradictions of the scopic regime which legitimates this visual practice. Broadly, Sartre views the world through rules of seeing which developed in the Renaissance and are codified in positivist science. This scopic regime, based as it is in positivism, valorizes the neutrality of seeing: the world is turned into a set of geometrical arrangements based on an abstract, fixed, universal, isotropic and material understanding of space (see Soja 1989: 124–5[12]); indeed, it is this 'space' which is properly presented in the generic map – a flat, supposedly all-seeing (if not all-showing), picture of (part of) the world. For Sartre, lines of power radiate out from his eyes over the geometrised field of vision: he captures the world in a fixed, disembodied and see-through gaze. It is hardly surprising, then, that he should feel threatened by the presence of an (inappropriate/d) other: Sartre is confronted by something this scopic regime denies, the (opaque and incommensurable) subjectivity of the (supposedly transparent and knowable) object. The lines of power that radiated out from his eyes now suffer interference from the lines of power that radiate out from the eyes of the other – Sartre can no longer presume the innocence of the gaze and this knowledge terrorises him. Lacan is also terrorised, but he starts off by examining the terror and he does not presume that the lines of power flow from the gaze of the centred-subject.

For Lacan, the gaze of the can and the fisherman captivate him, but this is a dynamic exchange of looks, which takes place within a scopic regime which defines them all, but this regime is ocularcentric in a different way. Sartre is the centre of his scene until he enters the gaze of another, Lacan is never the centre of his scene, he was always defined against another centre: the phallus defines the scopic regime.[13] Sexuality defines the field of vision (this position is most clearly outlined in Rose 1986). Lacan's ideas begin to undo a notion that space is somehow a passive backdrop against which bodies and subjectivity can be mapped – space looks back: space is dynamic and active: containing, defining, separating and naming many points of capture for power and meaning – for example, the can, the fisherman, regulatory practices such as language, the market and so on. Lacan's radical move is to place Desire as the reason to see, and to place phallocentrism as the structure of ways of seeing – in opposition to the presumption of neutrality and objectivity which Sartre cannot cope with – yet Lacan still presumes the Phallus as the signifier of power without explaining that power, and he presumes that the Phallus is one thing.

Lacan cannot account for other sites/sights of meaning and power: it is these that are suggested by the exchanges of looks in the other narratives. Those exchanges are all of a Lacanian order, yet Lacan's Phallus cannot account for the complexity of these situations. Instead, our analytical gaze must notice that lines of site/sight are oriented in each exchange of glances through lines of meaning and power, which can be specified in the hall of mirrors of the look. The little girl sees a dirty, working-class man; the child sees a black man; the black man sees a white mother and child; the lesbian woman sees a variegated and contested world, replete with desire and fear. Each dimension of seeing invokes differently a different kind of space between the person who looks and the object that looks back: there is a position, distance and an orientation to the look, which specifies a particular space of meaning and power: this space is neither isolated nor abstract; this space both contains and refuses an infinite number of invocations of meaning and power; this space is constitutive of the visual practice, it is staging and integrating the lines of power and meaning between the look and the look-back.

The quality of the look and the look-back can be defined still further; two sets of ideas will do for now: *first*, the fetish and the mirror, and, *second*, purity and the border. The first closure effects surround the desires that the viewer invests in the object that is looked at, while the second set of closure effects are marked by the fears that the subject feels when presented by the object.[14] First, then, the meeting of eyes may well be inscribed within the field of desire – hence the involvement of the fetish and the mirror in the mapping of the subject. Second, the exchange of glances may well provoke simultaneously unnamed fears (whether consciously or not). These ideas describe the closure effects of visual practices, such that power and meaning are understood as constructed neither on the firm ground of Truth nor on the flat ice-rink of Relativism. The final section of this introduction will suggest ways in which these closure effects fail, though in ways which tend to maintain that closure. Nevertheless, it is here that the possibilities of alternative topographies of the subject might be found.

The girl who sees the workman, the child who sees the black man, the lesbian who searches for home are caught into a look that wants to see something, something that is striated with geometries of fear. Paradoxically, the ambivalence and ambiguity, the incoherence and the conflict of the visual transaction constricts, integrates and names the exchange within the regulatory practice of vision. In each of the transactions, an object stands in the place of desire (as a fetish) and the object mirrors something of the viewer back (as a speculum): the workman, the black man, the home. The viewer wants to look at the object: for the children the desire is structured by visual codes of difference – dirt, work, skin – whereas for the woman the home stands for a place to live and love; mirrored back are fragments of these individuals' identity: kaleidoscopes of class, race, sexuality. Simultaneously, the fetish and the speculum are bounded and the object is 'purified' of association with the self, thus purifying the self (see Douglas 1966).

In a colonial context, Homi Bhabha talks ironically of 'the unknown territory mapped neatly onto the familiar' (1986: 73); but this mapping is partial. Colonial discourse creates 'a place for a "subject people's" through the production of knowledges in terms of which surveillance is exercised and a complex form of pleasure/unpleasure is incited' (1986: 75). The fetish and the mirror speak of the viewer as do the boundary and purity – it cannot speak of the alterity of the other. In order to ensure the safety of the viewer the object must be turned into something familiar, but this defence is radically unable to deal with the strange: the (un)seen other is placed as fetish and phobia.[15]

In the case of the children, they are instantly captivated by difference but come to recognise their confusion over the object – they have then to locate and name that difference in order to preserve that difference, where that difference must be absolute – or else they are in danger of dissolving into the other, so 'I am a girl, you are a man'; 'I am a child, you are an adult'; 'I (through identification with the mother) pay, you work'; 'I am rich, you are poor'; 'I do not tolerate dirt, you are dirty'; 'I am white, you are black'. The woman must tread carefully between parallel binary codifications of difference: she must blend in and not be seen as different, but she must also be able to be different – to be a lesbian; she must develop a sophisticated reading of the urban – its potential as a sight of pleasure, its danger as a site of intolerance – in order to lead her (never less than) double life.

Refractions of the body's location within the map of subjectivity are momentarily displayed, fixed and codified within the authorised map of meaning and power through the transaction of vision. Visual practices are regulatory, they demand that certain things are noticed, that other things are denied, and that other things are not seen at all. Codified in the aesthetics and ethics of meeting someone's eye/I, the scopic regime still remains a scene of ambiguity, uncertainty and conflict: transfixed by the interrogating gaze, people shuffle their feet and look away – to different places.

AESTHETICS/ETHICS – TO DIFFERENT PLACES

Let us conclude this chapter by returning to the map – this time as a fetish, a speculum, a bounded and purified re-presentation of mapper, mapping and mapped.

> 'Mirror', 'window', 'objective', 'accurate', 'transparent', 'neutral': all conspire to disguise the map as a ... *representation* ... of the world, disabling us from recognizing it for a social construction which, with other social constructions, brings that world into being out of the past and into our present.
>
> (Wood 1992: 22)

The practices of visual representation of the map serve to disguise the power that operates in and through cartography. Maps are not empty mirrors, they at once hide and reveal the hand of the cartographer. Maps are fleshly: of the body and of the mind of the individuals that produce them, they draw the eye of the map-reader. Maps are framed, marked with text, simplifications, fabrications. They raise to visibility, behind the map, around the map, in the map they consign invisibility. The map does not simply itemise the world: it fixes it within a discursive and visual practice of power and meaning; and, because it naturalises power and meaning against an impassive and neutral space, it serves to legitimate not only the exercise of that power but also the meaningfulness of that meaning.

This narrative can be extended to cover the practices of subjectivity and the body: the individual struggles to place themselves within regimes of power and meaning, but that struggle is naturalised in and through the spatial and temporal practices of the I/eye. The map and the subject portray truth: they seem to be what they seem to be – but they are always more than this. The map and the subject are neither a cover-up nor meaningless, but nor are they the truth of the matter or the centre of meaning. People incorporate and display maps of meaning and power into the practice of their body and subjectivity as a kind of masquerade, which (only) seems to be a self-grounding identity.

The individual is mapped as a subject through the practices of the body and subjectivity; practices which come to be seen as natural through spatial referents, such as position, movement, practice, encounter, vision (and aesthetics, as we will see); spatial referents which are 'natural' because space is understood as a passive, objective, neutral backdrop to thought, feeling and action. Space then appears to provide a self-grounding reality for identity.

However, the case studies of encounter show that space cannot be thought of in this way. People struggle to achieve the ability to make their appearance blend in, in different ways in different places, under the scrutiny of the gaze and graze of the other, under the self-scrutiny of the mirror. People mask the shards of their identities that threaten to expose them: revealed in those moments where people are not sure how to behave, or what other people think of them, or where people suddenly feel self-conscious, or alienated and so on. People map themselves into socially-sanctioned regulations of body and self – but they do so only imperfectly: people are not chameleons.

The mapping of the subject, then, continually reveals ruptures, tears, fraying, an inside-out. The map and the subject masquerade as something that they are not entirely: every day they put on their (brave) face to fit their bodies into those surfaces of power and meaning with which they are presented but which extend far beyond them. The mask/drag, that people use to get them through the day, is a veil which continually threatens to be torn away by the violence of the other, as Fanon found, as the lesbian fears. Identity is a fiction which must be continually established as a truth. Indeed, the practice of authority is revealed in the moment where identity is considered as a truth and forgets that it has been authored at all: hence, the attraction of identity politics as a way of establishing the legitimacy of alternative bodied subjects.

In a somewhat different practice of representation than mapping, the film-maker Trinh Minh-ha puts it this way:

> In short, what is at stake is a practice of subjectivity that is still unaware of its own constituted nature (hence the difficulty to exceed that simplistic pair, subjectivity and objectivity); unaware of its own continuous role in the production of meaning (as if things can *make sense* by themselves, so that the interpreter's function consists only of *choosing* among the many existing readings); unaware of representation as representation (the cultural, sexual, political inter-realities involved in the making: that of the filmmaker as subject; that of the subject filmed; and that of the cinematic apparatus); and, finally, unaware of the Inappropriate Other within every 'I'.
>
> (1991: 77)

We need only substitute 'the map' for 'the film', or to read the film as one simultaneously symbolic and imaginary exchange amongst many, or to change 'the subject' for 'the flesh' (in Merleau-Ponty's sense of simultane-ously body and mind), to see that some people's place in the world is more precarious than others. The map – as our allegory of power and knowledge – and the subject – as our allegory of the body and the self – reveal identity: its fluidity and fixity, its purity and hybridity, its safety and its terrors, its transparency and its opacity. The map – as allegory of space-time – and the subject – as allegory of place-in-the-world and limit-of-the-world – reveal that 'space' is actively constitutive of the practices of authority and resistance, of grounding meaning and re-placing meaning.

Mapping the subject, then, leads in three interrelated directions simultaneously: first, towards redrawing the old maps in ways which acknowledge their authority and authorship and in ways that delegitimate the claims to truth of those maps which rely on an unspoken universal and universalised subject; second, towards the resymbolisation, resignification and parodic repetition of the maps that we already have; and, third, towards new maps of the subject – and even throwing away maps altogether – in order to re-establish tolerance towards different practices of body and subjectivity.

We should not be under any illusions that just thinking new possibilities for practices of the body and subject will somehow undo the regulatory and oppressive maps of meaning and power. A new body politic will not be instituted the first time an English heterosexual male academic geographer turns up at the annual conference of the Royal Geographical Society wearing bright red lipstick, nipple clamps and a crotch-length lycra skirt (if this has not happened already). On the other hand, it is difficult to imagine new cartographies of the body and the self, of power and meaning, without continually revamping and re-placing the subject.

Mapping the Subject, then, is a determinedly partial activity, charged with subversion and resistance as well as meaning and authority. Mapping the Subject is a triangulation of power. Mapping the Subject is an ethics of wanting to know, not knowing and not wanting to know. Mapping the Subject is a contested ground, fixed through position, movement, practice, encounter, visuality. And Mapping the Subject is a masque. Mapping the Subject is a Necessary, Passionate Fiction.

NOTES

1   See Probyn (1990, 1993); Game (1991); and Rich (1986b).
2   See, especially, Carter (1987, 1992); Pratt (1992); and Blunt (1994).
3   See Crary (1990); Diprose and Ferrell (1990); Tagg (1988); and Lalvani (1993).
4   See, for example, Deleuze (1986, 1989).
5   See, for example, Virilio (1984, 1991, 1994).
6   See Springer (1991); Boddy (1994); and Doane (1993).
7   See Bukatman (1993b); Doane (1993); and Boddy (1994).
8   See, for example, Harré (1979, 1991, 1993); Harré, Clarke and de Carlo (1985).
9   See, for example, Gergen (1991).
10 See, for example, Shotter (1984, 1993a, 1993b).
11 See Callon (1986, 1991); Latour (1986, 1991, 1993); and Law (1994).
12 This argument draws on Soja's argument concerning the two illusions of space in contemporary social theory – 'the illusion of opacity' and 'the illusion of transparency' – but reorders this view by bringing both illusions under the same (contradictory) scopic regime. This shows that scopic regimes should be thought of as fields of conflict between different visual theories and practices (see also Jameson 1991).
13 Lacan's notion of the Phallus is extremely complex and contested within the psychoanalytic literature and elsewhere (see Lacan 1958a and b; Mitchell 1974; Irigaray 1977; Mitchell and Rose 1982; Elliott 1992). It is important to note that the Phallus is not the penis, but a signifier of power to which only men can claim to have access: thus, in Lacan's account of sexual difference, men have the Phallus though this is a myth, while women must be the Phallus, they must reflect men's (false) belief that they possess the Phallus. For Lacan, the woman does not exist, femininity is a masquerade. Thus, this notion links to Lacan's notion

of the mirror and masquerade, which will appear later in this chapter. Lacan has been accused of biological determinism (despite apparently separating penis and Phallus, they remain inseparable) and linguistic determinism (because all signification, and thus all meaningful exchanges between people, are centred on the Phallus alone), nevertheless this debate displays rather greater ambivalence and sophistication than can be found in the geographical literature (see Pile 1993; Pratt 1994).

14   Except where specified, the word object is used in the psychoanalytic sense to refer to anything which becomes the focus of the psyche, be it a person, an ideal, a fantasy, a word, an act, a thing and so on, where objects are never things-in-themselves (see Frosh 1987).

15   This account draws on Freud's notion of the fetish, in which the fetish object acts as a cover for anxiety, where the fetish disavows anxiety related to (sexual, racial and so on) difference, by protecting against the necessity of recognising alterity (see Freud 1927a). Fanon, of course, also drew heavily on psychoanalytic theories of identity and non-identity in subject constitution in what Bhabha describes as 'the grotesque psychodrama of everyday life in colonial societies' (Bhabha 1986: 71).

Part I

CONSTRUCTING THE SUBJECT

PART I

INTRODUCTION

The first part of mapping the subject takes as its stance the attitude that the subject is constructed and that, in order to demonstrate this contention, histories of the subject must be provided. The chapters in this part of the book therefore provide their own 'history' of the subject. Each chapter is simultaneously theoretical and empirical, general and specific. Each chapter wishes to place the subject, but this 'placing' is not presumed to be either merely and exclusively about someone's real location or free of metaphorical, imagined and symbolic significance.

Miles Ogborn takes a painting by Velázquez, *Las Meninas*, as his starting point. What interests him is that two of the key theorists of the modern subject both deploy an analysis of this painting in order to talk about the ways in which the subject can be known. Ogborn demonstrates that there are key similarities and differences between Michel Foucault's and Norbert Elias' theories of the subject. Interestingly, he argues that 'Foucault's concern with spatiality can be contrasted with Elias' concentration on temporality'. Both Foucault and Elias are committed to establishing a notion of subjectivity which is contingent on the power relations within which people are placed. In this sense, it is possible to argue that the sense of being an individual is an effect of these relations.

Perhaps one way in which 'we' think of ourselves as being an individual is the sense that 'we' are unique, that 'we' have something inside us that distinguishes us from everyone else. *Carolyn Steedman* is interested in the ways in which this 'interiority' became emblematic in the presentation of childhood subjectivity in the fiction of the eighteenth and nineteenth centuries. Theories of subjectivity often focus upon how subjecthood is achieved during the child's development, so Steedman starts by examining key theoretical discourses in this period. This examination provides a context within which it is possible to interpret the representation of 'the child's understanding of its own body and its own internal spaces' in fiction. Steedman takes three stories of childhood as her case studies: Goethe's *Wilhelm Meister* (1795–6), Andersen's 'The Snow Queen' (1844) and Charlotte Brontë's *Jane Eyre* (1847). She demonstrates that in each case the metaphor of 'the map' is used to describe the child's mind. The thinking behind this metaphor suggests, though, that the options available to children are limited because they are bound to follow the routes that the map provides. This placing of the child therefore permits certain kinds of development, but closes off others: thus, these stories 'describe an understanding of a self made in relation to the cold hard facts of life itself, which it is the task of the children in these tales to learn'. In practice, stories written for children provide adult maps of the child, which the child will do well to learn to read.

These landscapes of right living echo across time with those described by *David Matless*. While Steedman is concerned with the ways in which geographical and spatial metaphors are used to 'figure out' the child, Matless explores the ways in which geography, landscape and in-the-landscape activities are themselves figured 'out' in different, sometimes conflicting and competing, discourses: aesthetic, intellectual, spiritual, moral, physical, political. He explores the intersection of discourses of environmentalism, subjectivity and Englishness which together were intended to map out correct patterns of behaviour: thus, Matless clearly shows that 'versions of self [are] embodied in and made through historical geographical practice'. In particular, the writings of preservationists, planners, 'ramblers' and geographers are elucidated in order to show the kind of programmes that were being developed to promote a new kind of Englishness, to turn people into fit subjects.

Finally, *David Sibley* looks at the spaces of the child, examining the limits and boundaries that are 'placed' on their behaviour. This work resonates with the previous chapters in so far as it deals with the ways in which the child is meant to learn appropriate adult behaviour and thereby think of her or himself as an individual. Sibley predominantly uses adults' reconstructions of their past childhoods to look at children's social spaces and the ways in which they feel about being in certain places. To begin with, it is argued that 'children experience things acutely in a physical sense': thus, the world is constructed by thinking through the body and the body becomes a prime way of orienting the self in relation to both people and places. Rather than simply mapping feelings such as aversions, anxieties, pleasures and desires, Sibley is concerned

> to make sense of the personal geographies of childhood, focusing on the experience of boundaries, those demarcating the pure and the defiled (including the places that make you feel sick) and those markers we use to carve up time, like bedtime, playtime, getting-home-by time.

These boundaries permit/drive/help subjects to map themselves into social and geographical space, but in highly complex and dynamic ways. In particular, object relations psychoanalysis is drawn on to show how other people and the spaces of home and locality set limits on children's development, as the child tends to divide people and places into opposing categories of 'good' and 'bad'. The point is that these markers are then carried – often unconsciously – into adult life, providing a set of (usually implicit) patterns for subsequent behaviour.

3

KNOWING THE INDIVIDUAL

Michel Foucault and Norbert Elias on *Las Meninas* and the modern subject

Miles Ogborn

[E]s verdad, no pintura.
(Palomino 1724)

INTRODUCTION: LOOKING AT *LAS MENINAS*

Las Meninas,[1] the name that has been given to the painting that hangs in pride of place among the Velázquez collection in the Prado Museum, was not the name that was given to it by the artist. That name has been lost. The canvas, painted in 1656, shows (see Figure 3.1) a group of people in the Cuarto Bajo del Príncipe of the Alcázar Palace in Madrid.[2] Starting on the left it shows Diego Velázquez himself in the livery of a courtier with the red cross of a knight of Santiago on his chest.[3] He is in the process of painting a canvas stretched on a frame, the back of which takes up a substantial portion of the left-hand side of the picture (so substantial that many reproductions take the liberty of trimming it back). He has stepped back from the canvas, his brush is poised over his palette as he gazes at his model, engaging our eyes. To his left are composed several groups of people. At the centre is the Infanta Margarita, 5 years old when the picture was painted, her body turned slightly to one side, her head to the other, and her eyes meeting ours. On her right kneels María Augustina Sarmiento who is offering the Infanta a red jug of perfumed water on a silver tray. She offers us her profile. On her left stands Isabel de Velasco dipping as if in a curtsey and looking towards us with her head inclined towards the Infanta. Behind her, and in the shadows, stand Doña Marcela de Ulloa, the guardmujer de las damas de la reina and, beside her, an unidentified guardadamas. She is in conversation, he stands as if in prayer, his face indistinct. In front of them is a curious trio. The dwarf Marí-Barbola stares impassively at us, her left hand drawn up as if to nudge the midget Nicolasito de Pertusato who is rousing a supine dog with his dancing left foot. Behind all of these characters, silhouetted in a well-lit doorway stands José Nieto Velázquez, the aposentador, or palace marshal, to the Queen. He is looking back into the room and towards us. Between him and

Figure 3.1 Las Meninas by Diego Velázquez

Diego Velázquez, on a wall hung with dark paintings of mythical scenes,[4] hangs a mirror which, lit by the sunlight coming in from the windows on the right and through the doorway at the back of the room, shows, as if in a portrait, the reflections of King Philip IV and his second wife Queen María Ana. *Las Meninas* has been called 'Velázquez's claim to immortality' (Brown 1986: 259) and has prompted a huge range of reactions from artists and critics.[5]

This painting interests me here because it appears on the flyleaf of Michel Foucault's *The Order of Things*[6] and on the cover of Norbert Elias' *Involvement and Detachment*.[7] It is this connection, and their subsequent discussions of the painting, that I want to deal with here. This is not done in order to comment upon the painting itself or upon the many interpretations made of it by art historians, but to elaborate a concern with ways of theorising the subject, ways of writing its histories and its geographies, which concern both Foucault and Elias. Thus, *Las Meninas* provides a terrain upon which to debate the similarities and differences between Foucault and Elias in their discussion of a crucial figure: the modern subject – that subjectivity or selfhood characteristic of western modernity. Anthony Cascardi has described this as a 'vision of the self as subject, as ideally disengaged from the processes of nature and history, and as standing over both of these in a posture of confident self-possession' (Cascardi 1992: 63). This is a subject whose interrogations of interiority and subjectivity, and of the relationship between the self and the world, are conducted in terms of reason and non-reason, and lead it to feel divided between them. It is at once the 'fully integrated, cognitive and rational' subject which has dominated much of the discourse about the subject within the human sciences (Pile 1993: 122), and a feeling, emotive being. Crucially it is a subject which, despite being presented as universal, is gendered male and located in the West. This does not mean that it is simply understood as a rational *cogito*, but that it is understood as hierarchically split between the rational, which is gendered male, and the unthought, or unknown, which is gendered female and understood as non-Western (Rose 1993a; Torgovnick 1990). I shall return to these issues after setting out the positions taken by Foucault and Elias (although they should be borne in mind throughout what follows). In considering these positions Robert van Krieken has pointed out an initial similarity since both share a 'basically similar concern with the social history of subjectivity' (van Krieken 1990). Yet in setting out their projects I will demonstrate many differences before returning to similarities again.

THE REPRESENTATION OF REPRESENTATION: FOUCAULT ON *LAS MENINAS*

The Order of Things is an exercise in what Foucault called the archaeology of knowledge. It is an account of the transformations in ways of understanding since the Renaissance which proceeds via a 'spatialisation' of knowledge, setting out the 'epistemological space specific to [each] particular period' (Foucault 1970: xi), and, by degrees, showing what had gone before and what came after. As such it is an analysis governed by attention to the differences in the 'rules of formation' of knowledge in different periods, and the discontinuities between what Foucault then called epistemes (Foucault 1970: xi). Three of these were identified. First, a system of knowledge characterised by 'resemblance' which lasted until around the end of the sixteenth century. Here knowledge proceeded by reading the 'one vast single text' (Foucault 1970: 34) that nature presented through 'signatures' which made manifest relationships of connection, emulation, analogy, sympathy and antipathy. Second, the Classical episteme where the sovereign place of resemblance was usurped by an analysis of the identities and differences between things, and the possibility of mapping words onto things to produce a perfect classification. Third, the Modern episteme – beginning at the end of the eighteenth century – in which language becomes opaque as 'Man' takes centre stage and knowledge is organised in terms of historicity. Each of these epistemes is discussed in terms of the knowledges of language, living things and economic activity that they made possible, and the reader is jolted through the rapid and massive shifts which lie between their incompatible fields. Indeed, Foucault did not set out to explain these changes in the manner of more conventional histories of ideas or social scientific studies of science. By demonstrating their contours he aimed to illuminate their differences, to set each off against the other, revealing their contingencies and certainties to be as ludicrous as the impossible epistemological spaces of Borges's Chinese encyclopaedia.[8]

This analysis of knowledge opens with a 'bit of bravura, undoubtedly added at the last moment . . .' (Eribon 1991: 155), Foucault's discussion of *Las Meninas*. This is used to map out the epistemological space of the Classical episteme which lies at the heart of the book. Thus it sets a first marker of the differences between the three modes of knowing. For Foucault the Classical age is characterised by a new relation between words and things which is best apprehended through the notion of *taxinomia* and the construction of classificatory tables which are, quite literally, new epistemological spaces frequently made real in 'the general grid of differences' set out, for example, in herbariums, collections and gardens (Foucault 1970: 145, 131). Things are to be known and identified not through resemblances and signatures but through the 'identities and differences' (Foucault 1970: 50) which set them alongside but apart from other things. Things are, within the grid-lines of the classificatory table, 'what the others are not' (Foucault 1970: 144). Representation is crucial to this since it defines the new relationship between words and things such that 'language has withdrawn from the midst of beings themselves and has entered a period of neutrality and transparency' (Foucault 1970: 56). Its crucial role is naming, because 'to name is at the same time to give the verbal representation of a representation, and to place it in a general table' (Foucault 1970: 116). This is, therefore, an age of utopian visions of perfectly transparent languages able to perfectly represent the world. Particular forms of representation are also crucial since what is named is now only the visible. Medicinal or magical properties are no longer part of knowledge: it is a knowledge of 'surfaces and lines' rather than 'functions or invisible tissues' (Foucault 1970: 137). What is crucial about this episteme is that the only thing that cannot be included within the table is the classifier, or the act of classification. It is this notion that lies at the heart of Foucault's discussion of *Las Meninas* which he takes to be a rendering of the impossibility of the representation of the act of representation.

All the spaces of *Las Meninas* play out the implications of this epistemological impossibility. Following Foucault, I want to discuss this in terms of three relationships. First, the things depicted on the plane of the painting.

Second, the positioning of Velázquez's portrait of himself. Third, and most crucial here, the relationship between what lies inside the frame and what lies outside.

For Foucault the plane of the painting acts as a grid, or table, depicting representation. Indeed, he refers to the painting as the 'representation of ... Classical representation' (Foucault 1970: 16). This is constructed in terms of a spiral which loops clockwise around the elements of the picture. We are taken from the painter's gaze, via the back of the canvas, the paintings hung on the rear wall, the mirror, the man in the open doorway, the paintings on the right-hand wall which are visible only in sharp perspective and, finally, the light from the window which reconnects us to the painter's eyes. This he understands as depicting 'the entire cycle of representation' (Foucault 1970: 11): the material tools – the gaze, the palette and brush, the blank canvas. The representations – the paintings, the reflection in the mirror, the real man. The dissolving of representations – the pictures visible only in terms of their frames, and the light from outside which is also the condition of all representation as it touches the paintings and the brow of the painter. Thus the cycle is complete and never-ending.

What happens on the canvas is only part of the story. The rest is concerned with the impossibility of representing the act of representation within the frame of Classical representation. The positioning of Velázquez's self-portrait is read in these terms. For Foucault the figure in the painting is caught between standing back from his canvas allowing us to see him, or stepping forward to paint and slipping from our sight: 'As though the painter could not be seen at the same time on the picture where he is represented and also see that upon which he is representing something. He rules at the threshold of those two incompatible visibilities' (Foucault 1970: 4). This introduces the problematic spatial relationships between what lies within the painting and what stands outside it. Foucault identifies two 'centres' within the painting – the eyes of the infanta and the mirror on the back wall. Together these define a 'symbolically sovereign' (Foucault 1970: 14) point outside the picture which is asked to contain three observing functions: the model's gaze, the spectator's gaze and the painter's gaze (as he paints the picture being contemplated). This is 'the starting point that makes the representation possible' (Foucault 1970: 15) and it must always be both invisible – in that it is the point from which the representation is 'con-structed' – and visible in terms of the ways it is 'projected' within the representation (Cosgrove 1985). As Foucault says, 'It is an uncertain point because we cannot see it; yet it is an inevitable and perfectly defined point too' (Foucault 1970: 13). It is a point defined by the representation and defining it – 'A condition of pure reciprocity' (Foucault 1970: 14).

Foucault traces the epistemological implications of the relationship between these 'observing functions' and the gazes, reflections and silences present within the picture to reveal this 'uncertain point' as a site of instability. He comments upon the invisibility of the spectators. We are not represented in the painting and we cannot see ourselves. The artist in the picture only looks at us in that he is looking at his model, and we cannot tell our role since the canvas on which he is painting is invisible to us: 'subject and object, the spectator and the model, reverse their roles to infinity' (Foucault 1970: 5). In addition, the mirror is clearly crucial to connecting the 'inside' and the 'outside', the visible and the invisible. It makes the sovereigns the centre around which the depiction is organised, but only in so far as they are invisible. Their visibility (captured in the mirror) is a frail and indistinct form of reality, and we must also recognise that the position outside the painting that is allotted to them is also 'that ambiguous place in which the painter and the sovereign alternate, in a never-ending flicker' (Foucault 1970: 308). The mirror is also duplicitous in other ways. At first it promises 'that enchantment of the double that has been denied us' – a straightforward representation (Foucault 1970: 7). Yet, although it is the only representation which 'fulfils its function in all honesty and enables us to see what it is supposed to show' (Foucault 1970: 7) it is not looked at by anyone in the picture and, in turn, it represents nothing that is in the picture.[9] Second, the mirror acts to restore what is lacking in various gazes – the painter sees his model, the king sees his portrait, the spectator sees the real centre of the scene which 'he' has usurped. However, the mirror hides as much as it reveals – why is there no reflection of Velázquez painting the picture? Why is there no reflection of us, the spectators? These sagittal lines will always be incomplete, just as the cycle of representation is complete. It is in the nature of Classical representation that something is missing:

> It may be that, in this picture, as in all the representations of which it is, as it were, the manifest essence, the profound invisibility of what one sees is inseparable from the invisibility of the person seeing – despite all mirrors, reflections, imitations, and portraits ... [R]epresentation undertakes to represent itself here in all its elements, with its images, the eyes to which it is offered, the faces it makes visible, the gestures that call it into being. But there, in the midst of this dispersion which it is simultaneously grouping together and spreading out before us, indicated compellingly from every side, is an essential void: the necessary disappearance of that which is its foundation – of the person it resembles and the person in whose eyes it is only a resemblance. This very subject – which is the same – has been elided. And representation, freed finally from the relation that was impeding it, can offer itself as representation in its pure form.
>
> (Foucault 1970: 16)

It was only in the elision of the viewing, classifying, representing subject that representation could appear as if transparent and that Palomino could say of *Las Meninas*, 'this is truth, not painting' (Palomino, in Moffitt 1983: 271).

Las Meninas is, however, not simply the epitome of Classical representation, it also hints at what is to come. In tracing the differences between the Classical and Modern epistemes we are brought face to face with the painting once more in a manoeuvre which serves to show us the nature of the threshold between the Classical age and Modernity, and something of the nature of the modern subject:

> [M]an appears in his ambiguous position as an object of knowledge and as a subject that knows: enslaved sovereign, observed spectator, he appears in the place belonging to the king, which was assigned to him in advance by *Las Meninas*, but from which his real presence has for so long been excluded. As if, in that vacant space towards which Velázquez's whole painting was directed, but which it was nevertheless reflecting only in the chance presence of a mirror, and as though by stealth, all the figures whose alternation, reciprocal exclusion and interweaving, and fluttering one imagined (the model, the painter, the king, the spectator) suddenly stopped their imperceptible dance, immobilised into one substantial figure, and demanded that the entire space of representation should at last be related to one corporeal gaze.
>
> (Foucault 1970: 312)

Thus representation loses its position as the locus of truth – that position is taken by 'man's' consciousness. 'Man' becomes both subject and object: 'he' takes the place of the king and combines that role with that of model, spectator and creator. Things are no longer to be understood in terms of the identities and differences that they manifest in representation but in terms of 'the external relation they establish with the human being' (Foucault 1970: 313). In turn, 'Man' 'soon realises that what he is seeking to understand is not only the objects of the world but himself' (Dreyfus and Rabinow 1982: 28). This position – as both the foundation of knowledge, and as what is to be known – produces 'Man' as what Foucault calls a 'strange empirico-transcendental doublet' (Foucault 1970: 318), a figure which, within the search for truth, is the site of the tensions and connections between the empirical and the transcendental, of 'subject' and 'object'. This, in turn, is mapped by Foucault onto the relationship between the modern *cogito* and the unthought whereby each is seen to be dependent upon the other. For Foucault modernity is marked by a figure that stands at the centre of knowledge, who is both knower and known, and whose being 'is deployed in the distance between' the thought and the unthought (Foucault 1970: 327). It is *Las Meninas* that he uses to show us this radical new arrangement.

THE DETOUR VIA DETACHMENT: ELIAS ON *LAS MENINAS*

Norbert Elias' *Involvement and Detachment* chimes in many ways with Foucault's *The Order of Things*. In it Elias sought to set out a theory of knowledge, particularly the knowledge of human societies, which was dependent upon understanding its history in particular ways. As with all of Elias' work the argument is bound into his wider corpus, into the arguments developed in the two volumes of *The Civilizing Process* and in *The Court*

Society (Elias 1978, 1982, 1983). As in all his work the history and theory that Elias develops is profoundly developmental and profoundly processual, a tale of societies moving from one stage to another. *Las Meninas* fits in as an illustration of this story. Yet we start a long way from seventeenth-century Spain, since Elias' immediate concern is with the Cold War. He asks why it is that although humans have generated a large volume of knowledge, and specific ways of knowing, which give them a high degree of control over nature, they are unable to generate similar levels and types of knowledge in terms of the relations between human societies. This is a question that he answers via a discussion of the relations between knowledge and danger which serves to introduce the two key concepts of the book: the 'double-bind figuration' and the 'involvement and detachment balance'. As he says:

> The stronger the hold of involved forms of thinking, and thus of the inability to distance oneself from traditional attitudes, the stronger the danger inherent in the situation created by people's traditional attitudes towards each other and towards themselves. The greater the danger the more difficult it is for people to look at themselves, at each other and at the whole situation with a measure of detachment.
>
> (Elias 1987: xiv–xv)

The 'double-bind figuration' is a vicious circle. Elias identifies the relationship between the two super-powers in these terms, but more generally it refers to the situation of human societies which are operating with 'involved' knowledges: because of the danger levels inherent in their situation they are unable to create the detachment which would produce knowledge able to help them reduce the danger levels inherent in their situation! This, he argues, was previously the situation with regard to people and non-human nature. It is now the situation with regard to people and human societies.

Elias' concern for 'detachment' is not simply a plea for the human sciences to be more like the natural sciences, indeed he specifically rejects this aim. It is, instead, part of a concern which fills Elias' sociology: the need to think in terms of long-term developmental processes concerning the interdependencies between people and the power balances which they involve. This focus gives Elias' sociology a strong normative sense. In this case he argues that knowledge processes have a *direction* (regression or advance) that can be indicated by the involvement–detachment balance. Indeed, much of the book is dedicated to showing that the 'advance' from non-scientific to scientific knowledge about the non-human world is part of a change in the involvement–detachment balance. Prescientific societies are seen to be in a situation of greater 'involvement' where 'no ontological differences exist in human experience between the relation of human groups with each other, with animals and plants, or with earthquakes and thunderstorms ... [T]hey perceived the world as a society of spirits' (Elias 1987: xxvi). Elias argues that such people's questions about the world take the form: what does it mean for us? Such knowledge does not allow the danger levels posed by nature to be reduced which, in turn, does not allow more 'detached' knowledge to be produced.

In contrast, taking the scientific 'detour via detachment' means beginning to ask questions like 'What is it?' and 'How are these events connected with others?' (Elias 1956: 229). This, Elias argues, is less emotionally satisfying, because less ego-centred. It does, however, extend the 'security zone' of control over larger areas of life. Elias understands this change as an improvement, an advance. His framework is unashamedly developmental:[10] societies move from 'a smaller and less consistently reality-orientated fund of knowledge' (Elias 1987: 57) towards a 'more realistic and detached fund of knowledge' (Elias 1987: 52). For Elias all knowledge is to be understood as part of an historical process, a learning process. This process is, however, far from inevitable. There is no necessary advance but, instead, we must investigate a complex history which involves the relationships between selves, societies and nature.

Quite what is at stake here can be seen in Elias' discussion of a major 'spurt' of detachment in the relationship between people and nature which he identifies in the Renaissance. This act of detachment, particularly the shift from a geocentric to a heliocentric vision of the universe, 'requires very special conditions and a social attitude in individuals which includes a relatively high level of stable self-restraint all round' (Elias 1987: xxxviii). The self-restraint necessary for self-distancing or detachment makes the link between the 'involvement–detachment balance' and modes of subjectivity or selfhood. As Elias says, in using these terms

> one is referring to human beings including their movements, their gestures and their actions no less than their thoughts, their feelings, their drives and their drive control. One is referring, in short, to their self-regulation, including that which is regulated. Basically the two concepts refer to different ways in which human beings regulate themselves.
>
> (Elias 1987: xxxiv)

This self-regulation is, in turn, tied up with wider societal processes of state formation and the monopolisation of the means of violence which are explored at length in *The Civilizing Process*. Here Elias stresses the inseparability of self formation and state formation via a conjoint discussion of manners and the formation of monopolies of taxation and violence (Elias 1978, 1982; Ogborn 1991). In this historical sociology it is within the European court societies that these links between knowledge, selfhood and societal figurations and processes are most developed. These locations are seen as crucial to the formation of modern subjectivities and modern state forms (Elias 1983) and it is in this theoretical and historical context that Elias interprets *Las Meninas*.

For Elias the notion of a 'developmental' perspective is as important in art as in any other arena of life. He rejects what he calls a 'necklace model' (of knowledge or art) which would mean that 'each culture has to be considered as a human manifestation in its own right' (Elias 1987: xli) with no sense of advance or regress. Instead he argues for a 'staircase model' where there is 'a clearly recognisable sequential order of ascent or descent' (Elias 1987: xl). There is, he argues, no necessity driving societies up the staircase, but they do have to pass through each floor. This does not mean that any work of art can be considered as better or worse than any other, but Elias does suggest that they can be treated as social facts and that a 'staircase' can be constructed (according to the logic of the involvement–detachment balance) in which representations can be judged 'more realistic' (Elias 1987: xli) or, deploying a term he uses in relation to science, more 'reality-congruent' (Elias 1987: xix).

In this way he constructs a history of art, which is at the same time a history of knowledge, according to the notions of involvement and detachment. Thus, preperspective painting is not an attempt at realistic depiction but aimed to directly involve people in religious experiences (Latour 1988). The development of perspective in sixteenth-century Italy through the mathematical calculations of Masaccio and Uccello was, therefore, a move towards greater detachment. It 'contributed to the feeling of a really existing distance between the viewer and the painted event, between subject and object' (Elias 1987: xlvii), substituting the experience of gazing upon aestheticised objects for the emotional involvement which had gone before. Its further development involved using mirrors to convert three dimensions to two dimensions and to show people as they are normally seen by others, another 'spurt' of self-distancing. Yet all of this is not simply an exercise in increasing detachment. There is what Elias calls a 'secondary involvement' (Elias 1987: lii) which refers to a re-engagement, to the joy of painting and the arts (or artifices) of composition. It is within these processes that *Las Meninas* is understood as 'a particularly striking illustration of the complexities of the involvement–detachment balance' (Elias 1987: lii).

For Elias, *Las Meninas* certainly demonstrates a move of the balance towards detachment:

> In the development of European painting, it is one of the earliest pictures in which a painter paints himself painting a picture. It is thus a good illustration of a step on the road towards greater detachment ...
> He took a step towards perceiving himself more clearly as he might be perceived by others, a step towards distancing himself from himself.
>
> (Elias 1987: lxi)

Yet this does not simply make the painting in some way more 'realistic' than those which had gone before. Elias argues that 'a further act of detachment' had moved artists like Rembrandt and Velázquez beyond the reproduction of 'objects as they knew them to be or as they appeared to be if they were always seen in the same light' (Elias 1987: lix) towards the depiction of variations of light and shade in order to capture the 'animation' or 'openendedness' of a face or figure. Thus the picture is not a series of posed portraits. People, including Velázquez himself, are painted as if they were unobserved. It is this enigmatic openendedness that attracts Elias to *Las Meninas*: the secondary attachment which accompanies detachment.

Elias' understanding of *Las Meninas* centres on Velázquez' depiction of himself. He interprets the painting as a private moment between Velázquez and Philip IV. It is a sign of the painter's devotion to the monarch, a picture painted for the royal apartments and not for 'an anonymous public' (Elias 1987: liii).[11] Velázquez, he argues, has depicted himself as he would have wanted to be seen by others, particularly the King, as a member of a particular group of people: the small inner circle of the court. Elias is keen to set this group in historical context to make plain its specific character: the separate households of king and queen; the elevation of the sovereigns above all others; and the rules of etiquette which governed their contacts with those below them (Elias 1983). He indicates how Velázquez' seemingly informal picture is organised according to the strict hierarchies of court society. The mirror is read as a device which enabled the artist to resolve a problem of representation peculiar to court society: how to depict members of the court so as to show that their lives revolved around the royal couple when he could not depict the royal couple together with persons so inferior to them in rank (the mirror is a solution which required a substantial act of detachment). Velázquez also used light and scale to represent the hierarchy of persons in the picture. His own full-length self-portrait, which indicates his high standing in the favour of the King, is matched by the smaller but better-lit figure of his opposite number in the Queen's household. It is clear that Velázquez 'knew his place': 'The full light plays on the figure of the Infanta, while his own figure stands more in the shade' (Elias 1987: lxv). Thus in painting *Las Meninas* Velázquez was

> occupied with the problem of the painter's peculiarly divided con-
> sciousness, as someone who stood outside, who observed the world
> and formed pictures of it in his own mind, and who, at the same time,
> was also very much part of this world – who was, in a word, detached
> and involved at the same time.
>
> (Elias 1987: lxviii)

While Elias presents this 'balance' as the painter's position he also wants it to be understood as historically specific. As has been noted before he sees the crucial site of the formation of such a consciousness as the court society whose intricate interdependencies and power balances, all negotiated by the King, were part and parcel of the development of new forms of subjectivity. The intense observation of others to gauge their social status, to pick up any hints of advance or regress in social position, are coupled with 'a specific form of self-observation' (Elias 1983: 105) – seeing oneself as others would see you. This, in turn, is part of the transformations in the self-controls extended over affective impulses – the civilising process – also characteristic of this period and understood alongside the processes of the making of state territorial power within which the court played a crucial part. This self-observation is precisely the manoeuvre of detachment that Elias sees

Velázquez taking in producing *Las Meninas*. Moreover, what is important about this for Elias is its direct link to more philosophical understandings of the subject. He argues that the Cartesian *cogito* is also to be understood in these terms:

> This shift towards an increased consciousness of the autonomy of what is experienced in relation to the person who experiences, towards a greater autonomy of 'objects' in the experience of 'subjects', is closely related to the thickening armour that is being interposed between affective impulses and the objects at which they are directed, in the form of ingrained self-control.
>
> (Elias 1983: 252)

This, for Elias, is the modern subject.[12] It experiences the world and itself through a series of separations and distances. The separation between 'subject' and 'object', between the 'inner world' of the individual and the 'external world', a gap which is experienced and understood as filled with reason (or loss) and which leads this subject towards particular forms of knowledge of the world. The separation between the individual and society, 'ego' and 'other', which subjects experience, for good or ill, as autonomy in a disenchanted world (Elias 1978: 258; Cascardi 1992). The separation between drives or affective impulses and mechanisms of self-regulation, experienced as the differentiation of 'the true self, the core of individuality' (Elias 1978: 258) from what is learned, socially imposed or produced by self-regulation through reasoned reflection. These separations and distances, each dependent upon the other, are shown us by Elias' reading of *Las Meninas*.

THE DIFFERENCES OF THEORY

At the most basic level there are similarities between the methods that Elias and Foucault deploy. Or, at least, there are similarities between the rhetorical strategies through which their 'histories' are presented. They both look back into the past in order to demonstrate its differences from the present. Apart from the works under direct consideration here, this can clearly be seen within Elias' work on time or manners, and Foucault's work on punishment, madness or sexuality (Elias 1978, 1992; Foucault 1967, 1977, 1979). From that basis, however, their methods diverge. Their ways of explaining how we got from 'then' until 'now' are very different. For Elias the question revolves around what long-term social process, or interweaving sets of social processes, connect then and now. In contrast, Foucault poses the question: when and where was the discontinuity (or discontinuities) between then and now, and what did it look like? Their material is also very different. Foucault concerns himself with the discursive construction of a variety of worlds, Elias stresses the materiality of the changes that concern him. Finally, Foucault's concern with spatiality can be contrasted with Elias' concentration on temporality. Each of these differences is outlined at greater length below.[13]

Continuity and discontinuity

Their differing uses of continuity and discontinuity can be traced between the work of these two theorists, as well as within their works. Foucault made many important and dramatic shifts in position and methodology across his work, most notably from the earlier archaeologies to the later genealogies, although it should be noted that his interest in notions of 'the subject' continued, albeit in different form (Foucault 1982). Any discontinuities in Foucault's work are not, for him, a concern. As he says in a much quoted passage: 'Do not ask who I am and do not ask me to remain the same: leave it to our bureaucrats and our police to see that our papers are in order' (Foucault 1972: 17). Elias, on the other hand, remained remarkably consistent over an exceptionally long working life. His intellectual biographers and synthesisers have been able to stress that his later works were part of the same project as his earlier ones (Mennell 1992).

This difference is also clear in their forms of analysis. Elias' discussion of the modern subject weaves together a series of long-term processes: the civilising of manners; the extension of webs of interdependence across wider spans of space and time; the extension of human control over nature; and the formation of territorially monopolistic states. These have no distinct beginnings but, although reversals may be identified, there are trends, or directions of change, that can be established. Where there are discontinuities – the end of feudalisation, or the end of a double-bind figuration – these are not abrupt moments and Elias seeks to understand them in terms of other long-term processes. Thus no point of emergence can be identified for the modern subject, only its gradually coming into being suspended and interwoven within Elias' processes and figurations. In contrast, Foucault rejects continuity – particularly the continuity of reason:

> The order on the basis of which we think today does not have the same mode of being as that of the Classical thinkers. Despite the impression we may have of the almost uninterrupted development of the European *ratio* from the Renaissance to our own day ... all this quasi-continuity on the level of ideas and themes is doubtless only a surface appearance; on the archaeological level, we see that the system of positivities was transformed in a wholesale fashion at the end of the eighteenth and beginning of the nineteenth century.
>
> (Foucault 1970: xxii)

His mode of understanding presents the rapid and complete transformations of ways of knowing (epistemes). The modern subject emerges abruptly with a reconfiguration of the field of knowledge. It only gives the impression that it has been around for a long time.

'Reality-congruence' and 'discursive construction'

Some of the differences evident in the interpretations of *Las Meninas* presented by Elias and Foucault are part of a wider distinction between 'historical' and 'philosophical' approaches to the modern subject (Cascardi 1992) which are, in turn, mirrored in debates over *Las Meninas* between art historians (Brown 1986; Alpers 1983). Elias' contextual reading, in terms of questions of courtly status, contrasts with Foucault's concern with modes of representation (although this ground is also approached by Elias). What underpins these is a theoretical difference between Elias and Foucault. Elias, as is made clear in his sociological theory of knowledge, has a definite scepticism about any claims that there is nothing independent of conscious-ness. Instead he stresses the irreducible interdependence of people and things, argues that knowledges can be judged in terms of their 'reality-congruence', and makes definite claims about what exists (Mennell 1992). Foucault, while not reliant on a construction of the world in consciousness, seeks to show us the ways in which the world we know, the 'reality' that can be said to exist, is made and remade in very different ways within different discursive formations.

Temporality and spatiality

Elias' concern with long-term processes and, in the case of *Las Meninas*, the shift in the involvement–detachment balance, gives more attention to temporality than spatiality. Foucault's concern to identify epistemes, and to describe their contours and their implications, prioritises 'spatiality' – at least as metaphor – over temporality. For example, he offers no explanation for changes over time, and pursues no analysis of temporal transformations within epistemes. The two forms of analysis may be found compatible. A concern for spatiality can be found in Elias' notion of the figuration (Ogborn 1991), and there is a concern for temporality in Foucault's later genealogies. Foucault's epistemes might also be ranked on the basis of Elias' measuring stick. This, however, would run against the grain of each analysis since the thrust of Elias' case lies in demonstrating evidence of 'advance' or 'regress', while Foucault's lies in a critical relativism, setting epistemes and notions of the subject side by side like pages in the Chinese encyclopaedia.

These differences could be set alongside the differences that exist in their notions of the modern subject – Elias' argument that it is problematic because of a continuity with Cartesian notions as opposed to Foucault's sense that it represents a break with Descartes, and Elias' notion that the subject and object are understood as separated versus Foucault's more complex sense of the 'empirico-transcendental doublet' – to argue that, despite the similarity of their initial concerns, their projects are opposed and unconnected. There is, however, a substantial area of intersection that can be identified, and one which demonstrates the similar critical intentions of their quite different projects.

COMMON GROUND: THE SOCIAL THEORY OF INDIVIDUALISATION

Both Elias and Foucault are destroyers of myths. Their work operates by taking what may seem to be universal or commonsensical (for example, imprisonment or the handkerchief) and revealing it to be formed within a complex history and a dense network of social relations. In many ways their work revolves around attention to an enduring notion within western modernity – the notion of the individual – upon which are built political systems, philosophical schemas, social theories and social relations. In many ways, and despite the differences outlined above, they treat this figure in the same fashion. What I want to do here is to show how Elias and Foucault critique the notion of the modern subject – or, in their own terminologies, 'man' or *homo clausus* – through a conjoint historicisation and contextualisation which serves to undermine the legitimacy of this figure and its domination of the field of understanding by showing the conditions of its creation within discourse for Foucault and figurational processes for Elias.

Norbert Elias and the illusion of *homo clausus*

In *Involvement and Detachment* Elias presents his sociological theory of knowledge in direct opposition to philosophical theories of knowledge (which he associates with Descartes, Husserl and Sartre among others) which assume that the human acquisition of knowledge is always and everywhere the same, and suggest that it can be understood in terms of the ways in which a single adult knower relates to the world. This asocial understanding, and denial of the learning process, is, for Elias, ridiculous. As he says of Descartes: '*Cogito ergo sum*. What can be more absurd!' (Elias 1987: xviii). This challenge is also important because of the dominance of such notions in social theories, both academic and everyday, about the relationships between the individual and society. He argues that these are dominated by the figure of *homo clausus* – the closed man or the closed personality – an 'image of the individual as an entirely free, independent being, a "closed personality" who is "inwardly" quite self-sufficient and separate from all other people' (Elias 1978: 247). An image of the individual as separate from society.

This figure is inimical to Elias' figurational, or process, sociology and he sets out to undermine it by tracing its history in order to show that what is experienced as real is the product of particular sociological circumstances:

> The detachment of the thinking subject from his objects in the act of cognitive thought, and the affective restraint that is demanded, did not appear to those thinking about it at this stage as an act of distancing but as a distance actually present, as an eternal condition of spatial separation between a mental apparatus apparently locked 'inside' man, an 'understanding' or 'reason', and the objects 'outside' and divided from it by an invisible wall.

> (Elias 1978: 256)

The creation of a world of 'subjects' and 'objects', and the creation of Reason and the reasoning subject, are part of the same social relations as the civilising processes of court society. Indeed, 'it is these civilisational self-controls, functioning in part automatically, that are now experienced in individual self-perception as a wall, either between "subject" and "object" or between one's own "self" and other people ("society")' (Elias 1978: 257). The historical, or sociological, determinants of this mode of subjectivity can be revealed, and the philosophies and social theories which take it as universal can be rejected, by showing how

> this new form of self-consciousness was linked to the growing commercialisation and the formation of states, to the rise of rich court and urban classes and, not least, to the noticeably increasing power of human beings over non-human natural events.
>
> (Elias 1991: 97–8)

This returns us to the central theme of *Involvement and Detachment*, the need for 'reality-congruent' knowledge about human societies. For Elias we are living under the illusion of *homo clausus* and misunderstanding the workings of the world. We are like the members of court society who 'perceive[d] their self-constraint, their armour and masks and the kind of detachment corresponding to them, not as symptoms of a particular stage of human-social development, but as eternal feelings of unchanging human nature' (Elias 1983: 243). The feeling 'may be entirely genuine' (Elias 1983: 253) but it only exists as a feeling, a perception. If we live our lives by it then certain consequences follow under conditions of increasing social and spatial interdependence:

> No one is in charge. No one stands outside. Some want to go this way, others that. They fall upon each other and, vanquishing or defeated, still remain chained to each other. No one can regulate the movements of the whole unless a great part of them are able to understand, to see, as it were, from the outside, the whole patterns they form together. And they are not able to visualise themselves as part of these larger patterns, because, being hemmed in and moving incomprehendingly hither and thither in ways which none of them intended, they cannot help being preoccupied with the urgent, narrow and parochial problems which each of them has to face ... Thus what is formed of nothing but human beings acts upon each of them, and is experienced by many as an alien external force not unlike the forces of nature.
>
> (Elias 1987: 10)

People are not, despite the myth of *homo clausus*, independent rational beings. It is, Elias argues, necessary, particularly within modern societies, to remember that 'individualisation has its limits, that every human being is almost continuously dependent on others ... that individual identity is closely linked to a group identity' (Elias 1987: lxvi). Elias' aim is to show us precisely that, through a contextualised historical understanding of the conditions for the formation of that form of subjectivity, and to suggest other ways in which we might think our situations, our selves and those of others.

Michel Foucault and the modernity of 'Man'

Foucault's intention in *The Order of Things* is to reveal the conditions of existence, and the modernity, of 'Man' as the subject and object of knowledge in order to provide a critique of the human sciences and of those philosophies which rely on that figure. In arguing that 'Man' is 'a strange empirico-transcendental doublet' (Foucault 1970: 318) Foucault wants to undercut this notion in two ways. First, he reveals the instabilities, contradictions and difficulties inherent in this mode of understanding (Dreyfus and Rabinow 1982). Second, like Elias, he deploys a critical contextual historicisation to reveal 'Man' to be an effect of discourse, no less and no more 'real' than any other 'object of knowledge', and with decidedly fixed limits. What is crucial is that 'Man' appears old, offering in that sense an illusory history:

> He is a quite recent creature, which the demiurge of knowledge fabricated with its own hands less than two hundred years ago: but he has grown old so quickly that it is only too easy to imagine that he has been waiting for thousands of years in the darkness for that moment of illumination in which he would finally be known.
>
> (Foucault 1970: 308)

By revealing this history Foucault seeks to avoid the difficulties of the fluctuations between positivism and eschatology, the tensions and inter-dependencies of the empirical and transcendental, and the difficulties of Marxism and phenomenology by ceasing to play the game by the rules of 'Man'. Instead he asks a rather different question:

> Does man really exist? To imagine for an instant, what the world and thought and truth might be if man did not exist, is considered to be merely indulging in paradox. This is because we are so blinded by the recent manifestation of man that we can no longer remember a time – and it is not so long ago – when the world, its order, and human beings existed, but man did not.
>
> (Foucault 1970: 322)

Again, the sense of the illusions that that particular form of subjectivity makes dance before our eyes is strong.

This concern to reveal the histories of the subject, to demonstrate the conditions of existence of forms of selfhood and their connection to forms of knowledge, is one of the threads that runs through Foucault's work. There are alterations in the ways in which these questions are approached, shaped by the changes in methodology (which alter many of the periodisations set out in *The Order of Things*) and by political concerns. Yet the sense that the contextualised historicisation of 'the subject' can reveal something about ourselves and others remains. Within this attempt 'to create a history of the different modes by which, in our culture, human beings are made subjects' (Foucault 1982: 208) *The Order of Things* takes its place. It must be considered alongside the genealogies which, with their concern for institutions and power, set out to reveal the formation of modern forms of individualisation and individuality within the discourses and institutions of modern punishment and modern sexuality. Thus, *Discipline and Punish* was written as 'a genealogy of the modern soul' (Foucault 1977: 29) which would reveal the history of its production within relations of power and knowledge which connected punishment and the human sciences (an explicit link to the concerns of *The Order of Things*). The first volume of *The History of Sexuality* was, in turn, to reveal the modern construction of the inner self around notions of sexuality through the mechanism of the confessional (Foucault 1979). Later works, on sexuality and 'government', placed less emphasis on 'the technology of domination and power' and turned towards historically specific 'technologies of the self' to write 'the history of how an individual acts upon himself' (Foucault 1988: 19; Foucault 1985, 1990). All of these works used the differences of history to hold a mirror up to ourselves, to make us realise the implications of his analysis in *The Order of Things*:

> Strangely enough, man – the study of whom is supposed by the naïve to be the oldest investigation since Socrates – is probably no more than a kind of rift in the order of things, or, in any case, a configuration whose outlines are determined by the new position he has so recently taken up in the field of knowledge ... It is comforting, however, and a source of profound relief to think that man is only a recent invention, a figure not yet two centuries old, a new wrinkle in our knowledge, and that he will disappear again as soon as that knowledge has discovered a new form.
>
> (Foucault 1970: xxiii)

We are encouraged to see (our)selves as historically contingent, as formed within discourses and power relations, not as unimportant but as non-universal. This, in turn, is to govern our ways of understanding others.

CONCLUDING COMMENTS

The differences between the projects set out by Elias and Foucault, and revealed by their discussions of *Las Meninas*, mean that there can be no simple clipping together of their works to produce a critical history of the subject. There is also no certain basis on which to choose between them (van Krieken 1990; Burkitt 1993) since they operate, in many ways, within different orders. In this respect we might follow Cascardi when he says:

> What is required in describing the culture of modernity is thus not to ascertain the veracity of an abstract order of concepts or to establish the validity of a series of autonomous historical 'facts' but to comprehend the way in which the subject is positioned between these two orders.
>
> (Cascardi 1992: 10)

What we can do is to understand how they both depict the subject as historical and, in part, geographical – it is strung out across multiple figurations and discourses – formed within many sites and institutions. The subject is multiple and fractured. It is revealed to be part of the social relations of language and the 'incessant and irreducible intertwining of human beings' (Elias 1991: 31) despite its appearance of closure, independence and universality.

In short, then, they are both concerned to reveal what Gayatri Chakravorty Spivak has referred to as a 'subject effect' (Spivak 1987: 204). However, any reference to Spivak's work here signals a return to the issues flagged at the beginning of this discussion and makes necessary the bracketing of the critical nature of the intersection between the work of Elias and Foucault. Their claims to establish the contingent nature of 'Man' and *homo clausus* do not include revealing these conceptions of the subject to be both gendered and part of the power relations of colonialism. Feminists and postcolonial theorists have shown that the practices of self-distancing, the assumptions of individualistic autonomy, the 'subject'–'object' relations between people and both nature and other people, and the ways of seeing which Foucault and Elias ascribe to the subject positions characteristic of western modernity must all be understood in terms of the power relations between men and women and between colonisers and the colonised as they change over time. While there is, I think, a very similar intent in the critical projects of feminists such as Donna Haraway or Michelle le Doeuff and those of Elias and Foucault[14] – they all seek to undermine the modern (or master) subject's claims to universalism and autonomy from social relations – we need to go beyond Foucault and Elias to understand the modern subject as a 'master subject' and to understand all forms of subjectivity as gendered (Rose 1993a). In this context we should call to mind Svetlana Alpers' reading of *Las Meninas* where she interprets the painting as compounding two contradictory modes of representation, and installing an ambivalent representation of the Spanish court. At the heart of this representation stands the Infanta. As a woman she is the 'possession of the European painter's art', yet she is also a princess and a little girl: 'most marvellously self-possessed in bearing, but ... herself possessed by the court and by the royal lineage' (Alpers 1983: 39).[15] This begins to speak to Foucault of the genderings of modes of representation, and to Elias of the gendering of forms of self-regulation and human interdependency in ways which rework their critiques of the modern master subject.

ACKNOWLEDGEMENTS

I would like to thank Antonia Gross with whom I saw and discussed *Las Meninas* at Christmas 1991, and Chris Philo and Stephen Mennell who commented on an earlier version of this paper. It was also discussed at a

Social and Cultural Geography Study Group Conference in Manchester (February 1994) and I would like to thank the participants for their comments.

NOTES

1   The name means 'the maids of honour'. See note 11.
2   All the figures depicted were identified by Antonio Palomino and subsequently verified by others. A narrative account of the picture is given by Vahlne (1982), and the precise location is identified by Moffitt (1983).
3   This may have been added after Velázquez' death (Brown 1986).
4   These pictures have been identified as copies of Rubens by Juan Bautista del Mazo, pupil and son-in-law of Diego Velázquez (Moffitt 1983).
5   The art historical debate is outlined in Brown (1986). *Las Meninas* also prompted an Israeli artist to spend seven days sat on the benches before it, and led Picasso to paint numerous variations upon its themes.
6   Originally published in French in 1966 as *Le Mots et les Choses*, Paris: Editions Gallimard.
7   The book had been published in German in 1983 as *Engagement und Distanzierung*, Frankfurt am Main: Suhrkamp Verlag. However, several of the essays within it had been originally written in English. My attention was first brought to this connection between Foucault and Elias at the inaugural meeting of the Figurational Sociology Study Group at the British Sociological Association conference in Edinburgh in 1988.
8   *The Order of Things* is prefaced with a discussion of a Borges short story which concerns a Chinese encyclopaedia that presents an unthinkable system for the classification of animals.
9   There is some controversy here. Some interpreters argue that reconstructions of the perspective show that the mirror reflects the canvas which is being painted – see Searle (1980), Snyder and Cohen (1980) and Moffitt (1983). However, Brown (1986) argues that there are distinct ambiguities in the perspective which must leave this point unresolved.
10 Elias takes up a position strongly opposing what he sees as the relativism of Claude Lévi-Strauss (Elias 1987: 117). There is a discussion of criticisms of Elias' 'developmentalism' in Mennell (1992).
11 Elias argues that much of this is revealed by the naming of the painting. It only required a name after the deaths of Velázquez and the King, and only acquired the name *Las Meninas* when the painting left the royal collection for a public museum (Elias 1987: liii–liv). See also Varey (1984) and McKim-Smith *et al.* (1988).
12 Stephen Mennell has pointed out that Elias' entire academic life, from his 1922 Thesis to his last, unfinished writings, can be understood as a critique of Descartes' *cogito* (Mennell 1992 and personal communication).
13 This 'list' is not exhaustive. See van Krieken's (1989, 1990) argument that the key difference between Elias and Foucault can be understood in terms of intentionality. Whereas Elias talks of blind 'processes' of civilising/disciplining, Foucault talks of 'projects' devised to change lives. The same point is made by Burkitt (1993), but radically different conclusions are drawn.
14 There are also feminists who find little in common with Foucault's work, see Grosz (1990a), and while I am unaware of any feminist commentaries on Elias it is apparent that his notion of the desirability of being able to 'see ... from the outside' (Elias 1987: 10) is in danger of reproducing a masculinist schema for the production of knowledge (Deutsche 1991).
15 I have concentrated here on the feminist critique rather than the postcolonial critique because I wanted to return in this conclusion to Velázquez' painting and I have not found an interpretation of *Las Meninas* which considers it as a work produced within a court society which was also the centre of an extensive empire.

4

MAPS AND POLAR REGIONS
a note on the presentation of childhood subjectivity in fiction of the eighteenth and nineteenth centuries

Carolyn Steedman

INTRODUCTION

I have recently finished writing a history of subjectivity – one aspect of the history of subjectivity – attempting to describe the way in which, between the end of the eighteenth century and the early part of the twentieth, childhood (the cluster of ideas and beliefs connected to childhood: the 'idea of childhood') became representative, or emblematic, of adult interiority. 'Interiority' is a term quite widely used in modern literary and cultural history, to describe an interiorised subjectivity, a sense of the self *within* – a quite richly detailed self (Miles 1993: 124–42). I have tried to show that from about the end of the eighteenth century, what was felt and known about the self and its individual history was most easily articulated around the idea of the child, most obviously because so much information about growth, development and change was expounded in relationship to children as objects of scientific inquiry.

Towards the end of the nineteenth century, and in the early decades of the twentieth century, Sigmund Freud drew on many heterogeneous nineteenth-century understandings, derived from popular fiction as well as neurological physiology, in order to depict childhood as the individual historical past within the adult that haunts the present, and that is – almost irretrievably – lost and gone. Freud's delineation of the unconscious can be understood as the theorisation of 'childhood' in this sense, that is, as an abstract account of the way an individual past – its history and its vicissitudes – provide the aetiology of the adult self (Roth 1987; Rose 1984: 12–41).[1]

If Freud's achievement was a theorisation of questions of childhood sexuality and repression in the very notion of the unconscious itself (Roth 1987: 95–118), then it was done out of a very general nineteenth-century perception and delineation of childhood as a component of the adult self. In one of the texts that this chapter will discuss, in *Jane Eyre* (1847), the adult Jane dreams of a nameless child hanging round her neck, impeding her movements. This dream child, and actual children in the text (including Jane herself) are clinging burdens of responsibility, 'the emblem of my past life', as Jane, now working as a governess, remarks at one point of her charge Adèle. William Siebenschuh has argued most persuasively that 'the image of the child absolutely haunts the text of *Jane Eyre*' (Siebenschuh 1976). He shows how through the novel, we are 'repeatedly asked to perceive [the adult Jane] as a child seeking comfort, lodging, friends, security and love' (Siebenschuh 1976: 313). In Siebenschuh's reading, the lost is found, the past is restored, when Jane's dream child rolls from her knee as a presentiment of her marriage to Rochester, of her achievement of a home and children of her own, and of her redemptive rescue of Adèle from the school at which her former pupil is unhappy (Siebenschuh 1976: 315–16).

In my recently completed work I described the many *adult* uses of the idea of childhood that were attendant on understanding it as the epitome of personal history and interiority. I paid no attention at all to depictions of *childhood* subjectivity, indeed paid no attention to *Jane Eyre*, which interestingly narrates both the experiences of a young child, and the uses made of those experiences by the adult character the child becomes. Childhood subjectivity is the first topic of this chapter, and its persistent fictional presentation in terms of coldness, polar regions and maps (real maps, and figurative, cognitive and emotional maps), by which the fictional child, who is cold, might – not get warm – but rather, learn what kind of journey it is on.

In pursuing three fictional children (of Johann Wolfgang von Goethe, Hans Christian Andersen and Charlotte Brontë) through the cold regions, the psychological terrain in which Freud learned so much, and then abandoned when he formulated the unconscious, will be of some use. In his theorisation of childhood as the lost yet retrievable individual past, Freud in fact used 'the child' as evidence and epitome in the same way as did the recapitulatory child psychology in which he schooled himself in the 1880s and 1890s.

This establishment of a modern child psychology in the second half of the nineteenth century has frequently been described, usually in terms of its use and transmission of Darwinian and non-Darwinian evolutionary thought. Darwin himself was interested in the evidence that children presented, and made connections between evolutionary progress and the development of the faculties in young children (Darwin 1873: 13, 147–67; Darwin 1877; Sulloway 1979: 243–51; Morss 1990: 11–23; Cunningham 1991: 196–7). Freud put this perception of childhood to new uses; but the psychoanalytic unconscious is beyond the scope of this chapter. What must concern us here is not the unconscious self, but the self that 'thinks, suffers and wills' (Sully 1897: 70, 71) and to which child psychology of the period drew so very much attention.

To add to the historical evidence that the child was understood to embody were the striking findings of contemporary philology, which allowed language itself to be understood as an 'unconscious record of the growth and decay of ideas ... as the stratified deposit of thoughts' (Romanes 1888: 238). The historical evidence that language carried was added to testimony of the child mind: both offered confirmation of the processes of evolution, and were used figuratively to outline the project of a scientific child psychology. The evolutionary theory that was used by psychologists of the child-study movement involved an inherent teleology, with the idea of progress being embedded in the idea of development: the child's developing body and mind could be understood as an embodiment of a more general historical progress (Cunningham 1991: 197–9; Steedman 1982: 85–7, 230; Muirhead 1900: 114–24; Cavanagh 1981: 38–47; Wilson 1898: 541–89; Monroe 1899: 372–81; Stevens 1906: 245–9).

In some late nineteenth-century psychological accounts, the child's understanding of its own body and its own internal spaces was used as a form of historical evidence. James Sully thought that the child's ideas of 'origin, growth and final shrinkage', mirrored 'the development of the idea of the soul by the race', for among the ancient peoples 'its seat was placed in the trunk ... long before it was localised in the head' (Sully 1897: 68–9). When the child is able to grasp the idea of 'a conscious thinking "I", the head will become a principal portion of the bodily self'. This conscious self comes to be 'dimly discerned' by the end of the third year (Sully 1897: 70, 71). As it came into being, this self historicised itself, by constructing 'the unreachable past'. Sully observed how 'very curious are the directions of the first thought about the past self', for the child had to encounter the 'terrible mystery, time'. Sully described how children seem at first 'quite unable to think of it as adults think of it, in an abstract way. "Today," "tomorrow," and "yesterday" are spoken of as things that move.' When he pointed to the child's inability to grasp 'great lengths of time', he gave poetic expression to the great sadness that evolution and history had bequeathed, and he made a curious elision between adult and child, in suggesting that 'possibly [a] sense of immeasurable lengths of certain experiences of childhood gives the child's sense of past time something of an aching sadness which older people can hardly understand'. In Sully's description the subject feeling loss was at once adult and child (or neither; both were ageless subjects of time and history): 'Do not the words "long, long ago," when we use them in telling a child a story carry with them for our ears a strangely far-off sound?' (Sully 1897: 73–5. See also Preyer 1890a, b, Volume 1: 107–8, 209–10; Volume 2: 209–10).

This chapter concerns the story that was told about childhood in Goethe's *Wilhelm Meister* (1795–6), Andersen's 'The Snow Queen' (1844) and Charlotte Brontë's *Jane Eyre* (1847). These tales are only fragments of a long meditation on the questions of childhood subjectivity, how it is to be described and what to tell the children about it. All three involve cold places and cold children, and detailed mappings of the cold places, either made by the topography of the child's imagination, or by the child's journey through them. This chapter then, concerns formal and informal theories of childhood and their representation through spatial metaphors.

The suggestion is that since the middle years of the nineteenth century the map has been employed as metaphor, in order to say (to many professional and non-professional observers of childhood, and, much more covertly, to children themselves) that *something has already happened*, that a process of development has already taken place, and that children are already *in* a particular terrain: are already in a story. The idea of mapping in the psychological depiction of childhood was first formalised in the mid-Victorian years; part of the effect of telling the story of childhood subjectivity in professional circles, as happened in the European and North American child-study movement, as has briefly been described, was to make available to children covert accounts of the life-story they were already deemed to have set out on, and which they were bound to continue. Andersen's 'The Snow Queen' opens with the promise that the story means something: 'Well, now let us begin! When we have got to the end ... we shall know more than we do at present' (Andersen 1900: 139).[2]

COLD CHILDREN

The nineteenth century made manifold use of a late eighteenth-century child-figure: Mignon, from Goethe's *Wilhelm Meister* (1795–6).[3] In the century after Mignon was first written she passed through an extraordinary number of transmutations, crossed boundaries of language, of genre and of form. Her shade is there – attenuated, etiolated, but still there – in other child-figures and in depictions of actual, historically extant children of the nineteenth century. Through all the transmutations and reworkings of her (some of them by people who had never even heard her name) one thing went on being known about her. A sensibility shaped by a thousand renditions of Mignon's most famous song, her infinitely sad song of yearning for Italy, 'Kennst du das Land', permitted Victorian observers of outcast childhood to know that the child on the street was *cold* – whether it was performing its tricks, begging, importuning, or just stoically going about its business.[4] In Goethe's text of 1795–6, Mignon finishes her song and looks meaningfully at young Wilhelm Meister, by whom she is now employed as a servant. '"Do you know the land?" "It must be Italy that is meant," answered Wilhelm; "Where did you learn the little song?" "Italy!" Mignon said in a significant manner; "if you go to Italy, take me with you, I'm freezing here"' (Goethe 1977, Volume 1: 126). She repeats her desire to be warm more than once. In G. R. Sims and Clement Scott's melodrama of 1885, *Jack in the Box*, which was written as part of the British campaign against the exploitation of child acrobats, one of the Italian children trapped 'at Toroni's ... an Italian padrone's den in Saffron Hill', laments thus:

> 'I want to go back to Italy – To my mother. My father sold me to the padrone & my mother kissed me & cried, & said I should never see her again – and I shan't. I shall die & never see dear Italy again. I was so warm there. I'm so cold here.'
>
> (Sims and Scott 1885: 55)

Earlier in the century Henry Mayhew had made particular note of the cold appearance of the children clustering around the watercress dealers' stands in Farringdon Market, observed the snow sometimes falling on their 'numbed fingers' as they sat on neighbouring doorsteps, stringing their bundles (Mayhew 1850). But when he asked the child he interviewed about this aspect of her work, she said '"I bears the cold – you must ... No; I never see any children crying – it's no use"' (Mayhew 1861–2, Volume 1: 151–2).

The coldness – Mignon's feeling cold – remained; it was rewritten and reworked throughout the nineteenth century. What was lost was the extreme disturbance that Mignon manifests in the text of *Wilhelm Meister*. It is her silences, her hysteria, the complex of her *inabilities* that make her attractive to young Wilhelm Meister. The child speaks a poor German, mixed with French and Italian, and indeed, on some days 'she was wholly mute' (Goethe 1977, Volume 1: 97–8). However, when she sings 'the broken speech [is] made consistent and what was disjointed linked together' (Volume 1: 127–8). (In the century that follows what she sings at this moment, her song of yearning in 'Kennst du das Land', *is* Mignon.) The child finds written language as difficult as spoken language, and her 'letters remained unequal and the lines crooked' (Volume 1: 118).

Mignon desperately wants to understand how maps work, what form of representation they are, but she cannot perform this cognitive task (Volume 2: 73–4). Wilhelm notes the child's 'incapacity ... to represent anything'. He means by this her strained expression when she is asked to recite: 'in a few plays ... her small parts had been so dryly, so stiffly done that one might say that they were not acted at all.' But it is a much wider aphasia that the young man notes. There is something wrong with Mignon, 'something strange about the child, in all her comings and goings' (Volume 1: 97); 'her body seemed at variance with her mind' (Volume 1: 118). Meister means by this the way in which she finds it both extraordinarily difficult to control her body, and yet at the same time, can use it with uncanny deliberation and precision. 'She did not go up and down the stairs, but leapt; she climbed up the corridor bannisters, and before you knew where you were, she was sitting high up on the wardrobe' (Volume 1: 97). But the oddness is more general than this, and permeates all aspects of his observation of her. 'Wilhelm is constantly haunted by the feeling that there is something wrong with Mignon, something that he cannot put into words and that he cannot describe' (Eissler 1963, Volume 2: 757).

Her inabilities in map-reading are very marked. When encountering maps for the first time, the child is greatly astonished, asks innumerable questions about them, and 'her desire to learn seemed to ... [become] much livelier as a result of this new knowledge'. Later she pledges her silver shoe-buckles with a picture-dealer in order to get hold of a small atlas (Goethe 1977, Volume 2: 55). Her obsession persists: later she is found trying to explain the contents of her little atlas to a younger child,

> though in doing so she did not make use of the best method. For actually she did not seem to have any special interest in the different

countries apart from whether they were cold or warm. She could give a very good account of the polar regions, of the terrible ice there, and of the increasing warmth the further away one went from them. When anyone was going on a journey, her only question was whether he was going north or south, and she made efforts to find the way on her little maps.

(Volume 2: 73–4)

In literary-critical terms, and in many psychoanalytically informed readings of Goethe's work, Mignon is understood as an embodiment of the poet's longing for the South, a longing given the most piercing expression in the most famous of her songs. Her division of the globe into North and South and warm and cold to the exclusion of all other features, is an evident prefiguring of Goethe's achievement of warmth, in his Italian Journey of 1786–8. The lines that describe the cognitive inabilities of his child-figure were, in fact, first written in the 1770s in the first draft of *Wilhelm Meister*, long before the journey south became an actuality and when, in the text of *Wilhelm Meister's Theatrical Mission*, the child does not die as she does in the published version of 1795–6, and the narrative closes with her about to leave with Wilhelm for the journey across the Alps (Goethe 1913). In the later version – the version that the nineteenth century knew – Mignon's little maps are given more poignancy by the fact of her death, and her never learning what it is they represent.

In Charlotte Brontë's *Jane Eyre* (1847) we shall see another girl-child contemplate the same polar regions, though like Mignon, Jane never experiences them. This is not the case with Hans Christian Andersen's 'The Snow Queen' in which Gerda makes the journey to the North, travelling all the while without maps. 'The Snow Queen' (1844) is among Andersen's best-known tales; according to Bruno Bettelheim it is the only one of his corpus that comes close to being a 'true' fairy-tale because of the consolation it ultimately conveys (Bettelheim 1978: 37). Two children (they are unrelated, but have grown up together) spend the summers in play on the roof-space between the two garret apartments where they live, in some northern city, somewhere, sometime. There is a vegetable garden planted in a box and roses and other flowers grow here too. When winter comes the children are confined indoors, where the grandmother of one of them (they do have parents, who are described making the roof-garden, but then never mentioned again) tells them stories of the Snow Queen, who flies where the swarm of white flakes is thickest. Kay actually sees her, through the window: 'a full-grown woman ... very beautiful and graceful, but ... made of ice, dazzling glittering ice' (Andersen 1900: 142).

There is yet another story outside this narrative. Andersen begins the tale with a mirror, fashioned by the Devil to make the beautiful appear hideous and the worthless enhanced. The mirror has been broken into millions of pieces that now fly about the world, some of them no bigger than a grain of sand, but with the capacity to get into an eye and distort everything that is seen, and into the heart, which was 'the most terrible of all; those hearts became like lumps of ice'.

This is what happens to Kay, the following summer. 'Poor Kay had got one of the fragments right into his heart. It would soon become a lump of ice. It did not cause him any pain, but it was there' (143). When winter comes again, he goes out sledging and is abducted by the Snow Queen, who completes the work of the mirror-fragment by kissing him. Then she says ' "I shall give you no more kisses! . . . or I should kiss you to death".'

> Kay looked at her; she was very beautiful; a more intelligent or lovely face he could not imagine . . . In his eyes he knew she was perfect, and he did not feel the least afraid of her; he told her he knew mental arithmetic even in fractions, and how many square miles and inhabitants there were in all countries . . . But he felt he did not know enough after all, and he looked up into the great space above . . . They flew over forests and lakes, across the ocean and many countries; below them the cold blast scoured the plains, the wolves howled and the snow sparkled, and over them flew the black screeching crows, while the moon shone bright and clear, and by its light he beheld the long, dreary winter's night; by day he slept at the feet of the Snow Queen.
>
> (Andersen 1900: 145)

Gerda searches for her lost companion along all the edges of the world, setting her course for the North. The people and animals she meets help her in her quest, though some try to delay her passing and keep her with them. In a century of heart-wrenchingly determined little girl saviours, Gerda is the most resolute, and the most moving in her resolution. The Finnwoman has maps and directions to help her on her journey, but knows that she cannot give the child

> 'any greater power than she already possesses! Do you not see how great it is? Do you not see how men and animals must serve her, how she, barefooted, has got on so safely through the world? She must not be told by us of her power; it is seated in her heart . . . it consists in her being such a sweet and innocent child. If she cannot obtain access herself to the Snow Queen, and remove the bits of glass from little Kay, we cannot help her.'
>
> (Andersen 1900: 163)

This is what Gerda does, finally finding the boy in the vast halls of the Snow Queen's castle where, almost black with cold, he is trying to form 'some very intricate figures' out of flat pieces of ice he drags about the floor. He is playing 'the ice game of reason':

> He formed complete figures which represented a written word, but he was never able to form the word he most wanted. It was the 'Eternity,' and the Snow Queen had said 'If you can solve that figure, you shall

be your own master, and I will make you a present of the whole world and a pair of new skates'. But he could not.

Kay is at first insensible to Gerda's entry into the great, empty cold rooms, insensible to her cry of '"Kay! dear little Kay! So I have found you at last".' Then Gerda weeps, and her hot tears penetrate to his heart, and thaw the lump of ice, and consume the little piece of glass (166).

After some brief further adventures, in which Gerda's footsteps are retraced, and all those who have helped her are revisited, the children return home.

> They entered the town and found their way to their grandmother's door, – they went up the stairs, and into the parlour, where everything was in the same place as before . . . but as they passed in at the door they discovered they had become grown-up people.
>
> (Andersen 1900: 168)

Plot, themes and figures in 'The Snow Queen' allow a wide variety of interpretations. The most systematic recent analysis has explored it from a Jungian perspective and found in it an allegory of man's redemption by woman (Lederer 1986). Whilst this interpretation is as useful to its purposes as any other, it relies heavily on the notion that the child-figures' sexual identity is fixed and certain when the story opens and that Kay and Gerda are indeed separate persons of the opposite sex, rather than aspects of each other, or of the same component child-figure. Perhaps we should note Andersen's careful description of infantile depression: the child Kay is lost, and gone away from itself to a cold, distant place. The child can scarcely move its head, or lift its hand, it is so cold. The child is pressed into itself, as well as being absent from itself. In some way, in Andersen's telling of the story, this is the same child as the one who possesses the capacity to find itself, who will not stop searching along the margins of the world – a busy child, with a purpose – until the frozen self is woken, and the lump of ice dissolved.

At the very end of the tale, when Kay and Gerda sit down under the roses again and listen to the Grandmother read from the Bible '"Except you become as little children, ye shall in no wise enter into the Kingdom of God"', Lederer finds a denial of the adult sexuality that they might be assumed to have brought with them through the door, and a precise reading of Andersen's extreme difficulties in dealing with adult women and female sexuality. No doubt all these things are true, but it is in fact by ending the tale with the return of Kay and Gerda to the timeless space of childhood, that the story achieves its redemptive qualities. Sexless as they are at the end of it all, Kay and Gerda will not produce children of their own; no futurity shadows the end of this story.

In the opening passages of *Jane Eyre*, an 8-year-old contemplates the polar regions. The child takes up a volume of natural history, and her eye is drawn from the little engravings to the text that speaks of the bleak shores of Lapland, Siberia, Spitzbergen, Nova Zembla, Iceland, Greenland, with 'the vast sweep of the Arctic Zone, and those forlorn regions of dreary space – that reservoir of frost and snow, where firm fields of ice, the accumulation of centuries of winters, glazed in Alpine heights above heights, surround the pole, and concentrate the multiplied rigours of extreme cold'.

(Brontë 1966: 39–40)[5]

Jane's imagining of these 'death-white realms' is placed in conjuction with a complex and peculiar relationship to the ideas of cold and home. In effect, the narrative voice tells us that the child dislikes going out because she hates coming home. The novel opens with her pleasure at the cold winter wind that has driven her and the Reed children indoors:

There was no possibility of taking a walk that day ... I was glad of it; I never liked long walks, especially on chilly afternoons: dreadful to me was the coming home in the raw twilight, with nipped fingers and toes, and a heart saddened by the chidings of Bessie the nurse.

(Brontë 1966: 39)

The book that the fictional child hides in the window recess to read is Thomas Bewick's two-volume *History of British Birds* (Bewick: 1794, 1804) which contains a series of illustrated descriptions of the principal land- and sea-birds of the British Isles. At the end of each entry are smaller engravings unrelated to the text, the 'vignettes' that the child connects to the introductory passages of Volume 2 that describe the 'death-white realms'. From the precise description that Brontë provides, each engraving can be located, and as a group of images they have been subject to much discussion, seen particularly to provide an aetiology of the adult Jane's visual imagination, particularly as manifested in her later watercolour paintings (Stedman 1966–70; Kelly 1982; Hennelly 1984). Moreover, throughout the novel cold and ice are aligned with feelings of exclusion and isolation. In *Jane Eyre*, to be lonely is to be cold and desolate, a precise configuration learned in a childhood of extreme loneliness and emotional deprivation.

The presence of Bewick's engravings in the text of *Jane Eyre* has often been discussed in this way, as personifying the child's unloved state of loneliness at Gateshead ('the rock standing up alone in a sea of billow and spray ... the broken boat stranded on a desolate coast ... the cold and ghastly moon glancing through the bars of cloud at a wreck just sinking' (40)). These arguments have been particularly important for literary critics seeking to claim Jane Eyre as 'the first heroine in English fiction to be given, chronologically at least, as a psychic whole', an adult character shaped by the experience of childhood and childhood reading (Coveney 1967: 105). Certainly Brontë constructed the inner life of her character around Jane Eyre's childhood, which is constantly referred back to, in Jane's paintings, dreams and visions. But if the child is shown to unconsciously identify with the icy content of Bewick's little pictures in order to connect her inner feelings with the distant and mysterious exterior world he depicts, it is not at all clear why it is the wild Arctic zones that are the medium for the psychological act thus described.

WHAT TO TELL THE CHILDREN

In Charlotte Brontë's novel of 1847 the servant Bessie tells the child Jane how to behave like a child. In the cold mansion of containment and restriction, the advice has been given before. Mrs Reed, the orphan-child's guardian, has forbidden Jane the company of her own children until, she says, the 8-year-old has acquired 'a more sociable and childlike disposition, a more attractive and sprightly manner – something lighter, franker, more natural as it were' (Brontë 1966: 39). Jane knows that had she been 'a sanguine, brilliant, careless, exacting, romping child – though equally dependent and friendless – Mrs Reed would have endured [her] presence more complacently' (47). But it is Bessie who tells her exactly why her behaviour is so very off-putting. '"You're such a queer, frightened, shy little thing. You should be bolder ... Don't start when I chance to speak rather sharply: it's so provoking."' Bessie also has advice about Jane's behaviour at the boarding school she is about to be sent to, telling her that she must not go in fear of the new people she will meet: '"if you dread them, they'll dislike you"', she says (71).

The connection of childhood, service and servitude in English fiction is a complex and interesting one, and seems to have been made since at least the beginning of the eighteenth century. In 1978 Juliet Mitchell observed that in *Moll Flanders* (1721) Daniel Defoe inaugurated a new tradition of writing, in which Moll starts as a child, and 'what happens to [her] as a mature woman, indeed, who she is as a woman, depends on the conditions of her birth, her infancy, childhood and adolescence'. In this way, claims Mitchell, the first written 'individual' of fictional realism is Moll, a woman, or rather, 'capitalist woman' (Mitchell 1984: 217). Little Moll's utter refusal at the age of 8 to be put to service is what we are bound to see in retrospect as the determining factor in her life history, as well as the clearest expression of her personality, and the very point at which she is set on her path of ruin-and-riches. '"What doest Cry for?"' asks the child's foster-mother of Moll's failure to see the job of maid in a local household as a golden opportunity.

> 'Because they will take me away [says Moll] and put me to Service, and I can't Work House-Work ... and if I can't do it, they will Beat me, and the Maids will beat me to make me do great Work, and I am but a little Girl and I can't do it.'
>
> (Defoe 1989: 47)

This perception, of individual childhood history exercising a shaping force on the adult woman, profoundly shapes Mary Wollstonecraft's writing of Jemima, the prostitute-become-wardress and later servant, of her posthumous novel *The Wrongs of Woman, or Maria* (1798). Wollstonecraft gave the child Jemima fewer pages than Defoe gave his Moll (though it is important to note that Jemima's story of childhood is longer than that of the bourgeois heroine Maria, and is told first).

Comparisons between these two early working-class 'childhoods', written seventy years apart, would bear much more analysis, and indeed, a more thorough survey of the English novel of the intervening years might well turn to assertion the suggestion that the first extensively written childhoods in English literature belong not only to women, but to working-class women to boot. That these fictional women also achieved some measure of their fictional identity through a childhood or adolescent relationship to the idea of service and servitude may, of course, be revelatory of no more than realist assumptions at work in the early novel and among its modern readers: it is our conventional understanding as much as it was Daniel Defoe's that there was not much else for an unmarried woman of the poorer sort to *be* but a servant or a whore.

But the relationship between servants and children in the eighteenth-century novel has another dimension that is not entirely explained by statistics of female employment, nor the complex relationship between the idea of character in the novel, and the servant's letter of recommendation, or 'character'. This other dimension, which is to do with the terms in which household relationships between children and servants were discussed in eighteenth-century literature, was also explored in some detail by Mary Wollstonecraft, and echoed in Charlotte Brontë's fiction of 1847. Mary and Caroline, the child figures of Wollstonecraft's *Original Stories from Real Life: With Conversations Calculated to Regulate the Affections and Form the Mind to Truth and Goodness* (1788) are literally characterised by having been 'in their infancy, left entirely to the management of servants, or people equally ignorant' (Todd and Butler 1989, Volume 4: 361). Here, a long tradition of anxiety about the influence of servants (most widely given expression by reference to John Locke's strictures in *Some Thoughts Concerning Education* of 1693) is given new emphasis by Wollstonecraft. Mary's impertinence and hauteur to the maid Betty is used by Mrs Mason, the children's guardian and instructress, to impart one of the most important lessons of this book, which is that 'children are inferior to servants, who act from the dictates of reason and whose understandings are arrived at some degree of maturity', who have indeed in many cases, attained 'a virtuous character' by having 'done their duty [and] filled an humble station, as they ought to fill it, conscientiously' (Todd and Butler 1989, Volume 4: 412–13). In Brontë's work of 1847, the Gateshead Hall servants are in complete agreement with this view. When Jane Eyre passionately rejects the idea that the cruel and tyrannical son of the house is in any sense her master, and asks furiously '"Am I a servant?"', either Bessie or Abbott the lady's maid (the text does not make it clear who speaks at this moment) admonishes her by asserting that she is '"less than a servant, for you do nothing for your keep. There, sit down, and think over your wickedness"' (Brontë 1966: 44).

In 1787, through her character Mrs Mason, Wollstonecraft withdraws the servant's help to the child, as a means of imparting a sense of her own weakness and dependence on another's labour.[6] This pedagogic device, and the whole body of educational literature that Wollstonecraft translated and reviewed in the 1780s and 1790s, allows speculation about a new role for servants in enlightened, intellectual, middle-class households of the period, as a kind of teaching device, an audio-visual aid to empathy (Myers 1989: 61–5). The opinions of the servants in *Jane Eyre* suggest that these ideas were widely spread, at least across the years, and into other fictional realms. With this observation in mind, of a tradition of contemplating questions of childhood and child-likeness in relationship to servants, we should not be surprised that it is through Bessie's family that Brontë allows the first independent judgement of Jane's situation to enter the text. '"You are rather put upon, that's certain"', agrees Bessie at one point. '"My mother said, when she came to see me last week, that she would not like a little one of her own to be in your place"' (Brontë 1966: 71).

The dynamic of the early chapters of *Jane Eyre* is some kind of argument, or textual struggle, over the question of what children are like, and what children ought to be like. Reading them from this perspective, their interest lies in the fluency of the debate, and Brontë's attribution to her fictional child of a full knowledge of different modes or fashions of childhood. Claudia Nelson's recent *Boys Will Be Girls* could be taken to suggest that mid and late nineteenth-century fiction for children was one long advice manual for children, presenting them with pictures and models of different types of childhood (Nelson 1991). From this perspective, *Jane Eyre* could be read as a movement of prescription into fiction for adults, with the novel employing an individual childhood to map out the trajectory of an adult sensibility and subjectivity, that of Jane Eyre herself.

WHAT THE SERVANTS SAID

Why is knowledge about childhood conveyed to fictional children by fictional servants? I think that there are two possible and interconnected answers to this question. The first answer evokes much recent work on the history of subjectivity in western societies, and the way in which modern literary history and theory has found the image of modern personhood expressed in female figures, in a wide variety of literature produced since the end of the seventeenth century. Depth and space within, eloquent sensibility and a wide range of affect have been located in the woman-in-the-text. 'Domestic Woman', the 'Autobiographical Subject', the 'Proper Lady' – all the figures that now stand in for modern identity in the way that Economic or Capitalist Man once did – are subjects because they are subject to scrutiny (Armstrong 1986; Nussbaum 1989; Poovey 1984; Mitchell 1984: 195–218). As 'the sex', they are contemplated, regarded, written about and theorised because they are both subjects (they are inferior persons, who are to be ruled rather than ruling others), and non-subjects (for as inferior and ruled-over, they are not full legal and political persons).

Female servants might be expected to slip outside the net of the modern scholar's theoretical investment in writing a history of subjectivity, for whilst they were certainly women, they were also people of the poorer sort, and most recent investigation and theorisation has embodied accounts of *bourgeois* personhood in figures like Domestic Woman. And yet all the labouring poor who undertook service needed a 'Character', the piece of paper or verbal recommendation that enumerated their abilities, capacities and traits. In *The Servant's Hand* Bruce Robbins has revealed extraordinary apertures made by servants in a wide variety of texts (Robbins 1993). Ostensibly present as synecdoches, as mere hands, they erupt briefly as detailed, written personalities. In fact the female servant shares twice over in the attributes that have allowed modern scholars to name the first modern individual, for the servant possesses personality, and she is an inferior. And sometimes, like Bessie in *Jane Eyre*, she conveys knowledge to a third kind of figure, that can also be detailed by its unimportance, inferiority and insignificance: she conveys knowledge to a child. In the nineteenth century servants of both sexes and children (and married women) were still subject to the old common-law doctrine of *potestas*: to be held under the protection and rule of a master (Davidoff 1974: 406–7). All of them subjects in this sense, they became subjects in the other sense: ideas of selfhood, character and sensibility were written in their name.

If children (the idea or image of the child) really are a primary means of symbolisation, if they are – as I have claimed elsewhere – a first metaphor for people, whether those people have their own children or not, 'whether they are literate and in the business of constructing literary metaphors or not: a mapping of analogy and meaning for the self' (Steedman 1990: 259), then William Siebenschuh's is a redemptive account of this process, in which Jane is seen to achieve a resolution of her own story through a symbolic reshaping of her own childhood, particularly in the dream where the clinging, nameless child falls away from her. But to make that reading, you must forget what the servants said. On the eve of Mrs Reed's death, when the adult Jane is summoned back to Gateshead, she has another of her dreams about the child: 'during the past week scarcely a night had gone over my couch that had not brought with it a dream of an infant.' She remembers that

> When I was a little girl, only six years old, I one night heard Bessie Leaven say to Martha Abbott that she had been dreaming about a little child; and that to dream of children was a sure sign of trouble, either to one's self or one's kin.
>
> (Brontë 1966: 249)

She remembers too that what Bessie said was true, for the nursemaid had been sent for the next day, 'home to the deathbed of her little sister'.

If we heed the servant's words, then the glimpse of Jane as a mother that the penultimate page of the book provides, the glimpse of her holding a child with large, brilliant, black eyes (the others to come foreshadowed by the use of 'first-born' to describe him) is no redemption at all (476). Rather, this baby's existence is a presentiment and foretelling of *repetition*; of all the meanings that the child-figure has activated in the story just finished; of the idea that *it will go on like this*, that children will go on being born, and making their journey through life. Indeed, the baby's eyes are those of Jane right at the beginning of the book, when, locked in the Red Room, she sees in the 'visionary hollow' of the looking-glass a 'strange little figure there gazing at me with a white face and arms specking the gloom, and glittering eyes of fear moving where all else was still'. For the child Jane watching her own image, her reflection had the effect of 'a real spirit: I thought it like one of the tiny phantoms, half fairy, half imp, Bessie's stories represented coming out of lone ferny dells in the moor' (46). This is a changeling child in both guises; the story says so, at its beginning and its end: a sure sign of trouble rather than redemption.

TORRID ZONES AND POLAR REGIONS

The idea of the map has become a symbol of great motility in western writing; from a glance at the Contents page, readers of this volume will be aware of the service that the word is currently doing as metaphor, in modern accounts of selfhood and personhood. A first observation about maps and metaphors is that in literary and cultural analysis the map has been used more frequently to evoke cold places than hot ones; has been used to imagine the polar regions rather than the torrid zones. This is to say nothing at all about the degree of mapping that actually took place during the period of European colonial and European expansion, when it is quite evident that maps were made of all types of appropriated terrain, in many climates. And whilst it is also clear that the appropriation of the torrid zones was a source of metaphor, such metaphors work in a different way from those that evoke the death-white realms. Mary Ann Doane discusses sources for metaphors in the imperial conquest of Africa, by commenting on Sigmund Freud's evocation of the dark continent to denote female sexuality, and she indicates the work that these allegories of the self have done, in twentieth-century psycho-analysis and film. In this argument the torrid zones play an important imaginative role in the structuring of white identity and female sexuality across a variety of representational forms (Doane 1991: 209–48). What is striking however, is how blank and undetailed the hot places are made, in the kind of representational work discussed by Doane. If there is a journey to the torrid zones, it is mostly a journey without maps, to any old heart of darkness. In journeying to the polar regions on the other hand, the fictional children discussed here create inventories of ice floes, Alpine heights, icy billows of foam, the cold and ghastly moon.

These frozen topographies, made by the child's journeying through them, are an articulation of something that isn't actually very much to do with maps, as indeed more generally, the topic of the map is rarely to do with the ostensible subject of discussion. Hugh Brody has described the settlement of

British Columbia, and a long history of appropriation of Indian land by treaties, traplines and maps. Whilst the white settlers annexed vast tracts of land by these means, they simultaneously envisioned the harsh terrain even further north than the land already taken, as a place of limitless commercial potential and as a rich and fertile terrain. Considering these visions, Brody calls maps 'dreams'. In the kind of mapping he describes (which is a mapping of desire) 'the North conveniently moves north' (Brody 1981: 115). He is able to mobilise the idea of the map as a kind of dream because dreams are always about something that has already happened, as indeed in the history he is relating, the seizure of Indian lands had already taken place.

The fantastically detailed terrain that children map out in the fiction that has been described above is the story they are already in. The icy topography of the child Jane Eyre's imagination, the death-white realms that Gerda traverses, and the journey that Mignon never makes, are all what has already happened, both specifically to these figures, in these texts, and in general to all children, set on a course of development.

At the end of 'The Snow Queen' there is no servant to tell us what otherwise would be bound to come to pass (the production of more children; the unending story); perhaps it is because Andersen's story truly finishes, with Kay and Gerda made eternal children, that it does actually achieve its effect of great comfort and consolation. Yet as a tale 'The Snow Queen' has also suggested with the utmost and chilling clarity what might be a child's only solution to finding itself in the story it's in – which is existence itself – that is, in the infantile depression so powerfully described by Andersen. The story may serve to remind us too, that the very stuff of metaphor allows particular epistemological work to be done, and prevents other kinds of use. These childish journeys do not describe the construction of the self in relation to some arrogantly figured Other, some blank hot zone of darkness; they rather describe an understanding of a self made in relation to the cold hard facts of life itself, which it is the task of the children in these tales to learn.

NOTES

1 *Strange Dislocations: Childhood and the Idea of Human Interiority* will be published by Virago and Harvard University Press in 1995.
2 References in the text are to Volume 1 of the Braekstad translation (Andersen 1900).
3 Subsequent references in the text are to the Waidson translation of 1977.
4 I have counted over eighty nineteenth-century settings of 'Kennst du das Land'. The best known are by Beethoven (1809), Schubert (1815), Liszt (1842), Schumann (two versions of 1849), Tchaikovsky (1874) and Wolf (1888).
5 Subsequent page references in the text are to this Penguin edition of 1966.
6 This was a pedagogic device used in Salzmann's *Elements of Morality for the Use of Children*, which Wollstonecraft translated from the German in 1790 (Todd and Butler 1989, Volume 2: 142–4). Wollstonecraft reiterated the importance of imparting to children 'a sense of their own weakness ... [to make] them feel the natural equality of man' in the *Vindication* (Todd and Butler 1989, Volume 5: 262–3). Mme Cambon's *Young Grandison*,

which Wollstonecraft translated from the Dutch in the same year as *Elements of Morality*, shows a young bourgeoise, Emilia, who is able to demonstrate her civility through empathy with servants. Emilia knows that 'we ought not to give the meanest of our fellow-creatures trouble when we can avoid it, if we desire to be truly great' (Todd and Butler 1989, Volume 2: 220, 235, 237).

5

'THE ART OF RIGHT LIVING'
landscape and citizenship,
1918—39

David Matless

If our belief ... in the permanent importance of right leisure using ... has any content and meaning, then we shall see in the English countryside not only a possession of beauty which, having inherited from the past we are morally bound to hand down undefiled to posterity, but an instrument, the most important we possess, for the training of the citizens of the future in the art of right living.

(Joad 1934: 157–8)

INTRODUCTION

People, male and female, of all classes, moved in the country. Between 1918 and 1939 open-air leisure in England took on new scale and scope. A particular landscaped version of English citizenship emerged in the work of preservationists and planners; morally, spiritually and physically healthy, alert to the landscape. This essay takes the work of these preservationists, who were by no means resistant to the modern world but sought to order its progress into a distinctly English modern form,[1] as a way into matters of environment and self. The aim of the essay is twofold. First, it seeks to bring out an assertive and progressive vision of English landscape and citizenship hitherto downplayed.[2] Second, it explores the formulation of an environmental way of being, very much bound up with the geographical thinking of the period (Livingstone 1992; Matless 1992), which continues to govern many taken-for-granted assumptions regarding environmental behaviour and the benefits to be gained from landscape.

Matters of being in landscape were not simply the concerns of an élite preservationist group. Senses of selfhood and citizenship emerged out of the practice of such diverse groups as communist ramblers, scouts and guides, health campaigners, charabancers, modern dancers, youth hostellers and nature mystics. The preservationist ethos indeed braids into cultural fields which preservationists might themselves consider 'dangerous'. This essay steps into this disparate stream of environmental modernist sensibility, aiming to catch some of its cultural flow in a manner resonant with more recent discussions of self, subjectivity and environment (for example Rose 1993a). The aim here is to meet the self outdoors by tracing the culture of particular environmental practices. Cultural geography here meets the self not through mapping action onto interior psychology but through documenting the versions of self embodied in and made through historical geographical practice.

Walking, cycling, camping, map-reading made up for preservationists an 'art of right living' whereby an individual or nation might give form to itself environmentally.[3] This essay proceeds through the political, intellectual, spiritual and physical cultures of country movement, following the preservationist ethos as a central thread. Groups and practices threatening to take environmental practice beyond the preservationists' moral and political pale are also considered. We begin though with people making noise and leaving litter.

CONDUCT UNBECOMING

We have chosen browns for the dominant urban categories and for roads, using a broken line for rights of way and a brown stipple for public open spaces. Thus we may think of the urban population issuing forth in bulk along the roads, trailing over the footpaths and spreading themselves out in dots upon the grassy commons and heaths.

(Fagg and Hutchings 1930: 54)

Urban excursions into the country were not new in the interwar period; people of all classes had made their way before by rail and cycle. However as car ownership extended into the middle classes, and bus travel and communal charabanc trips were offered for the urban working class, the rural spaces of urban leisure were transformed. The railway, focusing passengers into the orbit of a station, still played a key role, but, as rambling activist and preservationist Cyril Joad put it, the 'motor's capacity for ubiquitous penetration' had 'created a new situation' (Joad 1934: 97).

Preservationists welcomed such extension as offering lessons in 'the art of right living'. But it was also seen to display conduct unbecoming citizens of a new England. Litter, noise, flower-picking, 'disobedient bathing', offered the contrast against which the right leisure user could be upheld. In 1930 Harry Peach, craft entrepreneur, socialist-of-sorts and design propagandist (Kirkham 1986), published *Let Us Tidy Up*:

The problem of litter ... has grown out of our laissez faire of the nineteenth century ... We have temporarily lost that sense of fitness and order which helps to make the beauty of the remaining unspoilt bits of eighteenth century England ... This lack of order applies to all sides of our life.

(Peach 1930: 3)[4]

ANTI-LITTER CAMPAIGN

Member of Picnic Party (just leaving) "Better have a look round among our litter and see we haven't left anything behind."
—*By kind permission of "Passing Show."*

Figure 5.1 Anti-litter postcard issued by the CPRE, 1928

The nation was seen to have a behavioural problem (Figure 5.1). Various stock litter-dropping noise-making figures emerge; thoughtless upper- and middle-class 'motor-picnickers' not clearing their empties, loud working-class charabancers. Offenders are often labelled 'Cockney', regardless of their precise geographical origin (Stedman Jones 1989). This cultural figure is picked upon as a grotesque, to be celebrated in its natural urban habitat but labelled out of place in the country. The Cockney, as Stedman Jones argues, could signify a commercial rather than industrial working class, styling its leisure life around consumption and display, and typically issuing from a city, London, dramatically expanding in the interwar years on a basis of financial rather than industrial capital. Such long-running moral geographical themes took on fresh meaning in the twentieth century as a 'Cockney' working-class leisure could be contrasted to the morally solid rambler or cyclist. Harry Batsford advised on *How To See The Country*:

We have most of us enough city-dreading Anglo-Saxon blood to feel a rejuvenating transformation at cutting adrift from the huddle of human habitations. But by contrast, take the case of the large party, presumably from the East End, who, a friend of mine said, disgorged themselves from motor-coaches under the Duke of Bridgewater's column on Berkhamstead Common above Oldbury. They produced a gramophone, and started fox-trotting on the turf. 'Why couldn't they have done that at home, Daddy?', said my friend's little boy.

(Batsford 1945–6: 6)

Such events become fables of interwar England, allegories of something more than a song and dance. Joad diagnosed litter as 'a grimy visiting-card which democracy, now on calling terms with the country, insists on leaving after each visit' (Joad 1938: 72; also Joad 1948).

A moral vocabulary of landscape emerges, a language for harmonious human–environment relations. 'Loudness', whether in buildings or people, is condemned. Planner Patrick Abercrombie wrote of 'a special ... tone in different countrysides ... the honk of the motor-car, the sound of the gramophone ... do not enter into the chord: their dissonance is seriously felt and of singular pervasiveness' (Abercrombie 1933: 243–4). Gramophones especially disturbed this precious music of place, Joad singling out black and American 'alien' noises: 'the atmosphere vibrates to the sounds of negroid music. Girls with men are jazzing to gramophones in meadows' (Joad 1934: 171). The Anti-Noise League, seeking the 'conservation of nervous energy', encouraged 'the well-mannered citizen to become noise-conscious', and urged prosecution of the loud (Horder 1938: 180–1).

A Country Code begins to develop, a 'special code of townsman's manners in the country' (Joad 1934: 176), presented as emerging from walkers' practice: 'they have a passion for the closing of gates, hunt litter like sleuths ... They even appoint voluntary officers ... to see that other walkers obey these ordinances which they have imposed upon themselves' (Joad 1934: 175). The call was for regulation and education, the former allowing the latter. People might be educationally and punitively 'taught better':

> It cannot be said that the squalid crowding of Haytor Rocks and Becky Falls, with the legacy of filthy litter, is an improvement in the recreation of the people. The people who do these things should, in their own interests, be taught better, and when they have been taught they will derive more pleasure from their visits, for some enjoyment of natural beauty will be added to that of mere jollification.
>
> (Cornish 1930: 45)

Regulated and educated, citizens might generate themselves anew in the country. As J. Wyatt, Chief Countryside Warden for Nottinghamshire put it, the cures for 'rowdyism' were 'Public enlightenment ... and litter receptacles' (Notts. County Record Office, File DDRC 8/1).[5]

MOVEMENT

People walking in the country, solo or *en masse*, were seen as part of a movement comparable in self-definition to modern movements of art and architecture. Open-air leisure was a part of England advancing morally, spiritually and physically. The material and metaphorical merge in this movement, invariably presented as upward and onward onto heights and futures before unreached or denied. The varying politics of movement can be traced through organised rambling and youth hostelling.

Mass movement

1930 saw the formation of both the Youth Hostels Association (YHA) and the National Council of Ramblers' Federations (later the Ramblers' Association). Geographer and preservationist Vaughan Cornish welcomed 'a new development of great promise ... the formation of clubs and associations for touring the countryside under definite rules of conduct' (Cornish 1932: 11–12). Some rambling bodies, however, asserted forms of conduct fitting uneasily into the preservationist vision of England. A landscape culture of mass rally and action emerged, ranging from the benign to the militant (Hill 1980; Lowerson 1980; Rothman 1982) (Figure 5.2). In 1932 country writer S.

Figure 5.2 Advancing leisure. The Ramblers' Association West of England rally at the White Horse, Westbury, Wiltshire, 20 June 1937; photography by Reuben Saidman

P. B. Mais led 16,000 to see the sun rise at Chanctonbury Ring in Sussex (it was cloudy). Movements of this number inevitably began, especially in the north of England, to come up against property. Open-air tensions of property and propriety arose when access was denied, and trespass was the response. Normally bombastic preservationists, wary of offending landed interests and worried by politicised mass movement, became reticent, except for Joad.

Joad addressed several mass outdoor rallies: 'It would be difficult to exaggerate the difference between these gatherings of ramblers ... and the ordinary indoor meeting, held in some stuffy hall, where the speaker's voice is punctuated by the coughs of the ailing audience' (Joad 1934: 59). Joad presented alert and increasingly 'militant' ramblers, 'ready to take the law into their own hands' (Joad 1934: 99), and recounted the most public example of militancy, the first April 1932 Mass Trespass on the Derbyshire moorland of Kinder Scout. Here was a popular politicised walking, making for a different country to the England of most preservationists. The trespass was organised by members of the communist British Workers Sports Federation. Unsympathetic media fixed on this as political manipulation, but in the north of England politics was deeply embedded in the organisation of the open air. Trespass organiser Benny Rothman recalls a land reclamation by the people, an assertion of a right to walk on moors closed by owners for shooting (Rothman 1982). This was a movement from below, both socially and topographically, workers ascending from industrial cities to the hills at weekends. Phil Barnes's photographic survey of 'Views of the Forbidden Moorlands of the Peak District' (Barnes 1934), arguing for the restoration of ancient walking rights, showed an empty denied land.

Such political walking troubled most preservationists, who sought, along with the leaders of many 'respectable' rambling groups, to direct a powerful movement away from a far Left politics. It is important though to stress that both preservationists and the most political ramblers often offered the same arguments for walking; a physical and spiritual escape from the city, a morally beneficial leisure taking the working class out of the pub and cinema. This was moral practice for all sides. A Sheffield socialist Clarion Club booklet cover declared how 'A rambler made is a man improved' (Hill 1980: 32). And ramblers, even those engaged in mass trespass, would insist on their own good conduct in the country, practising 'an ethic of nice trespassing' (Hill 1980: 53). Phil Barnes ends his trespassing photographic survey by criticising 'the behaviour of a small section of the public', and presents a code to be followed in the country (Barnes 1934). At both local and national level access and preservation bodies co-operated, while figures such as Joad and Tom Stephenson voiced the concerns of both movements.

Differences, however, remained. Many of the often wealthy preservationists preferred to trace problems to the abuse rather than the possession of landed wealth. Howard Hill blames preservation bodies for the dilution of rights of access in parliamentary legislation (Hill 1980: 76–82). When hundreds strode up Kinder, the preservationists' mix of élitism and populism hesitated. The political action of others trampled any expert middle ground underfoot. Here were not meek individuals seeking regulated education but bolshie groups who marched to urban folk songs written by the young communist Ewan MacColl: 'I'm a rambler, I'm a rambler from Manchester way/ I get all my pleasure the hard, moorland way/ I may be a wage slave on Monday/ But I am a free man on Sunday' ('The Manchester Rambler', reproduced in Rothman 1982: 9; on MacColl see Samuel 1989). This may not have been a vulgar jazzing of the moors, but it was a dangerously political walking tune. Did such a song 'enter into the chord' of the country? The preservationists found a more easily landscaped citizenship in the youth hostel.

Simple chains for a holiday

The YHA offered a less militant open air. Preservationist and historian G. M. Trevelyan acted as President until 1950, Abercrombie, founder of the Council for the Preservation of Rural England (CPRE), was a Vice-President. While the YHA had common ground with the rambling movement, notably through activists such as Tom Stephenson, the two could occupy different cultural spaces. The bolshie mass rambler hardly belonged in a youth hostel.

Youth hostels offered young people cheap overnight accommodation, supervised by a Warden. The movement began in Germany before 1914; the British movement emerged from a joint initiative of the CPRE, rambling clubs and the National Council for Social Service (Coburn 1950: 17–18). It produced a particular moral environment. Pleasure was to be taken in simplicity; comfort might detract. Landscape, basic nourishing meals and a bed at the day's close were the only necessities. A proto-Country Code appeared in the annual Handbook. Hostels operated on strict rules; single sex dormitories, no dormitory smoking, no intoxicants, no gambling, lights out and silence after 10:30, beds to be made on arrival, maintenance chores to be allocated between hostellers, the Warden able 'to retain the card of any member whose conduct is open to objection' (YHA 1939). As if advising on fieldwork, the Handbook listed necessary and sensible equipment and clothing. Rules were not presented as moral clampdown, but rather as devices to foster a new morality, indeed the YHA presented itself as running up against the 'guardians of public morality' (Coburn 1950: 3). Here male and female mixed, except in dormitories, and often in shorts. John Lowerson notes the YHA's 'achievement' in creating 'a sense of institutionalized respectability for activities all too readily seen as subversive' (Lowerson 1980: 270).

The doctrine of simplicity extended to hostel design. Country houses inherited by the YHA from private owners might infringe the principle, but new hostels were conspicuously lacking in ornament, whether built in local vernacular or modern functionalist form. Undecadent beauty was the order: 'daily life lived ... on a principle of self-service, a minimum of privacy, and

Figure 5.3 Youth on the march. The cover of *Youth Hostel Story* (1950),
a record of the YHA's first twenty years, designed by Conroy Maddox.
The image is predominantly red on the book cover, green on the dust jacket.
Individual hostel designs are picked out, including the modernist National
Demonstration Hostel at Holmbury St Mary in Surrey, designed by
Howard Lobb

absence of all kinds of upholstery, physical and mental ... Subject to that, I do not think it can be made too beautiful' (W. H. Perkins, quoted in Coburn 1950: 81). The arrangement of the hostels into regional and national networks also suggested a modern order. Regional groups sought to colonise their own areas and link into adjoining networks, aiming 'to cover England with a chain of hostels, each within a day's walking distance of the next' (Joad 1934: 16). YHA posters showed the hostels as red triangles on a large contoured map, stepping-stones across the land. By 1939, 280 hostels formed simple chains for a holiday (Figure 5.3).

Maps and scouts

Such movement did not occur on empty cultural ground. For many commentators existing youth movements, particularly the Boy Scouts, provided a model of outdoor citizenship, applicable to both men and women (Rosenthal 1986; Springhall 1977). A dibdobbery of walking emerges in three fields; the kit of the rambler, outdoor education and the role of the map.

Youth movements wore uniform, and a uniform was ascribed to the rambler. Efficiently and healthily attired in shorts and stout shoes, and carrying appropriate equipment, the rambler was, in the architectural language of the time, fit for his or her purpose. Tom Stephenson, editor of the YHA journal *The Rucksack*, pictured 'The Good Companions' of hobnail boots, map and well-packed sack (Stephenson 1946: 32) (Figure 5.4).

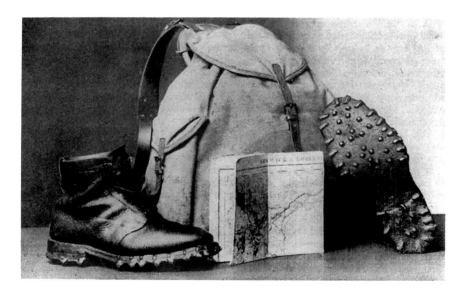

Figure 5.4 The Good Companions, photography by Tom Stephenson, from *The Countryside Companion* (1939), edited by Stephenson

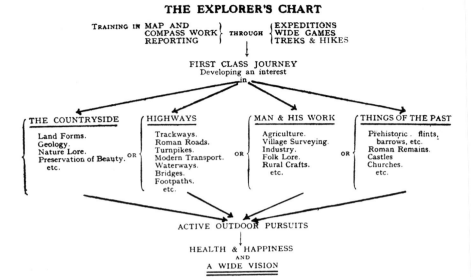

THE EXPLORER'S CHART

Figure 5.5 'The Explorer's Chart', from the 1930 'Gilcraft' scouting volume on *Exploring*, written by Francis Gidney. (Insert boy at the top of chart to produce citizen)

Citizenship, for Scout, Guide or rambler, could emerge from outdoor education. The 'Gilcraft' scouting booklet on *Exploring* presented 'Hiking' as 'the explorer's method above all others' (Gilcraft 1942: 72). *Exploring* reads like a handbook for elementary geographical fieldwork, outlining procedure for survey and reporting: 'not only so that they may enjoy a fuller life ... but also so that they may become informed citizens' (Gilcraft 1942: 56) (Figure 5.5). Education could be both behavioural and aesthetic. For behaviour the same conditions applied for adult or child. 'Good Manners', Joad wrote, were 'not Instinctive, but Acquired' (Joad 1938: 74): 'I would have every child ... pass an examination in country lore and country manners before he left school ... There is much to be said for requiring every townsman who had not succeeded ... to wear an "L" upon his back when he walked abroad in the country' (Joad 1938: 79–80; also Williams-Ellis 1928: 73–4). For aesthetic education different principles held. A sense of beauty was regarded as exposed in the child but buried in the adult: 'Children do not really love ugliness; on the contrary, they have a natural love for beautiful things' (Peach 1930: 26). Youth is presented as an innocent undifferentiated block, whose exposed faculties might be nurtured through education. For the contemporary adult such faculties remained latent, the task of education being 'to regain something of the direct appreciation of the child' (Osborn 1943: 11).

The central document of educational movement was the map: 'Maps are your charter to the countryside and its innermost recesses ... You need not fear to become a map-slave; the chains are light, and lightly worn' (Batsford 1945–6: 60). Map-sense could mark you as special; Batsford patronised the village pub:

> Country folk are generally not map-conscious ... You can astound the company of the village bar by telling them the message of the map for ... the surrounding landscape, and if you produce a map measurer you are likely to be suspected of black magic. They are ... unknown in country circles.
>
> (Batsford 1945–6: 63–4)

Map-consciousness also marked the good Scout. *Exploring*'s chapters on 'Map-reading' and 'Map-making', like Stephenson's 'Making the Most of the Map' (Stephenson 1946), are replete with diagrams of triangulation and traverse, key techniques for the walking, wondering and knowing citizen.

Musical analogy was often used to convey marvellous inquiry: 'The score of a musical composition would convey little ... to one unskilled in reading ... and it is so with a map ... we must become familiar with the notation' (Stephenson 1946: 33). A key text here was journalist, novelist, cyclist, Liberal and walker C. E. Montague's essay 'When the Map is in Tune'. Montague offered a ticket to wondrous and direct knowledge:

> The notation once learnt, the map conveys its own import with an immediateness and vivacity comparable with those of the score or the poem. Convexities and concavities of ground, the bluff, the defile, the long mounting bulge of a grassy ridge ... – all come directly into your presence and offer you the spectacle of their high or low relief with a vivid sensuous sharpness.
>
> (Montague 1924: 40–1, quoted in Gilcraft 1942: 41; Stephenson 1946: 41; on Montague see Elton 1929)

Magic comes not from vague dreaming but precision, from a mind trained to see. Montague sought to raise 'from the dead' to 'the sensuous imagination' the facts of geography: 'Geography, in such a guise, is quite a different muse from the pedantic harridan who used to plague the spirit of youth with lists of chief towns, rivers and lakes, and statistics of leather, hardware and jute' (Montague 1924: 43). Such delightful knowledge might seduce the reader, not away from sense and navigation but into a newly sensible world.

The map was often shown in action, most notably on Ellis Martin's covers for the new Popular and Tourist Edition Ordnance Survey maps (Browne 1992; Nicholson 1983) (Figure 5.6). Invariably the map user is stationed on a hill overlooking a valley. A church-gathered nucleated village nestles below. Unfolding the map unfolds the country. Mental overview only makes sense through concrete environmental practice; the field becomes the necessary place to exercise citizenly thought (Matless 1992). A sense of survey satisfies: 'I've stood on the edge of the Downfall/ And seen all the valleys outspread/ And sooner than part from the mountains/ I think I would rather be dead' (MacColl, quoted in Rothman 1982: 9). MacColl's Manchester Rambler

124

Figure 5.6 Advertisement for Ordnance Survey maps,
featuring one of Ellis Martin's cover designs

achieves the hard height of Kinder; Martin's covers tended to southern English valleys, places of less militant exertion, even of strolling. A standard design covered maps of any location in a southern Englishness.[6]

J. M. Tucker's 1936 *Hiking*, set at Avening in Gloucestershire, catches the ethos of composed English movement (Figure 5.7).[7] Three young women walk in the landscape. Kitted out for freedom with packs, shorts and accessories, their central enabling document is the map. Moving over the country, from one part of the national survey to another, finding localities within the national grid, coming upon things over hills, fixing their place by

Figure 5.7 Hiking, by J. W. Tucker. Laing Art Gallery, Newcastle upon Tyne, tempera on panel, 51 × 60 cm

the symbol for a church with a tower, taking refreshment in the village-in-the-valley and striding to the ridges, these are women fit for their purpose of discovery. Sun shines on the map, indeed the scene's light almost beams from the map, casting its language over the country.

PILGRIM WALKING

This essay has focused so far on intellectual, moral and aesthetic encounters with landscape. Country movement also however had spiritual and physical dimensions. The spiritual will be examined through moves to establish national parks, and through models of leisure as pilgrimage.

National park, spiritual space

Trevelyan, warning of 'breeding a race apart from nature' (Trevelyan 1929: 22), suggested that 'Without vision the people perish, and without natural beauty the English people will perish in the spiritual sense' (Trevelyan 1929:

19; Trevelyan 1931). For preservationists nature was a national and universal spiritual resource, with mystical potential for the walker (Matless 1991). The culture and politics of this mysticism is shown in interwar arguments for national parks, eventually established in 1949 (Sheail 1981). Questions of who might move in national spiritual space are central to debate.

Cornish, the leading parks campaigner in the late 1920s and early 1930s, presented parks as sanctuaries where 'the urban population, the majority of our people, can recover that close touch with Nature which is needful for the spiritual welfare of a nation' (Cornish 1932: 13). National socio-spiritual development required popular access to 'the untouched elemental prospects which are unrivalled in their power to impart a reverent conception of the Universe' (Cornish 1946: 78). Cornish's national parks would be one element in a new nation made under a new cultural authority:

> The National Parks which we constitute now will, it is reasonable to suppose, endure as such for centuries. The present careless indifference of the town tripper in his charabanc will, I believe, be replaced by a different mood in the succeeding generations. The faculty of appreciating beauty is latent in the generality and merely requires educating ... Within a time which will be short compared with the secular life of the National Park we shall be an educated people, the leaders of thought will lead all classes, not merely an educated minority.
>
> (Cornish 1930: 9)

As zones publicly regulated, though not publicly owned, national parks begged questions of property. A particular model visitor is central to the negotiation of private rights and public control. The key phrase, quoted by Cornish and others, is Wordsworth's in the *Guide to the Lakes*: 'persons of pure taste ... deem the district a sort of national property, in which every man has a right and interest, who has an eye to perceive and a heart to enjoy' (Wordsworth 1951: 127, quoted in Cornish 1930: 30; Williams-Ellis 1934: 227). To Cornish, 'Wordsworth's plea almost amounted to a proposal that the Lake District should be the National Park of Great Britain' (Cornish 1930: 32). The phrase appealed not just in its hint at public control but in its stress on taste. Wordsworth's implied model citizen tends to be missed in more recent discussions (Appleton 1986); Cornish however was quite attuned to such cultural implication. Given the principles of latency underlying his 'aesthetic geography' and nature-mysticism (Matless 1991), Wordsworth's 'eye to perceive' might be the property of all in an educated future, but for the present the national park and its visitors demanded regulation under the eye of an expert. As others were not yet finding beauty and the mystic through their common senses, Cornish & co. were still required to teach them better.

Pilgrimage

Any walk in the country, in or outside a national park, might be a spiritual trip if conducted as a pilgrimage. Nature worship and the iconography of landscape could combine to make walkers 'pilgrims of scenery'.

English travel writing of the time abounds with accounts of 'pilgrimage', suggesting, in content and form, ways of being in the country. H. V. Morton hit upon two senses of the term in his best-selling *In Search of England*. In Stratford he bemoans the tourist:

> I suppose the old religious shrines also received thousands of sheep-like pilgrims who had no idea why they were pilgrims beyond the fact that it was the right thing to do. How I detest the word 'pilgrim' in its modern sense. Also the word 'shrine'.

In the same chapter though Morton makes 'a real pilgrimage' to woods by the Avon which he is sure inspired Shakespeare's *A Midsummer Night's Dream* (Morton 1944: 258–60). Beyond the sheep, distinguished, Morton tracks an English cultural icon.

Preservationists also hit upon pilgrimage to capture desired contemporary movement. R. G. Stapledon, agriculturalist and ruralist, saw country visits as 'the most fruitful human tendency of this century' (1935: 268): 'If I mistake not the meaning of the great pilgrimage ... then the new age has in fact dawned' (Stapledon 1935: 277). Cornish termed himself, and anyone who might follow him, a 'Pilgrim of Scenery'. Pilgrimage here demanded reverence for both nature and nation. Cornish's *The Scenery of England* presents an iconographic gallery of English (and occasionally Welsh) environments; mystical experiences are recounted at Stonehenge, outside Buckingham Palace and by Ullswater. Pilgrimage suggested more than simple worship though. It proposed a goal and a progress towards it, carried a promise of ecstasy but with a rein of humility. English landscape could be a humbling arena, preventing 'our people' losing 'that sense of the true proportion between civilisation and the cosmos which is essential to the religious welfare of a Nation' (Cornish 1933: 323). Pilgrimage suggested disciplined devotion, a discipline generated from within, citizens walking in an embracing order.

BODIES OF ENGLAND

Landscape's physical culture

Interwar public events invariably featured displays of 'mass physical culture'; people jumping up and down, forming pyramids, etc. Country walking can be connected to these bodies of England. Preservationists were concerned for physical health; mind, body and spirit were held to be interdependent. Joad wrote of 'The Making of Whole Men and Women':

the culture of the body as well as of the mind must play its part ...
Whence can we derive ... an education alike of body, of mind and of
spirit, so happily as from Nature? The feeling of the air upon the skin,
of the sun upon the face; the tautening of the muscles as we climb;
rough weather to give us strength, blue skies and golden sunny hours
to humanize us – these things have their influence upon every side of
our being.

(Joad 1934: 150)

Cornish outlined three 'disciplines' in the 'cult of Scenery'; for the spirit
the 'cultivation of the state of receptive contemplation', for the mind the
'acquisition of the scientific faculty ... often the hardest of all for people of
emotional temperament', and for the body the 'athletic' discipline. Cornish
noted that in the 'trinity of eternal values' of Goodness, Truth and Beauty the
latter was often

suspect on account of the opinion that the aesthetic life often leads
towards sybaritic luxury rather than spiritual exaltation. But the
Pilgrim of Scenery is beset by no such snare, for a Spartan habit is
needed for the enjoyment of Nature in her sterner moods.

(Cornish 1935: ix)

Cornish offered the example of mountaineering, 'where the supreme
satisfaction of seeing the world spread out at one's feet is attained only by
those who keep the body in fine discipline' (Cornish 1935: ix). A notional
encouragement of an activity for any class or gender offers an élite and
masculine version of climbing as a model (Robbins 1987). Contradictions
emerge in calls for a mass of such individuals. Chief Scout Baden-Powell also
prescribed climbing: 'the best possible physical developer of nerve and
muscle and endurance. A good rock climber cannot be a C3 man. And it is
ripping good sport' (Baden-Powell 1922: 44). Attending a 1931 International
Rover Moot in Switzerland he admired the physique, conduct and climbing
of 2,500 17-years-and-over Rover Scouts, 'storm-troops of the larger army':

Their arms are alpen-stocks, their discipline that of good will from
within ... one saw the endless succession of these splendid specimens
of the young manhood of all nations setting out in comradeship
together with heavy packs on their backs and ice-axe in hand to tackle
the neighbouring mountains.

(quoted in Reynolds 1950: 151–2; on Rover Scouts see Warren 1987)

The storm-troop allusion will be returned to below.

The Spartan suggested environmental exposure, a sense of more-than-
observation, an elemental physicality placing great store on skin and lungs,
rain and wind. Stapledon lamented the low 'coefficient of rurality' in
England, 'in the main a function of the precise extent to which a people as
a whole have direct contact with nature'. Walking, as part of a dietary and
environmental bodily regime, could help:

The extent to which they breathe uncontaminated air, the extent to which they eat unprocessed foods, and, for example, the chances open to them of getting a wet shirt in either their work or their play. It is a depressing thought to contemplate that there must be millions of people in England to-day who have never experienced the exhilaration of a thorough good drenching, and whose individual coefficient of ruralicity must be practically nil.

(Stapledon 1935, 4; on Stapledon see Waller 1962; Bramwell 1989; Chase 1989)

A key rain-and-wind text was Trevelyan's essay 'Walking': 'The fight against fierce wind and snowstorm is among the higher joys of Walking, and produces in the shortest time the state of ecstasy' (Trevelyan 1930: 18). Trevelyan merged with the world:

Whether I am alone or with one fit companion, then most is the quiet soul awake; for then the body, drugged with health, is felt only as a part of the physical nature that surrounds it and to which it is indeed akin, while the mind's sole function is to be conscious of calm delight.

(Trevelyan 1930: 16)

The near loss of composure, drugged up to delightful calm, will be returned to below.

Trevelyan is conspicuously not in a mass rambling party. His language, full of Wordsworth and especially George Meredith, subject of a Trevelyan biography, is classically romantic. An individual romanticism is to somehow translate to popular walking, raising those of a lower physical and spiritual culture into whole men and women, yet without disrupting the order of things. A kind of mass romanticism is being called up, a youth hostelling to ecstasy. Political models which offered to resolve such class contradictions will be considered below.

National fitness

Open-air recreation, if allied to planning and education, might make 'A1' citizens, curing the physical and moral 'degeneracy' of urban industrial life. Such arguments gave a modernist twist to late Victorian worries, emphasising the potentials of mass society if expertly planned, with technology serving values 'higher' than the commercial. Hygienic and efficient homes and workplaces, orderly public spaces, a planned town and country, citizens in tune with their environment, national parks for required spiritual immersion; walking took place in a larger scheme of things.

Fitness was central to the vocabulary of this new England. Buildings were to show 'fitness for purpose', and people likewise. E. P. Richards wrote of the open-air movement as 'a coming chief antidote in Great Britain to city, office, shop and factory confinement; to cancer and constipation, nerves and tuberculosis ... to "THREE C-ISM" in all its senses – physical, mental and

Figure 5.8 Hovis as life-enhancing National Health Bread, reproduced from *The Nottinghamshire Countryside*, 1938

higher' (Richards 1935: 2). Regular bodies inhabited the new country, walkers moving in the national health service. V. G. Biller, writing in Stephenson's *Countryside Companion* on 'Camping and Caravanning', argued in medical newsreel tones: 'camping helps him to become a good citizen, and the health-giving powers of recreation ... are widely recognised as being of great assistance in the creation of an A1 nation' (Stephenson 1946: 365–6). Biller followed Richards' free-moving eulogy, and his warnings against 'disobedient bathing' (Richards 1935: 8), by recommending proper disposal of hiking's regularly induced waste. To avoid 'a sanitary nuisance',

Figure 5.9 A men's keep fit class in an unspecified Nottinghamshire
village, from the Notts. Rural Community Council journal *The
Nottinghamshire Countryside*, 1939, 3(1), page 26

dig a hole eight inches deep to 'cover the excreta and protect it from flies'
(Stephenson 1946: 377). The country toilet code seems emblematic of orderly
fit new walking England, body and landscape in functional harmony, moving
well together.

Such environmental regularity prefigures the more recent wholefood
vision of holistic environmental bodies and selves (Coward 1989; Bishop
1991). Such a vision was then, however, tied as much to nation as nature.
Thus Hovis marketed itself as 'the National Health Bread . . . Ask your Baker
definitely for Hovis' (Figure 5.8). Ironically, such holistic expert discourse
helped produce the faith in planning underpinning a post-war technological
medical vision of national health, against which the contemporary wholefood
vision reacts. Medicine moved into the science of the surgery; the outdoor
body re-emerged in anti-modern opposition. Many inter- and early post-war
national health arguments, however, emphasised both surgery and landscape.
The work of Harry Roberts, London East End doctor and Labour activist,
broadly Fabian socialist and journalist, is a case in point (Stamp 1949).
Roberts professed an expert humanism, grounded in day-to-day contact with
working people, and with education at its heart, aiming to plan society

'towards the highest attainable common standard – financial, cultural and hygienic' (Roberts 1942a: 48). In his *The Practical Way To Keep Fit*, open-air movement forms one element of a body culture of 'sensible' clothing and posture, sports and holidays, medicine, town planning for air and gardens, and emotional education. Roberts places walking in what now seems a part-radical part-reactionary complex frame:

> I look upon the enthusiasm for hiking as one of the most important things that has happened in England for many a year. From a sane philosophic and hygienic point of view, it is, perhaps, the most significant social phenomenon since the foundation on a world scale of the Boy Scout movement.
>
> (Roberts 1942b: 140)

Roberts' frontispiece, 'Towards a Healthy and Contented Life', shows a walker pausing in a bracing breeze on a hill over a valley, a church-gathered village below:

> Today in Britain ... there is a real, though often unexpressed, enthusiasm for physical fitness. The 'hygienic' revolution that took place in Germany a few years ago was not particularly Teutonic in its essence. Nothing but organizing zeal is needed to make the movement spread all over Britain.
>
> (Roberts 1942b: 316)[8]

In 1939, in a Nottinghamshire village, thirteen men, aided by a trestle table, balanced themselves into a curious human pyramid (Figure 5.9). What possessed them? Had they taken Hovis? '"The Nottinghamshire Country-side"! What an inspiring title, and how well it links up with the National Fitness Movement!', exclaimed Lord Aberdare, Chair of the National Fitness Council (NFC), in the journal of the Nottinghamshire Rural Community Council (Aberdare 1938). The village thirteen, consciously or not, were pyramiding in a national movement. The NFC, set up by the 1937 Physical Training and Recreation Act, instigated a National Fitness Campaign, aimed particularly at the young. As with the YHA, older people were not excluded; the aim was a metaphorically young country. Youth was a key word of the time, suggesting energy, vigour and the future. Inspired in part by a parliamentary report on the German 'Strength Through Joy' movement (Jones 1987), the Council made films, organised 'mass physical culture' demonstrations, and gave grants. Gymnasia, swimming baths, youth hostels, all modern sites for modern bodies, were favoured. Publications showed people striding outdoors or synchronised in exercise: 'It is everybody's duty as a citizen to be as fit as possible' (National Fitness Council 1939: 8).

Jones (1987) and Lowerson (1980) present the Campaign as reflecting state anxiety over citizens' capacity to fight a forthcoming war. While this was certainly a factor, the ideas of the NFC were not novel. The idea of national fitness, first propounded in the 'National Efficiency' drive following the revelation of poor physical capacity in Boer War recruits (Searle 1971), had been central to a wide environmental discourse of citizenship. The NFC sought to harness this discourse into state service, both by rooting national movement locally and embracing local movements in a national order. Which returns us to the 'inspiring' Nottinghamshire Countryside.

Opposite Aberdare's enthusings came an anonymous article headlined 'Strength Through Joy: Suggestion For a Rural Fitness Policy'. A Nazi slogan appears as a banner in a local English magazine. The author suggested a 'national rural fitness festival' in every village, with competitions including No. 7 Gymnastic Displays, No. 11 Folk Dancing, No. 12 Volleyball, No. 18 Relay Race for Parish Councillors, No. 19 A Series of Purposeful Games: 'Can Notts. villages give us . . . just that experience in practice that will enable us to plan a modern "Merrie England" along the lines of Strength Through Joy?' A Nazi ethos could be locally reworked: 'While continental countries achieve fitness by discipline imposed from above, Britain plans to succeed with fitness schemes that appeal because they come from a self-imposed discipline generated in the heart of the individual' (Anon. 1938). The author tries to detach the slogan from its country of origin, freeing it for Englishness.

This is half-sinister half-farce, Parish Councillors relaying to a higher plane of citizenship through discipline nicer than the nasty continental type. People are to impose themselves upon themselves. Are the thirteen in the village pyramid enthusiastically marking themselves out from slovenly others? Or merely put upon by Aberdare and his ilk? Or enjoying themselves, recreating their lives through new everyday practice, while poking fun at the huff and puff of the instructor? Jill Julius Matthews addresses parallel questions in her study of the Women's League of Health and Beauty (WLHB). The League, set up in 1930 (motto: 'Movement is Life'), organised many a demonstration of mass physical culture, providing through local groups and national gatherings a space for a particular kind of femininity. However, the emphasis on human 'Racial Health', the use of mass spectacle, the uniform and elements of leader worship and Germanophilia, fuelled an 'association between fitness and fascism' which would 'haunt the League for decades' (Matthews 1990: 40), and this despite the displays being as much Busby Berkeley as Nuremberg, and the 'rational dress' uniform expressing a feminism more than a fascism (Wilson 1985). We return to questions of fascism below.

Fascism was not the only 'dangerous' moral and political ground trodden by open-air and fitness movements. As Matthews notes, bodies like the League, in bringing to mass popularity activities 'which had formerly been relegated to the world of foreigners and health cranks–sunbathing and tanning, hiking, dieting and slimming' (Matthews 1990: 26), generated accusations and attractions of hedonism and voyeurism. It is time to look upon walking from this angle.

An everyday English leisure stressing mysticism, physique and physical pleasure could not help but court hedonism and voyeurism. In its effort to be all-encompassing, to provide a design for living, a discourse of authority reached into danger, threatened to undercut itself. The field of movement was hard to hedge; outdoor bodies might move beyond the pale.

Hedonism

To encourage the mystic in the everyday risked decontrol and self-abandon, a dissolving rather than a sharpening of categories of pleasure. Care is taken to insist that the mystical produces spiritual clarity rather than hazy bliss (Matless 1991). Loose mystic cannons should not fire in the youth hostel. Likewise the process of 'Educating the Emotions' involved direction into 'useful channels', such as mountaineering (Roberts 1942b: 113). Emotions are conceived of as fluid; from this proceeds their potential but also their risk. Canalisation is required lest dangerous floods overcome the self.

A heightened sense of one's own body could also suggest narcissism walking. Preservationists caution against transgression, Cornish emphasising 'the tactile sensation' only to immediately warn against decadence:

> Those who intend to get the maximum of pleasure from the world's scenery should cultivate an outdoor habit which will extend the range of pleasurable response to heat and cold. The Spartan not the Sybarite is the epicure of scenery.
>
> (Cornish 1935: 24–5)

Cornish remains a connoisseur of sensations, hot or frosty, but seeks to walk the tightrope away from lax association. The Spartan denotes an alert joy in the senses.

Associations of hedonism were hard to avoid. Foreignness helped place certain activities beyond respect, Germany in particular lending dubious cultural tones; in the thirties of fascism, in the twenties of dodgy modernism. Stephen Spender recalls a late twenties German modernist 'popular mass-movement': 'Roofless houses, expressionist painting, atonal music, bars for homosexuals, nudism, sun-bathing, camping, all were accepted ... It was easy to be advanced. You had only to take off your clothes' (Spender 1953: 92–3). This was a landscape of a different citizenship, having scarce regard for nation or state. Lights out at 10:30? Unshared beds? This was a different walking landscape, and hints of its striding atonality carried to England. Preservationists can be thought of as moving a Victorian morality into a restrained modernism, beauty and the body moving within bounds. The preservationist movement might be self-consciously different and new, but it was not subversive. Like the modern yet 'respectable' women of the League, the preservationists were: 'conservative progressives on the side of a beauty culture that was winning the struggle to dissociate the cultivation of

physical beauty from accusations of narcissistic vanity and sexual abandon'
(Matthews 1990: 48).

Winning the struggle, perhaps, but such bodies caught those English
cultural undertows which pull bodily exhibition into nudge-and-wink
territory. An illuminating parallel comes in the interwar emergence of
nudism, whose texts continually guarded against sauce. The leading naturist
magazine was the fit-sounding *Health and Efficiency*, while Maurice
Parmelee's *The New Gymnosophy* argued for nudism on grounds of 'Natural
Rearing of the Young', 'Man the Air and Light Animal', 'The Aesthetics of
the Human Body' and 'Gymnosophy and Humanitarian Democracy'
(Parmelee 1927). Nudism constantly stressed the restraint of pleasure; even
so it provided a ripe target for that saucy humour running out of the music
hall into the cinema. Thirties star George Formby, ukulele in hand, could sing
and wink how:

> I've got a picture of a nudist camp/ In my little snapshot album/ All
> very jolly but a trifle damp/ In my little snapshot album/ There's Uncle
> Dick without a care/ Discarding all his underwear/ But his watch and
> chain still dangles there/ In my little snapshot album.
> ('In My Little Snapshot Album', by Harper, Haines and Parr Davies,
> from the film *I See Ice*)

Moving to make a new Englishness through a new bodily morality, the
preservationists also moved in an England of Formby songs and seaside
postcards.[9] Whether nudism was *de rigueur* among preservationists is
unclear, but Joad recommended nude sea bathing on deserted beaches, where
'afterwards you lie on the floor of the cove, naked to the sun' (Joad 1946:
149–50), while League founder Mollie Stack advocated private nudism for
'skin-airing' (Matthews 1990: 29). Two kinds of Englishness rub against one
another, one looking down on common vulgarity, and seeking to raise people
from low humour to a higher body, the other popularly laughing from below
at the moral heights.

Voyeurism

Those who roared at Uncle Dick's watch and chain might also be tempted
to peer. Laughter, whether male or female, might pause by the hole in the
nudist camp fence. And hints of voyeurism appear too in those promoting
physical culture.

Fitness literature delights in displaying young flesh, while preservationists
relish the sight of youth moving, 'the spectacle of our youth making joyous
contact with nature' (Caine 1938). The walking and sporting subjects of new
England were objects to behold. Looking does not of course constitute
voyeurism, but a terrain emerges here where youth becomes the thrilling
object of the gaze of older, usually male authority. Scouting, led by one keen
to train young male bodies against dubious women and 'self-abuse' ('keep the
racial organ cleaned daily' (Baden-Powell 1922: 111)), shows a clear opening

for voyeurism. The rhetoric of catching and holding and moulding boys lays the ground for the scoutmaster joke. *Scouting For Boys*, Baden-Powell's most famous text, becomes a ready-made double entendre. Authority walks straight into mockery, inviting comedy from those it seeks to enlist. And such upright authority leaves itself open to other uses. In the confines of the troop and the camp, spaces of potential (though not necessary) voyeurism and more are opened up through the great authority vested in the boy-moulding master.

Martin Green has traced the common interwar upholding of the youthful figure, especially the male 'naif', for adulation (Green 1977). Such admiration tended to accompany a desire for radical change, especially on the far Right. Mountaineers such as George Mallory, aviators such as Lindbergh, adventurers such as T. E. Lawrence, mavericks such as Edward Prince of Wales, the last three all associated with the radical Right, became emblems of an attractive force for change (Green 1977; Cunningham 1988). Outdoorly active young men signified hope. Green ties such adoration into 'the worship of the male adolescent by older men that is expressed in the myths of Narcissus, Adonis, and such' (Green 1977: 27). The preservationists, generally men of middle age or older, can again be read as seeking to rework a Victorian imperial masculinity into an assertively modern form, though caring to avoid more 'dangerous' modern moralities.

Such danger was however diffused by a feature of the English class system through which most preservationists had passed, the public school. The idolisation of the sporting male, the hints of the homosexual, appear far less dangerous when this context is recalled. Jeffrey Richards has traced the various meanings of 'manly love' in the public school, at times a locus of scandal and fear, but more often a safe ground (Richards 1987). The public school playing-field was scouting's ancestral turf (Mangan and Walvin 1987; Rosenthal 1986). A mass of communist ramblers striding onto Kinder might shake this ground, but Cornish, secure in his class masculinity, could safely recall the thrill of sailors in the Great War:

> In the home country the aspect of Manhood at this period was less exhilarating, but there was a distant haven in home waters, Scapa Flow, where one could feel the full force that lies pent in the finest specimens of our Race.
>
> (Cornish 1946: 38)

CHOREOGRAPHY

Before concluding this essay it is worth discussing the body in the landscape via a notion of choreography. Where awkward Cockneys blare, loudness issuing from every clumsy movement, hikers and hostellers move in composed formation, in choreographed Englishness. Even in the howling ecstasy of a wet gale Trevelyan or Cornish would not emit undue noise. Their bodies are sound, their selves do not decompose.

Figure 5.10 'The Author Ski-ing in the Alps. Nearing the Summit.' J. P. Muller pauses in the cool fresh air, from *My Sun Bathing and Fresh Air System, c.*1930: 86: 'Even in winter a sportsman requires no more warmth than that supplied by the sun. I have often spent half-a-day running on skis in the mountains with only my boots and socks on, the thermometer registering many degrees below zero' (*c.*1930: 42)

Again some cultural detours can cast a different light on country walking. A sense that 'Life is Movement'[10] captures the outdoor ethos of many health books, especially those offering a male athletic aesthetic (Lewis 1985). Anti-decadent masculinity walks the land, choreographed in health. Thus ex-Danish Army Lieutenant J. P. Muller, author of a range of 'My System' books for men, women and children, offered a mix of hedonism, narcissism and voyeurism in his *My Sun Bathing and Fresh Air System* (*c.* 1930). The author argues for vitamins and town planning, and displays his body 'bathing' in snow, skiing up mountains aged 60 in shorts (Figure 5.10), performing 'rubbing exercises', etc. A blend of hygiene, recreation and voyeurism, 'Issued under the patronage of H. R. H. The Prince of Wales', is to choreograph citizens into landscaped health.

Outdoor choreography emerges too in Modern Dance. Martin Green's study of the early twentieth-century 'counter-culture' at Ascona in Switzerland finds dance as a key component in an aesthetic and political radicalism conducted through the body (Green 1986). Choreographers and dancers such as Rudolf Laban and Mary Wigman reacted against the theatrical ballet through a modernist movement of natural and bodily 'eurythmics', belonging outdoors in a culture of 'Life-body-gesture-movement' (Green 1986: 98). Laban was expelled from Germany in the late 1930s, and moved to England, where his 'Basic Classification of Movement Analysis' was adopted by the Ministry of Education in its physical education training courses. Laban later co-devised the Laban–Lawrence Industrial Rhythm, assessing the kinetic quality of industrial labour.

Laban's biography brings out the connection of landscape's physical culture to a kind of Taylorism of the body, 'an ethos of mechanized capitalism ... an ethos based in a principle of simultaneity of the forms of work and art and leisure' (Matthews 1990: 43). There are tensions in such bodily planning between expression and standardisation, but the connection is an important one. Landscape's culture of the body could feed into a state corporatist public culture, a culture furthered through the post-1945 town and country planning established by preservationists such as Abercrombie. It is less than fanciful to connect, though not to equate, Laban's Modern Dance and Harry Roberts' call for more open-air dances, outside of stuffy unhygienic halls: 'There is a pleasing, and not unintelligent, abandon, which provides exactly the ideal condition for the efficient working of those unconscious processes of metabolism which make up the whole basis of healthy life' (Roberts 1942b: 68–70). Tucker's hiking women perhaps perform an open-air dance of England, a folk-modernist dance around the map, their bodies in trim with the rhetoric of preservation.

There is another pertinent culture of the body here though. Laban fled Germany only after falling out of favour with the Nazis; his dance had been well-favoured before. 'Modern Dance' had become 'German Dance', and Laban was entrusted with the dance component of the Berlin Olympics. Goebbels' disapproval however led to arrest and subsequent exile (Green 1986: 109–12). Again the association of fascism crops up in this essay; we must now consider in more detail the place of this politics in the English culture of landscape.

LANDSCAPE, AUTHORITY AND PSYCHOLOGY

For preservationists the resonance of fascism, whether Italian or German, lay in its concern for landscape and citizenship, and its model of cultural and political authority. It should be stressed that none of the preservationists figure in Richard Griffiths' survey of British *Fellow Travellers of the Right* (Griffiths 1983); indeed many, as has been indicated, professed a Liberal or socialist position. This did not though prevent fascism offering example or temptation.

Fascist Germany's environmental and planning policy is increasingly well documented (Bramwell 1989; Groning and Wolschke-Bulmahn 1987; Shand 1984). Preservationists expressed admiration for specific policies; the autobahn programme, nature conservation measures, etc. Nazi Germany was also enthusiastic about youth hostelling, the thriving existing movement being redirected for fascist ends (Coburn 1950). Baden-Powell's storm-trooping Rover Scouts would have found a fine network of purpose-built hostels. While preservationists might applaud such policies, however, they seldom express pro-fascist political argument. Occasionally though there is a hint that there is more than policy content to admire, that the cultural authority embodied in fascism might also tempt a movement keen to reorder England's town and country.

The preservation movement took firm shape around the turn of the 1920s and 1930s, a time of great debate within the élite over the ability of existing forms of government to stave off social unrest and economic collapse (Skidelsky 1967). New forms of authority were floated, most notably by Oswald Mosley, who left Labour to form the New Party, for which Joad acted as Director of Propaganda. Joad left when Mosley began to move to fascism. Preservationist discourse was formed in the context of such debates on government and authority. Planning, order and discipline were the watchwords, whether for economic policy, political organisation or country walking. While Cornish mused on national parks under new cultural authority, Clough Williams-Ellis, introducing the key preservationist text *The Face of the Land*, offered a choice between 'Laissez faire or Government', chaos or discipline:

> We need direction and leadership now as never before, because now, in this generation, a new England is being made, its form is being hastily cast ... If there is no master-founder, no co-related plan, we may well live to be aghast at what we have made ... What then must we do to be saved from this future state of chaos, ugliness and inefficiency?
>
> (Peach and Carrington 1930: 20)

Answers could be sought abroad, in the planned advance of Stalin's Soviet Union, whose architectural and planning policies Williams-Ellis admired (Williams-Ellis 1971: 184–6), in Mussolini's Italy, and later the America of the New Deal. The theme of new kinds of authority for a planned society was not exclusively fascist, but the concern of preservationists to transform individuals through environmental practice and to create a specifically national planned environment perhaps found their clearest echo in Germany after 1933.

Some preservationists, notably Stapledon, who argued for Britain to follow the land-reclaiming example of Mussolini by reclaiming The Wash for agriculture as a national morale-booster (Stapledon 1935), were linked to that strand of English ruralism which found much to admire in German agricultural policy (Chase 1989, 1992; Bramwell 1989). Most however looked more to planning and environmental than specifically agrarian policy. Before 1933 we find odd links between English preservation and German fascism. Harry Peach, Fabian socialist, looked in his litter campaign to well-designed German bins with rhyming notices (Peach 1930). He encountered these through his correspondence with Walther Schoenichen, long a proponent of racial doctrine, who would join the Nazi party in 1932, act from 1933 to 1938 as Director of the 'Governmental Agency for Preservation of Natural Monuments', and in 1943, still anticipating victory, propose a system of national parks for occupied Central Europe (Groning and Wolschke-

Bulmahn 1987). In late 1931 and early 1932 Peach corresponded with Schoenichen over a tear-off calendar the German had produced, showing landscape scenes for each month. Peach urged the CPRE to produce something similar for England (Council for the Preservation of Rural England Archive, file 254). One cannot infer here that Peach, who died in 1933, had fascist sympathies, but one can I think connect such preservationist thinking to that German 'reactionary modernism' of the twenties which had 'turned the romantic anticapitalism of the German Right away from backward-looking pastoralism, pointing instead to the outlines of a beautiful new order replacing the formless chaos due to capitalism in a united, technologically advanced nation' (Herf 1984: 2). It is perhaps less important here to pin a 'Left' or 'Right' label onto preservationists – current political categories do not easily project back (Potts 1989) – than to stress their consistency in calling for authority, planning and order.

The temptation of fascist authority is clearest in Lord Howard of Penrith's essay in Williams-Ellis' collection *Britain and the Beast* on 'Lessons from Other Countries'. Howard covers Swedish and Swiss measures, but gives greatest attention to Germany, quoting as 'characteristic and essentially true' the German legislation on nature:

> The protection of objects of natural interest which has been growing for centuries could be carried out with but partial success, because the necessary political and cultural conditions were lacking. It was only the transformation of the German man which created the preliminary conditions necessary for an effective system of protection of Natural Beauty.

Howard wrote that 'Whatever we may think or feel about Nazi political philosophy', such efforts should be applauded, and he regretted the unlikelihood of 'a British Government' introducing such measures (Howard 1938: 284–5). Such a respectful and envious citation of fascist action, a citation which connects landscape and citizenship, nature and 'the German man', suggests that for some preservationists Germany provided an example not only of policy content but cultural authority. 'Lessons from Other Countries' might suggest a radical political means to a new England. The most common preservationist tack though was not to applaud fascism but to learn lessons of authority from it. 'Strength Through Joy' might be turned to English ends. J. M. Keynes, arguing in *Britain and the Beast* for the provision of public festivals, warned 'western democracies' to tend communal feeling: 'These mass emotions can be exceedingly dangerous, none more so; but this is a reason why they should be rightly guided and satisfied, not for ignoring them' (Keynes 1938: 6).

Keynes' words show the importance placed by preservationists on the moulding of individual and collective psychology. It is worth briefly attending to this psychology in conclusion. The preservationist concern for the selves of England often ventured into contemporary psychological debate. Regional surveyor and preservationist C. C. Fagg pursued a serious interest in psychoanalysis (Matless 1992); Stapledon, advising on 'The Non-Material Needs of the Nation', urged contact with nature as the chief psychological need implied by psychoanalytic findings (Stapledon 1943). An intriguing story remains to be written on the links between environmental and psychological theory at this time; Arthur Tansley, for example, was the leading British proponent of both ecology and *The New Psychology*, which to him implied a move away from individualism to a new kind of national and international collective life (Tansley 1920). G. H. Green of the Health and Cleanliness Council provides a final example, on the one hand writing *The Healthway Books* for children on hygienic lifestyle ('Where there's dirt there's danger'), orderly housing and town planning (Green 1939), and on the other providing teachers with a guide to deploying *Psychanalysis* (sic) *in the Class Room* (Green 1921). The most modern psychology is to be harnessed for orderly progress:

> the object of psychanalysis (sic) is to investigate the unconscious regions of the mind, and to make possible the removal of the obstructions which dam or divert the stream [of energy], so that the freed 'libido' may flow singly, as a powerful river, from unconsciousness into consciousness, there to be diverted into interests of value.
>
> (Green 1921: 175)

The values of the preservationist might channel the unconscious. For Green the Scouts and Guides were exemplary, impressing upon children through ritual their position in 'the circles of the family, the school, the country and the empire' (Green 1921: 88).

Whether or not preservationists directly engaged with psychoanalysis, questions of psychology were integral to their programme of landscape and citizenship. The individual and the collective mind were to be made through geographical and environmental practice. That way might lie a new England. This essay has sought to show that such concerns of psychology, culture and geography were at the heart of a highly influential discourse of self and environment whose assumptions often remain taken for granted in everyday life. Should noise be made in the country? Why is it good to be in the landscape? Which practices and people are fit for the English countryside? This essay has sought to understand the contours of such questions by tracing a part of their genealogy. A story of landscape and citizenship has entailed excursions into litter, noise, country walking, maps, trespass, youth movements, fitness, mysticism, hedonism, voyeurism, fascism and psychoanalysis. To tell such a story should at least begin to unsettle some everyday ground.

NOTES

1   The modernism in this preservation is sometimes missed in contemporary discussion (for example Jeans 1990). It is increasingly recognised, though, that the polarities of tradition

and modernism commonly used in discussions of landscape and heritage are difficult to sustain, especially for the interwar period. On this see Luckin (1990), Matless (1990a, 1990b), Potts (1989) and Wright (1992). Alison Light (1991) makes a similar point in her discussion of the 'conservative modernity' embodied in much women's writing between the wars.

2   For a briefer consideration of this theme see Matless (1990c). It should not of course be assumed that all country leisure was conducted in such a spirit. The period also sees the development of a highly nostalgic and conservative middle-class motoring pastoral, expressed in many of the English guidebooks of the time. The car itself becomes a tamer, less futuristic mode of transport, used to potter around rather than speed through the landscape (see Morden 1983).

3   For a more theoretical discussion of these themes of conduct around the concept of 'moral geography' see Matless (1994).

4   On the preservationists' narrative of landscape history, broadly one of eighteenth-century peak, nineteenth-century fall and modern revival, see Matless (1990a).

5   Wyatt was addressing the Notts. CPRE County Committee on 14 September 1937. The committee included Nottingham geographer K. C. Edwards, also active in local rambling groups.

6   Martin did not only produce standard cover designs. For his range of designs for individual special area maps see Browne (1992).

7   Tucker's picture has since been regarded as typifying its era, appearing in exhibitions of the period and on the cover of the best general interwar social history text (Stevenson 1984).

8   It is unclear as to whether Roberts' German reference is to the Nazi or pre-Nazi period. Perhaps only Roberts' inner-urban medical practice distances him from Orwell's label of 'outer-suburban creeping Jesus . . . who goes about saying "Why must we level down? Why not level up?" and proposes to level the working class "up" (up to his own standard) by means of hygiene, fruit-juice, birth-control, poetry, etc.' (Orwell 1965: 163; on Orwell's own prescriptions on working-class diet see Bishop 1991).

9   Formby would also star in a late 1930s skit of the fitness movement, 'Keep Fit', poking gentle fun at 'Biceps, Muscle and Brawn'.

10 The phrase 'Life is Movement' may have been first deployed by body-builder and general physical culture exhibitionist Eugen Sandow, famous from before 1914, who used it as a book title around 1919. Sandow may have taken the term from other writers (Sandow c. 1919; Rosenthal 1986). The Women's League of Health and Beauty reversed the phrase for their motto.

6

FAMILIES AND DOMESTIC ROUTINES

constructing the boundaries of childhood

David Sibley

There are many childhoods. The essentialist discourses of developmental psychology suggest otherwise but it is important to recognise that childhood is a problematic concept, one which calls for a polytextual approach to understanding (Stainton Rogers and Stainton Rogers 1992). Thus studies of children in social space need to be complex. Feelings of children about their material and social environments, adult recollections of childhood and adult feelings about children in the family, the home, the neighbourhood and so on, are all important in building up a picture of children's places and spaces.

I am particularly concerned here with boundaries – the limits beyond which children feel they should not go and the excitement, exhilaration or anxieties associated with transgression. Happiness and fulfilment, as well as anxiety and misery, can be associated with boundaries within the family and home as well as those that separate and bring together others beyond the home. Boundary experiences affect the quality of interpersonal relations and the quality of the relationship between the child and the material environment. For this reason, boundary questions seem to me to be an important focus for research on childhood. They constitute a part of the larger problem of liminality as an aspect of social space, one which raises questions of identity, where we belong and how others are perceived (Shields 1991).

Some of these concerns about childhood spaces, relating specifically to the built environment, have been touched on before, for example, in the writing of Colin Ward (1977) and Robin Moore (1986) and there is a growing awareness of the complexity of the problem in human geography (James 1990). In psychological research, emphasis on boundaries is evident in work on family dynamics and family therapy by Minuchin (1974) and Olson, Russell and Sprenkle (1983). However, nowhere are children situated at the same time in the context of the family, domestic space and the larger spaces of the locality and the city. Incorporating all these elements of the problem is one of the objects of this study, although none are explored in as much depth as they need to be. The aspects of childhood which I discuss in this chapter are a few of many possible ones and they are mostly adult constructions of past childhoods. I attempt to locate children in social space, not using the conventional medium of mental maps but rather, drawing on children's feelings about people and places.

Children experience things acutely in a physical sense. Places, events, relationships with others, may be experienced as butterflies in the stomach, nausea, or may engender a pleasant physical sensation. According to Kristeva (1982: 2–3), this is how we as adults recognise the border between self and other, through visceral feelings. As a way of representing children's experience of their environment, the recovery and articulation of feelings about people and places seems particularly appropriate. My inclination to make use of sensations as a way of marking the spaces of childhood comes partly from my own experience, growing up in north London in the 1940s, and partly from reading texts like Kristeva's *Powers of Horror* (1982) and Perin's *Belonging in America* (1988). Although there were things that happened in my childhood which I recall as pleasant and gratifying, there was also plenty to be anxious about. As I remember it, pleasure was not unalloyed. There were people and places to avoid, like that part of the primary school playground where I had been hit by boys who were handier with their fists than I was, and Sproul's cat meat shop, one of several vomit-inducing retail premises and houses on the walk to school. Time was also a problem, particularly at school where the dreadful importance of punctuality was reinforced by the head teacher rapping his cane on the white-tiled walls on the way back up the stairs at the end of playtime.

Personal experience, while not an entirely reliable guide to significant problems, does suggest that there are some interesting issues connected with aversions, pleasures and desires which are marked out in the the spaces of the home and the localities of childhood. Fear is one emotion which is now recognised as an important constituent of social space, for example, in the writing of Gill Valentine (1990) and Jo Goodey (1993), in relation to women and girls, respectively, but I think that we can go farther in constructing maps of aversions, anxieties, pleasures and desires which delineate social space, a kind of geography anticipated by Corbin's olfactory history of French culture (Corbin 1986). It is possible to make sense of the personal geographies of childhood, focusing on the experience of boundaries, those demarcating the pure and the defiled (including the places that make you feel sick) and those markers we use to carve up time, like bedtime, playtime, getting-home-by time. These boundaries are elements of a geography which is partly experienced and defined by sensations – fear, anxiety, excitement, desire – which shape the developing child's relationship to people and places, and it is a geography which can be recovered by adults recalling their own childhoods and uncovered by children themselves.

In this chapter, I will examine boundary questions initially in relation to the self, as the concept has been developed in psychoanalysis, because psychoanalysis and, more specifically, object relations theory, connects the individual, the social world and the material environment in an interesting way (Pile 1993). I will then focus on those environments with which children will be most familiar, namely, the home and the near-home environment, where they experience space in both its oppressive and its liberating aspects. I am conscious of the fact that I will be referring almost entirely to western, and particularly British childhoods and that I fail to examine systematically differences between the experiences of girls and boys. Clearly, gender and culture differences are very important (Weisner 1984; Katz 1993) but I have only accounts of British middle-class and working-class childhoods to work with – some boys, some girls, but not enough of each. The material consists largely of recollections of childhood[1] but I also draw on accounts of childhood from the Mass Observation archive, some of which are children's own narratives. Narrow as the selection of narratives is, it is still a rather mixed bag in terms of social class, time and locale and one should be wary of generalising about British childhoods from this sample.

In particular, looking backwards presents problems. Memories are partial and selective and they refer to different pasts. Since the 1940s, there have been big changes in social values and in the material circumstances of childhood. Attitudes to space and time and boundaries have changed considerably during this period as, according to several observers (Katz 1993; Ward 1977; Hillman, Adams and Whitelegg 1990) the environment in which many children grow up has deteriorated, become more hazardous. It is important, then, to be clear about what pasts we are referring to and to avoid generalising about the spaces of childhood during a period of considerable social change.

THE BOUNDARIES OF THE SELF

I will first present a summary of object relations theory or, rather, aspects of object relations theory which seem interesting in relation to the problem of delineating the boundaries of childhood. The observations of psychoanalysts in relation to the emerging self provide an account of expanding relationships to others, not necessarily involving interactions with others, and of relationships to the places populated by family and others who contribute to a sense of self in the growing child. Object relations theory provides us with a map of the self in place, an integration of the spaces of the body, the space of the self and the other, and the mediating material environments of the home, the locality and the world beyond. As it was first articulated by Freud, the theory was more narrowly focused, concerned with the infant's relationship to people. Freud charted the earliest stages of development when a part of the baby's initially undifferentiated feelings are transferred to part-objects (the breast) and then to the whole person (initially the mother). He thus suggested how a sense of border, a differentiation of self and other, develops. However, subsequent, broader interpretations of object relations theory are more appealing, and more relevant to this discussion, because of their assertions about the relationship between the self and both the social and material world.

There are several strands of an expanded version of object relations theory which help to locate the child in social space. First, Mead (1934: 154) provided an interesting cue for a geography of the self in the world when he postulated a 'generalized other', consisting of inanimate objects as well as people ... 'Any thing – any object or set of objects, whether animate or inanimate, human or animal, or merely physical – towards which [the child] acts, or to which he responds socially'. This broader conception of the object world as both social and material has been developed at some length by Csikszentmihalyi and Rochberg-Halton (1981) in relation to domestic space and I will suggest that the bounded spaces which must be negotiated by the child should be incorporated in this generalised other. Second, Winnicott (1957) implied that it was necessary to have a contextual understanding of the self which was similarly broader than Freud's initial conception. As Winnicott put it,

> The family protects the child from the world. But gradually the world begins to seep in. The aunts and uncles, the neighbours, the earliest sibling groups, leading on to schools. This gradual environmental seeping in is the way in which the child can best come to terms with the wider world and follows exactly the pattern of the introduction to external reality by the mother.

Today, the external world enters the life of many children rather less gradually than it did when Winnicott was writing, partly through electronic media, so there is a wider range of external objects to which the child relates and which contribute to the development of a sense of border between self and other. Third, Erikson (1959) argued that the boundaries of the self change over the life course rather than being fixed in infancy and, if we accept this, it becomes possible to draw freely on object relations theory in discussing childhood and adolescence.

Ignoring wider contexts for the moment, we can note how, according to most theories of object relations, the initial sense of boundary emerges. In early infancy, there is a pre-Oedipal one-ness with the mother, what Davis and Wallbridge (1987) call 'primary unintegration or total merging with the environment'. This one-ness with the mother gives way gradually. Initially, the infant relates to 'subjective objects', parts of the body, like the fingers, which appear to be 'other than me'. The child acquires a conception of self through 'a series of semi-objects that stake out the transition from a state of indifferentiation to one of discretion (subject/object) – semi-objects that are called precisely "transitional" by Winnicott' (Kristeva 1982: 32). As the infant further develops a sense of border between itself and the other, it rejects the mother, the mother becomes the other. There is a fear of the other in the form of the mother who threatens the dissolution of the self but there is also a feeling of loss, a desire to re-establish the pre-Oedipal unity of mother and child. Aversion and desire, repulsion and attraction, play against each other in defining the border which gives the self identity and, importantly, these opposing feelings are transferred to others during childhood. At this point, Julia Kristeva and Melanie Klein both provide insights into the link between the individual and the wider social and material worlds.

Kristeva, like Freud, recognises that a sense of border and a sense of otherness are established partly through feelings of revulsion towards bodily residues – faeces, urine, sweat, scurf – which become symbols of defilement, distinct from a pure self. She describes the feelings towards residues as one of *abjection*, in one sense a visceral feeling, nausea or spasms in the stomach. However, abjection is also defined by Kristeva as a desire to expel but powerlessness to achieve it. As Gross (1990: 87) puts it, abject things 'hover on the border of the subject's identity, threatening apparent unities with disruption and possible dissolution'. Abjection is, thus, a perpetual state. The abject cannot be eliminated so fear of the abject becomes a part of object relations. There are two aspects of abjection which are crucial to an understanding of the role of boundaries in childhood (and adulthood), boundaries separating the pure and the abject. First, the abject is an expanding category which includes people and places through the elision of the biological and the social. Bodily residues *become* social residues. As Constance Perin (1988: 178) perceptively notes, 'Evil is embodied, according to Western beliefs, in excrement; Defilement, Deviltry, Disease, and Sin shape this conceptual system.' The second, related, feature of abjection is that it is learned. Abjection displaces the easy relationship infants have with bodily residues because of socialisation and, similarly, the recognition of the abject in the material and social world is learned. Thus, the young child exists in a world where conceptions of the abject differ from those of its parents and this different world-view is often a source of conflict. What enters the catalogue of the abject is also culturally dependent. A mixing or merging of things or people might be seen as defilement in one culture but accepted in another. Likewise, within cultures, some groups, like the residents of stereotypical suburban environments, may have a heightened sense of the abject while, in individuals, a concern with boundaries and cleanliness may be recognised by others as phobic, a form of problem behaviour or deviance. The borders of the self as defined by the abject are therefore socially/culturally constructed. As Lorraine (1990: 16) observes,

> in the sense that the interpretations that make up my consciousness are drawn from a public realm of the interpretive possibilities in my culture, the self that I am is not even 'my' self. Although my interpretations *may* be my own, they are constrained by the possibilities my culture makes available to me.

Kristeva's *Powers of Horror* is an essay on the abject so she necessarily dwells on fear and loathing. Melanie Klein had earlier presented a more balanced account of object relations, particularly distinguishing between good objects which strengthen the ego and bad objects, sources of pain, fear or loathing. An object can be simultaneously good and bad, so, for example, the mother, the first good object as the source of love and comfort, is also the first source of frustration and pain. The object, good or bad, can exist outside the self, or it can be internalised. Further, the good and the bad object can be incorporated in a generalised other, expressed in the benign and malign faces of a stereotype (Gilman 1985), something which could be usefully explored in relation to children's representations of strangers. Klein further managed to knit together the individual and the social through her development of Freud's concepts of introjection and projection. Thus,

> introjection means that the outer world, its impact, the situations the infant lives through, and the objects he encounters, are not only experienced as external but are taken into the self and become a part of his inner life. Inner life cannot be evaluated, even in the adult, without these additions to the personality that derive from continuous introjection. Projection, which goes on simultaneously, implies that there is a capacity in the child to attribute to other people around him feelings of various kinds, predominantly love and hate.
>
> (Klein 1960: 5)

Klein suggests that 'if the interplay between introjection and projection is not dominated by hostility or over-dependence, and is well-balanced, the inner world is enriched and relations with the external world are improved'. In some individuals, however, hostility or repulsion may dominate and this hostility will be projected onto abject others, manifest as racism, for example, as Hoggett (1992) has demonstrated in an interesting application of Klein's psychoanalysis. In such cases, boundary maintenance, keeping the abject at bay, will be a dominant concern, exacerbated by the introjection of negative stereotypes. However, Klein's main interest, stemming from her work with children, was in balance, a balance based on a range of relationships between the self and others and centred around attraction, affection, fear and repulsion, with the implication that most children are not overly concerned with boundaries.

I will now examine some of the questions raised by object relations theorists in the context of the home and the local area, the environments in which children spend most of their time. Specifically, I will consider how the self in childhood is bounded by others and by the spaces of the home and locality – people and things which variously constitute good and bad objects, shaping childhood experience. In the narratives which appear later in this essay, people talk about how the boundaries of the home, the locality and of time were experienced in childhood. This means also talking about the family. Adults relate to children partly through their attitudes to space and time in the home and one way in which children express anxiety and pleasure is through their connection with domestic spaces and objects. The peopling of these spaces and the ways in which families structure space and time in the home are crucial issues. The child, the family and domestic space need to be considered together in order to understand the role of boundaries in childhood.

HOMES

The home is one place where children are subject to controls by parents over the use of space and time and where the child attempts to carve out its own spaces and set its own times. The possibilities for conflict here are considerable. Children may find the domestic regime oppressive because of rigid parental control of space, the availability of space in the home may limit opportunities for children to secure privacy, adults may feel that children get in the way and so on. These problems clearly spill over into public spaces. For example, children playing out with their friends or walking to school are affected by controls exercised in the family. At the same time, regimes which are external to the home, like working hours and the school day, impinge on activities and relationships in the home.

Much of the literature on the home, particularly in environmental psychology, fails to convey the frustrations and anxieties that may be associated with home life. Human geographers have generally shied away from domestic interiors, restricting their investigations to public spaces, although a research agenda was sketched out in the mid-1980s by Williams (1986). The dominant message of environmental psychology is that the private domain of the home is a benign, controllable, personal space standing in contrast to the exterior, public domain which is uncontrollable, uncertain and riven with conflict. The house is haven, the dwelling place in western culture is a 'locus of sentiment'. According to Lee Rainwater (1966):

> There is in [American] culture a long history of the development of the house as a place of safety from both nonhuman and human threats, a history which culminates in guaranteeing the house, a man's castle (sic), against unreasonable search and seizure. The house becomes the place of maximum exercise of individual autonomy, minimum conformity to the formal and complex rules of public demeanor. The house acquires a sacred character from its complex intertwining with the self and from the symbolic character it has as a representation of the family.

Many studies of homes have focused on these qualities. There has been a particular interest in middle-class homes where individuals restore themselves and reconnect with a symbolically rich environment. Rainwater, in the 1960s, recognised a middle-class bias in academic literature on the home and this has continued. According to Korosec-Serfaty (1984: 304) the home provides opportunities for self-expression, so the living room, for example, 'bespeaks the dweller and is a part of the being's anchoring in space'. The home is a place for 'authentic living', presumably meaning that at home a person does not have to act a public role. Similarly, Dovey (1985: 46) argues that the positive experience of the home, because it contrasts with the negative experience of the wider environment, gives home life greater intensity and depth. Some writers on this theme become quite lyrical. Thus, Cooper (1990: 37) pictures the home as:

made up of histories and possibilities. So, the empty house is full of spaces for the imagination, of hopes and opportunities. There is a dreamlike quality in the momentary association of things in the process of change, the accidental relationships of light, space and clutter. Endless alternatives exist in walls almost without traces. The empty space slowly fills, a kind of order is imposed, disciplining, choosing, fixing. The wide view becomes a picture on the wall, a backdrop for the contents of the room – we look increasingly inward toward the detail. But while the limited possibilities in empty, pristine spaces are lost, the changes are the acquisition of a history, a mirror to life ... The home is a space replete with pasts and memories.

This evocation echoes Gaston Bachelard's view of the home as a womb, a place recalled in dreams, 'giving access to the initial shell which shelters the being' (Bachelard 1981). Such an appreciation of the home as a restorative, anchoring, protective and insulating shell clearly has meaning for some people but the realisation of the dream depends on wealth and, apparently, an absence of children. It bears no correspondence to the experience of many people living in families, where adults and children may have conflicting needs and where the appropriation and transformation of domestic space may be frustrated by a lack of money. Thus, Bachelard's happy phenomenology is not very helpful.

In order to understand how children experience the home, we need a perspective which focuses on power relations, the way power is expressed in family interactions and played out in the spaces of the home. While boundary issues, the separation and bringing together of children and adults in space, are my main concern, the route into this problem is through the family as a locus of power relations.

FAMILIES

Dichotomies are often crude conceptual implements, but a useful way to start thinking about the way power is exercised in families is to adopt a dichotomous categorisation, namely, to distinguish between what Basil Bernstein called *positional* and *personalising* families (see Atkinson 1985). Positional and personalising can be taken as poles separating a number of intermediate modes of control or forms of control which are mixed in varying degrees.

'Positional' means power is vested in position, so 'father', for example, may signify power, someone who relates to other family members in an authoritarian manner. This would be manifest in the imposition of arbitrary rules and in giving instructions without explanation. A typical exchange between a positional parent and a child might be: 'Do this' ... 'Why should I?' ... 'Because I say so.' The positional parent would fit the profile of 'the foreclosed personality', a person who is inflexible, rigid, intolerant of ambiguity. This implies a rigid attitude to space and time in the home and anxiety over spatial boundaries. The practice of keeping children out of rooms or spaces decreed as adult spaces and a concern with the temporal regulation of children's activities would be typically positional. Keeping control means maintaining clear, unambiguous boundaries. In Bernstein's terms (Sibley 1988) space in the positional family is strongly classified, that is, spaces are characterised by single uses, there are strong rules to maintain these singularities and any mixing of activities is seen as pollution. Corresponding to strong classification, strong framing may be used by the positional individual to maintain control, that is, within rooms, there is a highly ordered arrangement of objects and activities. Only father sits in a particular chair, for example, or there is a fixed seating arrangement for meals.

In the personalising family, all the distinguishing features of the positional family are reversed. Notionally, power is equally distributed between family members with the implication that the uses of space and time in the home are negotiable. Weisner (1984: 357), for example, asserts that North American families are characterised by 'parental warmth, personal attention to children, family democracy and negotiation and an absence of overcontrol (sic) in family discipline styles'; a gross generalisation, but this is what a personalising family would be like. In regard to domestic space, the mixing of activities is encouraged because the exclusive use of space infringes someone's rights. There is, therefore, no concern with boundary maintenance. Using Bernstein's terminology, domestic space is weakly classified and weakly framed so toys spread across the living-room carpet would not be a problem.

One problem in applying Bernstein's authoritarian/egalitarian dichotomy is that it cannot accommodate what some family therapists have recognised as the needs of the child for *both* relationships with others and separateness. Children need good relationships with others but, in western societies, they are also seen to need privacy in order to develop autonomy. This is evidently not the case in some cultures. There is an almost total lack of privacy in many Gypsy communities, for example, in many rural communities in Africa and elsewhere in the developing world, and there is no evidence that this lack of privacy is psychologically damaging. However, family therapists working in the United States, particularly Salvador Minuchin, have identified the lack of personal space as well as the excessive separation of children from parents as potentially problem-creating.

Minuchin (1974) identified two kinds of family regimes where problems may arise because parents have an oppressive or alienating relationship with their children, one where parents and children are *disengaged* or detached, that is, living in their separate, bounded worlds, and the other where family members are *enmeshed*, or living in each other's laps. In the first case, boundaries are strong and in the second, boundaries are weak or absent. Two comments on childhood/adolescence by adult problem drinkers, the first describing a disengaged family, the second an enmeshed family, suggest that unhappiness or conflict can be associated with both modes of control:

1   D's father was a bit unapproachable. Typically, he would return home from work, have his tea and then spend his evening in the living room, watching TV, listening to the radio or reading. He had his special chair near the TV and radio and the rest of the family had their own recognised seating places. The other living room was the best room, reserved for visitors but not attractive in any case because it was always freezing.

2   A was allowed to take his girlfriend into the best room but every ten minutes a head would pop round the door and father would ask if they were all right. A's eldest sister received a thick ear for talking back to Dad after she discovered him following her to see what she was up to.

Minuchin suggests that being 'connected', as opposed to enmeshed and being 'separated', not disengaged, are normal modes of interaction. Both the problem-creating behaviours are likely to be characteristic of positional families, where there is a strong parental urge to control, one through the erection of boundaries to separate children and parents (disengagement) and the other through the violation of personal space, ignoring the child's own boundaries. However, it is also feasible that an unbounded, personalising regime will prove oppressive if it does not make space for the child to develop a sense of autonomy. What I am suggesting is that the child's sense of boundary, anxieties about space and time or feelings of attachment to particular spaces, will be affected by the domestic environment as it is shaped and manipulated by family members. Clearly, the opportunities for control, or for giving children their own spaces, will be affected by the size of the home, the way space in the home is partitioned, and the relationship between private and public space. In relation to the private/public distinction, there is a serious issue in British society associated with the decline of exterior space as a space for children. The home then becomes more important as a place where the child develops a sense of boundaries and an awareness of others.

HOME SPACES AS EXPERIENCED BY CHILDREN

So far, I have concentrated on boundaries which are imposed on children and which may be oppressive. However, children's own narratives, or adult recollections of the use of space in the home as it was experienced in childhood, may express little concern with exclusions or regulation. Expressions of satisfaction with the way things are or were, which are common in the surveys of British childhoods which I draw on in this essay, may reflect the attainment of some autonomy by children who have grown up mostly in middle-class families where there is no shortage of space. It goes without saying that these stories of childhood would be alien to a child living in the *barriadas* of Caracas or in bed-and-breakfast accommodation in a British city. Their significance is that they demonstrate the limits to autonomy and the rather subtle ways in which boundary controls are exercised in 'ordinary' middle-class families.

Having one's own space is important in developing autonomy and this distinguishes the middle-class child who is part of a small family from one with many siblings or living in poverty. Particularly when a child has been given its own bedroom, then the space may be appropriated, transformed and the boundaries secured by marking that space as its own:

> My bedroom is very small. Our attic is a lot bigger but it is lovely (I think) ... On my far wall, I have my wardrobe. It is white and built into my wall. The catch on it is magnetic. It has gold (brass) handles on it. The dressing table is nearly the same. It is white with two drawers with gold handles on them. It also has a big round mirror on the facing wall. The things I keep on my dressing table is my jewellry box, my little black and white portable television set. I also have a two-way rag doll sitting in the corner. I have some soap out and some powder and a few items of make-up! I also have a rather fragile China doll on it about twelve inches high (thirty centimetres). On top of my television I have two old birthday cards, a little white kind of vase thing with a lid and a pink one and two ornament soldiers ...
>
> Then I have a white bookcase. It has five shelves and is very tall it is very narrow though. It has tons of books on it and a spider plant. Next to the bed there is my bedside table. On this I have a goldfish (in its bowl), a pink lamp, a photo of my brother, a picture I drew (I framed it) ... Oh yes, I almost forgot, on top of the wall near the top of my bed there is a life-size poster of Lady Di and Prince Charles.
>
> (Mass Observation, People's Homes, 1983)

The room, with its cherished objects, is an entirely personal space. In another account, much less embellished with detail, the bedroom is not described as if it were personalised and there is a hint of parental authority which in the past had rendered the child's space less secure:

> My bedroom is the medium sized room and I have quite a lot of furniture in it. I have a bed, bedside table, chest of drawers and a desk ... I have not very much space if I want to play with some of my toys and games. My bedroom is normally messy, my clothes are usually all over the room and I put everything that gets in the way under my bed. I used to be in the biggest bedroom but my Mum is in there now ... This bedroom was much bigger than my bedroom now. I didn't like the big bedroom because it was very hard to keep tidy and I was always getting told off because my room was always messy.
>
> (Mass Observation, People's Homes, 1983)

Similarly, a 21-year-old man, describing his bedroom when he was about 10 or 12, recalled the parental controls as well as the freedom:

> I remember having a blue ceiling which was something I particularly wanted. My Mum was quite happy to let me do that. Yes, it was very personalized. It was generally full of little bits and pieces and things

that I would get. But it wasn't untidy, though. My Mum would have come in and tidied it if it got untidy. That was the main thing between my sister and my Mum. That was the main point of conflict – the untidiness of her room.

Although many children in middle-class homes can secure some privacy and in their relationship with the rest of the family experience separation without disengagement, in Minuchin's terms, the autonomy which this represents may be very limited. Elsewhere in the home, children may still constitute a polluting presence, requiring regulation or exclusion. Parents commonly determine what are adult spaces and adult times, creating a mixed regime with elements of separation and little concern about the control of the child's space, combined with regulation and strong boundary maintenance. Thus, the boy who could paint his ceiling blue had much less freedom in the living room:

> It wasn't that rigid about where you sat although, if Dad came in and said 'That's my chair', then you moved. Apart from that, it wasn't that rigid. As soon as Dad came home, then it was more my parents' room than our room.

The timing of activities in the home combines with the partitioning of space to create liminal zones, spaces of anxiety both for the child and parent, the parent in extreme cases attempting to distance herself or himself from the child but often lacking the power, the child becoming anxious about being in a place when it should not be. The regulation of time by adults is particularly anxiety-creating and I will first illustrate this point with a case concerning the role of time in the childhood experience of a girl – an unhappy recollection – and then show how time regulation is expressed in ordinary family routines.

A woman recalls how, as a child in Glasgow,

> my brother and I used to race to the bus stop for the 1 o'clock bus back home, to travel the mile and a quarter to a too hot dinner, followed by a sprint down the drive to catch the 1.30 p.m. bus back to school. I remember on one of these racings to and from the bus stop falling onto a newly tarred road but, dead or alive, I had to get to school. Bells, of course, rang between periods (45 minutes or so) to end play time. I remember when my leg was broken by a girl falling onto it that I was horrified to hear the bell ringing between my wailing, that I could not get up and join the serried ranks of children waiting for the janitor to direct them into the building.

She continues:

> My parents were most wonderfully organized in running their hotel. The clock ruled. Its discipline was sacrosanct. The time I spent over homework caused a lot of trouble between my parents and myself. I realize now that the internalized whip made me anxious about learning

... Father would come back late from a freemason's meeting and stand by the electric light, demanding that I would stop wasting his money and go to bed ... Our relationship was a very guarded one though this was never talked about.

(Mass Observation, Summer, 1988, 'Time', woman correspondent)

In families where the regime is more relaxed than in this case, the time of day may determine when spaces change from child or family spaces to adult spaces and it is at these transition points that power relations are exposed. The living room which had been shared by mother and children becomes a regulated adult space when Dad returns from work, as in the example of the blue ceiling child, above. Similarly, a woman recalls that:

The only arguments used to be about what time I went to bed. I think it used to be just as the news came on at night or something. But I can remember it changing once I went up to the comprehensive school. I then decided that as I'd got homework, I was very important now, I'd got the right to stay up later. It was more left up to me once I got up to the big comp.

Here, we have positionality surfacing in an otherwise personalising family but also the child asserting herself and beginning to define her own boundaries.

Even in the most benign accounts of the home and family relationships, there are intimations of conflicting world-views, manifest in arguments over untidiness and bedtimes, in particular. In a sense, these kinds of tensions represent a conflict between order (adult) and disorder (child), between a preference for strong boundaries (adult) and weak boundaries (child) although, occasionally, a child's yearning for order may be frustrated by a parent's disregard of regularity in space and time. The more usual conflict over order draws on a developmental view of childhood, the child through socialisation becoming more responsible, more orderly, cleaner and better mannered. In this, there is a shadow of Rousseau's child of nature affecting adult attitudes to children (Prout and James 1990) with their assumed proximity to nature threatening the boundaries of adult society. This was evident in a 1992 television commercial for Persil in which children were portrayed as 'savage', romping through a jungle in American Indian head-dresses and then returning home to be cleaned and civilised by their Persil-packing mother. Through association with dirt and nature, the child is a source of abjection for the mother. As Perin (1988: 169) puts it: 'Wild until Tamed, Barbaric until Civilized, Beasts to be trained up as Angels – so we have been constituting children ... With more than an anxious slip of the tongue, we animalize children (and humanize dogs).'

HOME AND LOCALITY

For children in the most highly developed societies, the house is becoming increasingly a haven. Some children appropriate more of their own space in the home and constitute an important market for home-based leisure. At the same time, the outside becomes more threatening, populated by potential molesters and abductors, so the boundary between the home (safe) and the locality (threatening) is more strongly defined. Hillman, Adams and White-legg (1990) have documented a dramatic withdrawal of children from the street in Britain, and to a lesser extent in Germany, between 1971 and 1990. There are several reasons for this. One is the increase in the number of cars which increases the risk of injury and makes the street a hazardous place to play. Another is the fear of others which is heightened by 'stranger danger' campaigns and stereotypical images of threatening urban environments projected by the media. As recently as 1982, according to one 21-year-old man, 'The street was very important ... it was the meeting place. You'd go out to the street and there were always people out there and then usually you would decide where you went afterwards.' Now, however, the locality is more likely to be experienced from the car, necessarily in the company of adults, rather than alone or in the company of other children. The car then functions as a protective capsule from which the child observes the world but does not experience it directly through encounters with others. This was also the conclusion of Hillman, Adams and Whitelegg (1990: 90–1):

> More of our lives are now spent in the cocoons of house and car, and the outside world has become impersonal. As the streets fill with traffic, they tend to empty of people, and as street life retreats and public transport declines, the world outside also becomes menacing.

These authors note that, according to some psychologists, television constitutes an alternative to this physical space previously experienced through play in the local environment. However, it is a poor substitute and maybe a dangerous one because of the frequent stereotypical representation of others by the electronic media. If the environment in which the child grows up is being populated in the imagination by dangerous and deviant others, we may be producing more fearful and purified selves who contribute to the creation of strongly bounded and purified localities and homes.

CONCLUSION

As Hoggett (1992: 345) remarked: 'We all know fear, uncertainty, desire and envy.' These are feelings which are etched into space but academics generally remain silent about them. In childhood, sensations are particularly acute but they are often unarticulated. As adults, we forget what it felt like to be in a new place which looked and smelled different to anything experienced before, or to be late for school or to be sent to bed. To recover these feelings seems to me to be important in the task of constructing and reconstructing childhood. When talking about children's spaces, however, we necessarily implicate adults who themselves construct childhood in different ways. We cannot isolate children from their social experience with adults nor the particular places in which these social relationships occur, notably the home. Maps of the home, peopled by adults and other children, provide one context for childhood experience but the contexts of the home – the locality, the global images of consumption culture and so on – are also integral elements of the social space of the child.

Academic studies of childhood are fragmented, reflecting a disciplinary fragmentation and the lingering appeal of different paradigms among the social sciences. Recognising the bounded, incomplete and often incompatible representations of childhood, psychoanalytic theory appears attractive because it crosses boundaries and promises an integration. As Pile (1993: 123) argues:

> Psychoanalytic theory, in its theories of the unconscious, describes how the social enters, constitutes and positions the individual. Similarly, by showing that desire, fantasy and meaning are a (real) part of everyday life, it shows how the social is entered, constituted and positioned by individuals.

I have suggested that object relations theory provides one way of connecting the people, things and spaces which make up a generalised other and give the self a sense of border. Things and spaces, as they are appropriated, cared for, shared, traded, barricaded, disturbed or destroyed, conjure feelings, sensations. They are a part of the complex of relationships defining the boundary between self and other. The observations on childhood recounted in this essay, mostly drawn from 'normal' British childhoods, convey only a little of this complexity – there is a rich world of intimate spaces to be recovered. There is a danger that this may become an indulgence, however, if academic inquiry is focused only on more accessible, western middle-class childhoods. The street children in Brazilian and Colombian cities, who are viewed by the death squads as dirt to be swept from the streets, are also a part of the problem, and their representation as dirt or a defiling presence is a more pressing issue than mealtimes in Laburnum Crescent.

ACKNOWLEDGEMENT

I am grateful to Dorothy Sheridan, University of Sussex, for her help in using the Mass Observation Archive.

NOTE

1   These come from counselling sessions with adult problem drinkers (Lowe, Foxcroft and Sibley 1993) and from interviews about home life in childhood with 21-year-old students conducted by the author. Unless indicated otherwise, the extracts are from the latter study.

Part II

SEXUALITY AND SUBJECTIVITY

PART II

INTRODUCTION

In the very first sentence of their chapter, Bell and Valentine raise the pertinent question 'What does it mean to be a sexed subject?'; such a question implies at least that it means something. There is a paradox here: this question is not new, but then again it is hardly ever asked. Thus Sigmund Freud, in introducing a lecture on the problem (sic) of femininity, states that 'when you meet a human being, the first distinction you make is "male or female?" and you are accustomed to make the distinction with unhesitating certainty', but quickly adds 'anatomical science shares your certainty at one point and not much further' (1933: 146). This part of the book contains chapters which look into this 'not much further'; not so certain, they unhesitatingly inquire into the place of sexuality in subjectivity and subjectivity in sexuality.

David Bell and *Gill Valentine* start out by asserting that 'we can usefully think of sexualities, like genders, as performative constructions naturalised through repetition'. Such a position enables them 'to think about the way sexualities become codified – even stylised – and how that codification informs the subjectivity of our sexed selves'. With similarities to Sibley's position (in Part I), this requires an investigation of the complex, conflictual and dynamic processes through which the categories of sexual identity, and the borders between sexual identities, are created, sustained and changed. For Bell and Valentine, the essence of sexuality is to be found in its repeated and (re)stylised performance, within a context of social permission, regulation and prohibition. Their interest is in the ways in which people resist social norms and they provide three examples of resistance through the perform- ance of the sexed self in specific places and at particular times: first, the everyday tactics and strategies which lesbians use to manage other people's impressions of their sexual identities; second, the reclamation of other possibilities for a sexualised corporeality through body modifications such as piercing, tattooing, scarring; and, third, the anger in the AIDS activism of groups such as ACT UP, Gay Men Fighting AIDS, Queer Nation and OutRage!. This evidence demonstrates that 'sexuality is not merely defined by private sexual acts but is a public process of power relations in which everyday interactions take place between actors with sexual identities in sexualised locations'.

Indeed, sexuality is a component in the practices of social meaning; where people with sexual identities in sexualised locations make themselves intelligible to others and become intelligible to others through specific grids of meaning which are written (as it were) on the body. *Julia Cream* takes the case of the woman on the pill to ask why it is that this sexed subject has become understandable, familiar and acceptable, when the idea of a male pill is not and nor is the taking of other kinds of hormones (by almost anyone). Cream argues that it is the 'appearance of sex as prediscursive, prior to culture, that obscures and disavows the constraints that produce a domain of intelligible and unintelligible bodies'. If sex is 'unhesitatingly certain', then this can disguise the power-infused and discursively-constituted circumstances which lead women – rather than men or rather than men and women – to take the, or any, pill. Cream demonstrates the ways in which the woman-on-the-pill has been both constituted and made culturally intelligible by looking more closely at three 'bodies' that 'make sense' – the infertile woman, the Puerto Rican woman and the contracepted western woman. In a similar vein to Bell and Valentine, Cream concludes that 'the gendered performance of the woman on the pill offers opportunities for reworking our bodies and our social values'.

If gender is – even only in some small part – a repeated and (re)worked performance, then masculinity (and heterosexuality) can no longer be either assumed away or absented from analyses of sexuality (a term often used to describe every other person's sexual practices but straight men's) and power. *Vic Seidler*'s underlying position is that knowledge itself is constituted by sexuality and, more precisely, heterosexuality, in a context where the powerful (usually men) can usually identify themselves with things that are culturally valued and thereby denigrate the powerless (often women) by associating them with the things that are socially abhorred. Where rationality has been valued to the exclusion of emotionality (usually successfully), men have identified themselves with reason while others are (commonly) derided as being emotional or controlled by their bodies – but this antagonism can only be achieved and maintained where men sever themselves from their own emotional lives. It is important to realise, then, that 'heterosexuality exists not simply as a sexual preference but as a powerful institution within a patriarchal society' and that 'this has consequences for the ways we understand the space of intimate and personal relationships and *how* power operates within relationships'. The problem is that there remains little debate about how straight men relate to dominant patterns of masculinity and heterosexuality or about how this might be done differently. Seidler's search is for a positive response to feminist and queer critical politics and, in order to do this, he maps out the terrains of masculinity. He argues that straight men live largely in ignorance of their bodies, emotions, needs, sexualities and mastery: for example, 'there is so little that teaches us as boys that sexuality has to do with vulnerability and contact'. Seidler concludes that men must learn to think through their own experiences, if masculinities are to be redefined in less oppressive, less hurtful ways.

7

THE SEXED SELF

strategies of performance, sites of resistance

David Bell and Gill Valentine

What does it mean to be a sexed subject? It might mean simply 'being' sexed in a particular way – having, for instance, a 'gay sensibility' or a 'lesbian essence' (and we might note here that all too often it is only non- or counter-hegemonic sexualities which are thought of as marking their bearers in particular ways – no one talks of 'straight sensibility'). But thinking from various theoretical angles – from deconstructive feminism, from identity politics, from anti-essentialist viewpoints – has suggested instead that we can usefully think of sexualities, like genders, as performative constructions naturalised through repetition (Fuss 1989; Butler 1990, 1993; see also Julia Cream's chapter in this volume). This might help us to think about the way sexualities become codified – even stylised – and how that codification informs the subjectivity of our sexed selves. Social histories of sexual minorities have shown us how vital this performative vocabulary can be, both as a marking of difference (from heterosexual hegemonies) and as a marking of sameness (creating a cohesive group identity essential for the formation of recognisable 'communities' and so on). Stressing the relational nature of sexual identities, and interrogating the strategies for the perform- ance of our sexed selves, we can think through Diana Fuss' (1991: 2) questions about 'the complicated processes by which sexual borders are constructed, sexual identities assigned, and sexual politics formulated'.

Performativity, then, must be seen not as a singular act, but as a repetitive one, thus reminding us of its historicity (Butler 1993). As Elspeth Probyn (1993: 2) says, the self 'is not simply put forward, but rather it is reworked in its enunciation'. Within the tense arena of sexual politics, the performative choices available to those with non- or counter-hegemonic sexualities are in part an embodiment of the regulatory regimes which operate to constrain the possibilities of performance, and in part a claiming of the sexed self as a site of resistance precisely to those regulatory regimes. The tension between these discourses of regulation and resistance, as they are enacted through the sexed body performing in space, are articulated in different ways at different times and in different places. It is our aim here to examine three performative strategies of the sexed subject in space, each of which enacts its identity in relation to the boundaries set on this performative identification; each mobilises the performance of the sexed self as an embodied site of resistance. Gillian Rose (1993a: 5) has written that we 'position ourselves in relation to others'. The ways we choose to demarcate our position in relation to others speaks of the possibilities and impossibilities of the sexed subject – of exposing the limits of the performance, the rules of the game.

While it is not possible (nor, hopefully, necessary) to rehearse Judith Butler's arguments about gender performativity here, we might say a word or two about their deployment in this chapter. Her arguments, outlined in *Gender Trouble: Feminism and the Subversion of Identity* (1990), about how a parodic act like drag can destabilise and denaturalise any gendered subject position by revealing the constructedness of gender itself, has been applied to thinking about the construction of space as prediscursively heterosexual, and the disruptive or transgressive potential of what she calls 'heterosexual conventions within homosexual constructs' (Butler 1990: 31) performing in those 'heterosexualised' spaces (Bell, Binnie, Cream and Valentine 1994). More recently, she has turned to sex and to the body (Butler 1993), while others have appraised (and praised) the body of her theory for, among other things, 'placing theater and theatrical performance at front and center of questions of subjectivity and sexuality' (Sedgwick 1993: 1). In particular, Eve Kosofsky Sedgwick's paper on queer performativity has usefully considered the role of shame and shaming in the naming of 'Queer', and the gathering up of transformational energy from being shamed:

> There's a strong sense, I think, in which the subtitle of any truly queer politics will be the same as the one Erving Goffman gave to his book *Stigma: Notes on the Management of Spoiled Identity*. But more than its management: its experimental, creative, performative force.
>
> (Sedgwick 1993: 4)

In a sense, the three figures we invoke in this chapter all bear a relation to shaming, and all might gain some of Sedgwick's transformational energy from that sense of shame: through a reading of the performance of the managed, adorned and angry selves outlined below, we can understand a little of how performativity can build on shaming in creative and restorative ways – as Sedgwick (1993: 14) says, 'for certain ("queer") people, shame is simply the first, and remains a permanent, structuring fact of identity: one that has its own, powerfully productive and powerfully social metaphoric possibilities'. For if shame is a permanent structuring fact of what she calls a 'queer' identity (and we would like to note what a contested term this is), then the responses enacted by 'passing' lesbians, pierced perverts or AIDS activists mark out some of the ways this structuring fact can shape identities.

The configurations of the sexed (and shamed?) self described and theorised below perform very different 'enunciations': the managed lesbian self of elective publicity and privacy discussed in the first section is a very particular kind of 'subversive bodily act' (Butler 1990: 79). While it might be seen as primarily a strategy of survival – of the care of the self through the 'disguise' of 'passing' as heterosexual – it nevertheless opens up radical and trans-gressive possibilities which can work from *inside* hegemonic discourses. As Joseph Bristow has written, with reference to gay male 'passing':

> Stylizing particular aspects of conventional masculine dress, we can adopt and subvert given identities, appearing like 'real men' and yet being the last thing a 'real man' would want to be mistaken for: gay ... This type of gay identity, therefore, consciously inhabits a publicly acceptable one which is, in fact, its enemy.
>
> (Bristow 1989: 70)

The managed lesbian self therefore offers very different transgressive possibilities from either the adorned self or the confrontational body-terrorism of AIDS activism. The first of these owes much to theories of playful transgression. The adorned body – pierced, tattooed, scarified – represents a particular performance of sexual deviance or perversion, and one which confronts the regulatory regimes of state and law in complex ways. Its political potential relies on exposing the boundaries of precisely those regimes of power, and as such it provokes harsh censorship and close surveillance under the sign of 'obscenity' (Bibbings and Alldridge 1993).

The angry self of the 'AIDS activist aesthetic' (Crimp 1987) is an 'in your face' performance, borrowing much of its language from other realms of civil disobedience and political action but fusing it with what has been termed a 'queer subjectivity' (de Lauretis 1991). While not solely confined to the sexual, there is an intimate entwinement of AIDS activism with sexual politics which means that the activist aesthetic, while battling for much more than sexual citizenship, mobilises those labelled as sexual outsiders to 'act up, fight back, fight AIDS', as the slogan goes. A particularly corporeal set of spatial strategies of protest marks the angry self of AIDS activism as a powerful political transgression. By thinking through this and the managed and adorned self, we can begin to theorise some of the complex implications of each performative strategy.

I AM NOT WHAT I AM, OR AM I? MANAGING THE SEXED SELF IN EVERYDAY SPACES

Historically, women's relationships with each other, and in particular lesbian relationships, have largely gone undocumented (Faderman 1991; Jenness 1992). Despite the recent growth of lesbian and gay studies and the raised political profile of lesbian activists as a result of stunts such as abseiling into the House of Commons, lesbians still remain largely invisible in contempo-rary popular culture and everyday life. This marginality is evident in the problems experienced by researchers trying to locate lesbians to participate in academic research: 'My own process of gaining a group of [lesbian] teachers was tortuous and halting. Advertisements yielded few responses, not surprisingly so given the risks had my advertisement not been genuine' (Squirrell 1989: 89, quoted in Oerton 1993: 3). 'Any account of lesbians' employment experiences is at best a speculative effort to overcome a deep cultural silence and intentional invisibility' (Schneider 1988: 274, quoted in Oerton 1993: 3). As both these quotes hint, this invisibility is largely deliberate. In modern western societies heterosexuality is the dominant form of sexuality, with 'normal' sex defined as potentially reproductive. Lesbians and other sexual dissidents have been and continue to be (with different degrees of acceptability) perceived as 'unnatural' and deviant (Young 1990). In particular, lesbians are stigmatised in the media and popular culture as man-hating, butch, ugly, a danger to children and a threat to the family and hence the entire social fabric (Lonsdale 1993; Phillips 1993). They are therefore vulnerable to discrimination, prejudice and anti-gay violence (Herek and Berrill 1992).

Given this regulatory regime of stigmatisation, discrimination and violence, many lesbians manage their identities in order to 'fit' within the boundaries of the hegemonic heterosexual discourse. This is not merely a case of keeping quiet about what goes on behind closed doors in private space: sexuality is not merely defined by private sexual acts but is a public process of power relations in which everyday interactions take place between actors with sexual identities in sexualised locations (Burrell and Hearn 1989; Valentine 1993a). Marny Hall (1989) uses the term 'role suffusion' to describe the way in which so-called 'private' sexual identities thread everyday 'public' conversations and encounters between men and women. In particular she argues that women in the workplace are perceived to be inherently sexual in dress, appearance and behaviour and are expected to be sexually submissive to men and to engage in subtle sexual dynamics – being coy, flirtatious or motherly. In this way heterosexual women become the symbolic 'other' which Hall argues (heterosexual) men need for their own continuing process of self-definition. In turn women can use sexual banter and flirting to manipulate (heterosexual) men for their own ends.

By expressing a dissident sexuality at work, lesbians disrupt such heterosexualised dialogues: 'The sheer weight of dominant cultural attributions that lesbians must carry, if their orientation is known, renders them unavailable for the myriad and quickly shifting micro-projections necessary to maintain and elaborate the male narrative of self' (Hall 1989: 127). Lesbians who disclose their sexual identity at work therefore often find it difficult to operate within these heterosexual boundaries. In the same way they also encounter problems in most other everyday spaces – from the bank to the high street – where similar socio-sexual behaviour is also used by men and women to enhance their own power and to control others. As Gillian Rose (1993a: 37) writes, 'everyday space is not only not self-evidently innocent, but [is] also bound into various and diverse social and psychic dynamics of subjectivity and power'.

Many lesbians therefore adopt a mantle of heterosexually-defined femininity to fit in with everyday social environments, using signifiers such as dress, make-up and 'wedding rings' to publicly project an apparent 'private' 'heterosexual' identity across their bodies. To maintain this performance they must become highly tuned to the 'usually hidden matrices of behaviour, values and attitudes in self and others' (Hall 1989: 129). Similarly some lesbians create apparently asexual identities by avoiding reference to their personal life or playing the role of the 'career woman' or 'spinster'. But all the time gay women may still bear lesbian signifiers such as a discreet pink triangle or a 'pinkie ring' which can be read by those 'in the know'. By combining these two strategies as circumstances dictate, 'passing' lesbians shuttle through different appearances, putting on or taking off different 'masks', sometimes maintaining multiple identities in one space at different times or in different spaces at the same time (Valentine 1993b). But in the crucible of everyday life there is always a danger that people from one place or time will stray into another and so stumble on the performance of the 'wrong' identity, spoiling the 'impression' created. To avoid a rupture of their 'identity' many lesbians use time-space strategies to segregate their audiences. These include establishing geographical boundaries between past and present identities, separating different activity spheres and hence identities in space, expressing a lesbian identity only in formal 'gay spaces', confining their 'gay' socialising to homes or informal 'gay spaces', expressing their lesbian identity only in public places at specific times, and altering the layout and decoration of private spaces to conceal clues about their sexual identity from specific people (Valentine 1993b).

In his now classic book *The Presentation of Self in Everyday Life*, Erving Goffman (1959) used the notion of the theatre to describe the social processes through which actors execute different performances in front of different audiences. In his dramaturgical analysis people have no stable or essential self; the self is a fleeting image. In his words, 'this self is a *product* of a scene that comes off and is not a *cause* of it ... [The person] and his (sic) body merely provide the peg on which something of collaborative manufacture will be hung for a time' (Goffman 1959: 252–3). In this way the self is constantly created and recreated through interactions with other actors, whom Edgley and Turner (1975: 7) claim 'tend to act as members of a performative team, assisting each other in sculpturing their shows through ... impression management'.

An essential part of Goffman's notion of self-presentation is the concept of audience segregation. In other words, that 'those before whom one plays one of his (sic) parts won't be the same individuals before whom he (sic) plays a different part in another setting' (Goffman 1959: 57).

Goffman uses further theatre metaphors of stage and backstage to articulate this concept, which many sociologists and social psychologists appear to have taken literally (e.g. Schlenker 1985; Snyder 1987). Conceiving the notion of stage and backstage (and hence the self) as a 'public–private' duality, they argue that people play roles in public (front stage) in order to communicate the image of themselves that they wish others to believe. The private self (backstage) is often not revealed, in the same way that in 'real theatre' actors play a role whilst maintaining a private identity that the audience doesn't see (tabloid exposés excepted). The private self is therefore the source of the roles people perform publicly rather than 'the self' being, in Goffman's terms, fleetingly created through each performance. As Tseelon (1992: 116) says, 'this game is not an end in itself but *a means to an end* of gaining benefits. It is a game of *misrepresentation*', a game where only the private self is sincere or authentic, whilst the public self presents a false impression – I am not what I am.

Implicit in much of this Impression Management theorising is the notion of the individual as manipulative and of self-presentation behaviour as a disguise. The actor is seen as a social con-artist, floating identities as trial balloons or outright deceptions usually with the aim of (re)packaging information about the self and projecting a 'positive' identity in order to gain control or power (Baumeister 1986; Tedeschi 1986). In this game others are used as a social mirror to feed back information about the public (presented) self. This can then be used to perfect self-presentations to achieve a desired effect.

The management of lesbian identities in everyday spaces appears on the surface to fit this model of manipulation. Research suggests that gay women presenting themselves as heterosexual or asexual in environments such as the workplace think of their private lesbian identity as their 'real self' – the 'natural' identity they return to when not performing in public. This self has different friendship networks and a different lifestyle from the other selves which are presented (Davies 1992; Valentine 1993c).

Often, of course, gay women have to make little effort to generate a false impression of their identity; such is the hegemony of heterosexuality in most everyday environments that heterosexuals commonly assume all those in their company share their sexual identity. The lengths to which lesbians therefore have to go to project an appropriate self for different audiences depends on the shifting forms that ideologies and discourses of lesbianism and heterosexuality take in different spaces. In different environments gay women may adopt a tactical or short-term approach to their identity management (Arkin and Baumgardner 1986), presenting themselves as heterosexual whilst waiting to see how the land lies before 'coming out'. Others may take a more strategic or long-term view, consciously deciding to disguise their lesbian identity permanently because of the homophobia expressed in their home or workplace (Hall 1989; Oerton 1993). This is an approach commonly adopted by women in sensitive occupations, such as teaching (Squirrell 1989), where the risks of performing the 'wrong' identity could jeopardise their livelihood. Others manage their identities in a less conscious or non-strategic way: some women who have relationships with women do not conceive of themselves as 'lesbians' (Jenness 1992), so that their identity management is more passive or non-reflective; they merely habitually fit in with others as the script of everyday life unfolds around them.

But audiences are not always deceived. Even when suspicions about the 'realness' of self-presentations are raised, people appear to give others the benefit of the doubt (Blumberg 1972). Sometimes lesbians make slips in their identity management (Hall 1989) or colleagues guess a woman is a lesbian. But, to borrow another theatrical metaphor, 'the show must go on', and both actor and audience collude in maintaining the illusion of the identity the actor is trying to present.

Despite the fact that many gay women perceive their lesbian identity as their 'real self' this is often no more stable than their other publicly-managed identities. As Jenness (1992) and others have argued, the identities of lesbians are as fluid as any identities, with women having heterosexual relationships and then identifying as lesbians, or having lesbian relationships but then identifying as bisexual. Similarly, the notion of 'what a lesbian looks like' is also in flux, with women adopting political or lesbian feminist identities but then later adopting other roles, becoming lipstick lesbians or having butch–femme relationships. There is no essential or unified 'I'; the self is slippery, elusive. We may have a sense of self-consistency and not notice our moment-to-moment shifts, but every now and again we do recognise that we are not the same as we once were; we are more feminine, more confident, more political: I am not what I am, or am I? (Tedeschi 1986). The postmodern self is a 'discursive phenomenon, not an essentialist one' (Tseelon 1992: 120), in which, as Gergen (1990: 156) says, 'self is replaced by the reality of relatedness – or the transformation of "you" and "I" to "us"'.

This interconnectedness of the self to others suggests that lesbian identity-management strategies share the potential to destabilise the identities of others claimed by those lesbians and gay men who adopt hyper-feminine and hyper-masculine identities as deliberate transgressions (Bell, Binnie, Cream and Valentine 1994). The lipstick lesbian and gay skinhead are credited with the potential to undermine heterosexuals' confidence in their own and others' identities by appearing to be like 'real (straight) men' and 'real (straight) women'. In the same way, by 'passing' as 'normal' in everyday life, lesbians who manage their identities also have the potential to shake the foundations of the 'stable' temple of heterosexuality if their 'deviance' is revealed. By destabilising heterosexual identities they also have the power to destabilise the heterosexual space that those performing hegemonic hetero-sexual identities produce.

Managing a lesbian identity, or 'passing', is often seen, however, as collusion or compromise which allows all women to enjoy the privileges of heterosexuality. The recent campaign to 'out' public gay figures has highlighted the need to mobilise the destabilising potential of these managed identities as sites of resistance. It is only through revealing their performativ-ity that the link between certain identities and meanings is disrupted and the slipperiness of all selves is revealed.

THE ADORNED SELF: WEARING THE BODY POLITIC?

A celebratory sexualised body-aesthetic which bears the mark of the pervert is a strategic resistance to regulatory regimes and a performative statement of self-conscious 'othering'. Body modifications and adornments – piercing, tattooing, scarification – signal a particular relationship with the body and certain sexual practices which invoke a project of corporeal reclamation and, advocates claim, the tapping of deeper psychic forces which can impact on the broader social sphere in somehow mystical ways:

> Amidst an almost universal feeling of powerlessness to 'change the world', individuals are changing what they *do* have power over: *their own bodies*. That shadowy zone between the physical and the psychic is being probed for whatever insight and freedoms may be claimed. By giving visible bodily expression to unknown desires and latent obsessions welling up from within, individuals can provoke change – however inexplicable – in the external world of the social, besides freeing up a creative part of themselves; some part of their essence.
>
> (Vale and Juno 1989: 4)

The 'inexplicable change' to the social world which body modifications and decorations are deemed to provoke take us once more into the embattled terrain of 'transgression', the realm of 'subversive bodily acts'. The body, Butler (1990: 139) writes, 'is not a "being", but a variable boundary, a surface whose permeability is politically regulated, a signifying practice within a cultural field'. Within this context, modifying or adorning the body is taken as a sign of dissent (Curry 1993).

Further, the surficial permeability of the body-as-boundary has been characterised as a medium for playing out deep psychic desires, for unlocking the creative and wild possibilities of the self. As psychotherapist David Curry (1993: 69) has written:

> Body decoration lies at the interface between the private and the public. The skin is the actual membrane between what, on one side, is inside me and, on the other side, is outside me. It is superficially me and at the same time a surface onto which I can both consciously and unconsciously project that which is more deeply me. This property of the body surface is the basis for body decoration.

To a large degree, acts of bodily subversion or dissent through adornment function as markers of difference. They thus belong, as has been noted, to a whole set of performative strategies and practices sited around what Sue Golding (1993b: 25) has termed 'a place beyond a natural limit ... the elsewhere of sexual mutation curiosity'. As she writes, the identity of the pervert is 'a peculiar identity: one that must always bear an excess, the excessiveness of the game itself, the perverse and excessive game of self, of mastery and submission' (Golding 1993b: 26). The 'excessive game' of 'mastery and submission' is played out across one's own body when it is self-consciously modified. As Curry (1993: 71) says of tattooing:

> One of the core motives for being tattooed is to declare a peculiar relationship with one's own body, and this properly includes a responsibility for it ... There is a tension between the desire to be

tattooed and an anxiety about its permanence. This tension is one of the attractions.[1]

There is also a notable tension in all body modification between the public and the private. The surface of the body may be the membrane between self and other, but it is not necessarily our outward presentation of self, which is more often articulated through clothing, deportment, gesture and language. Body modifications, although marking the exterior of the body, are often concealed from 'public view', or at least can be electively rendered 'private' by clothing. A tattooed shoulder or a pierced clitoris can be hidden and displayed (in most circumstances) at the mastery of the bearer. Adornments to the face and hands are obvious exceptions to this, hence the reluctance of many tattooists to execute facial tattoos (piercing cannot be considered a permanent modification in the same way as tattooing, and so anxieties about ear, nose, eyebrow and lip piercing are less deeply-felt, although Curry (1993) argues that even a discontinued piercing has permanently changed the wearer's sense of self). The knowledge that under the business suit there is a full body-tattoo and pierced nipples is one of the most frequently-celebrated 'transgressive' pleasures offered by adornment.

The relationship between adornment and sexual practice is an ambivalent one. While there might be some broad connection between, for example, body piercing and certain kinds of sex acts – certainly those which eroticise or fetishise the piercing site – the signifying capacity of piercing cannot be necessarily read as a sign of any sexual practices (this might be even more so for tattooing, but may be less so for certain kinds of scarification). A recent court ruling has made explicit the gap between piercing-as-adornment and piercing-as-sexual-act: one famous London piercer, Mr Sebastian, was charged (as part of the infamous Operation Spanner trial of same-sex sadomasochists) with assault occasioning actual bodily harm for piercing his lover's penis, on the grounds that the piercing was for the purposes of both parties' sexual pleasure (Bibbings and Alldridge 1993). While an occasion like this offers up 'one domain in which the force of the regulatory law can be turned against itself to spawn rearticulations that call into question the hegemonic force of that very regulatory law' (Butler 1993: 2), the material consequences – in this case a fifteen-month suspended prison sentence and untold harassment and adverse media attention – show the often-precarious position of the pervert on the bounds of the law (Bell 1994).

This brings us back yet again to the whole question of transgression, which has become one of the most contested and emblematic notions in current discourse on, among other things, presentations and performances of the sexed self (e.g. Wilson 1993). The transgressive play-of-signifiers within cross-dressing, for example, has been subject to a heightened scrutiny (see for example Butler 1991; Garber 1992; Tyler 1991). In a telling statement from an interview, artist Nayland Blake says:

the community in which I live, for instance, is so quick to adopt transgression, that things very quickly cease to be transgressive. Believe

me, everybody has something pierced in San Francisco! And what we learn from that is that notions of piercing per se are not transgressive.
(Gange and Johnstone 1993: 60)

On the other hand, notable queer discourses have celebrated the transgressive body as implicated in new possibilities of radical democracy, singling out the pervert as the harbinger of a new creative urban politics (e.g. Golding 1993a). Questions of context, of 'author' and 'audience' for such bodily inscriptions, become important here. The tension between advocating the creativity of authorship (which we find, say, in Butler's and Golding's work) and seeking an understanding of the 'practical, historical, institutional reverberations' (Bordo 1992: 159) of subversion and transgression is one which resonates through much 'queer' theory and praxis (see also Bell, Binnie, Cream and Valentine 1994). Andrew Travers' recent essay on camp is notably explicit in this context; it is, he says, 'a sociological "My Way" that invites the reader to forget society and at the same time promises no further demand on the reader but that he or she enjoy the show' (Travers 1993: 128). Enjoying the show and the showing of tattoos, piercings and scars may be a transgressive pleasure (although greater enjoyment perhaps comes from watching reactions to the show), but if it is only reaped at the cost of forgetting society then its radical potential may be severely blunted.

Further, Vale and Juno's (1989) assertion that body modification is a sign of control over our own bodies does not question other regimes of power which regulate bodies and which limit our own 'control': as has been mentioned, law dictates how body modifications can be carried out, on and by whom, as well as censoring material *about* adornment and modification (including, for a time, Vale and Juno's book; see Califia 1993). In this sense, then, the adorned body is a *political* body, in the sense that its existence provokes reaction and counter-action from the state and from law. The act of wearing this decorated body politic, however, raises familiar questions about context and audience. Is it enough to know that one's cock has a ring through it? Can cock-piercing be a political act in itself – an act, as Curry (1993) writes it, of politicised dissent? Brought out at piercing parties and tattoo festivals, the adorned self is the *adored* self, adored by the self. Certainly, there is a strong polymorphously perverse politics associated with the SM community to which body modification activity is often appended, and which is mobilised more strongly than ever over issues of consent, rights and freedom – issues of 'life, liberty and the pursuit of sex (all in the name of democracy)' as Sue Golding (1993a: 183) calls them; but the ways in which body modifications articulate a politics of dissent other than through well-worn notions of the refusal to conform remain unclear. Occasions when adorned perverts are on parade in 'straight space'[2] – on carnivalesque Pride marches, for example – might signal the political potential of the transgressive body operating a kind of 'shock tactic' (although the *regulation* and *containment* of such 'public' displays must always be remembered); outside of this, the pleasures of the text – the inscribed body – operate through

Jonathan Dollimore's 'perverse dynamic', an articulation of 'perversity, proximity, paradox, and desire' (Dollimore 1991: 230) in which the 'passing pervert' (tattooed and pierced beneath her or his business suit) can achieve a transgressive politics of location – location in 'straight space' – which carries the potential for disruption more meaningful – more *dangerous* – than the bared scarred buttocks and pierced nipples of shock tactics.

EMBODYING AIDS ACTIVISM: THE ANGRY SELF

Resistance to the dominant mythologies of the AIDS epidemic and to the state's *in*activity and feeble responses to the current health crisis has been articulated through an embodied geography of raging activism. Groups like ACT UP, Gay Men Fighting AIDS, Queer Nation and OutRage![3] have intervened in the AIDS panic by using a radical toolbox comprising techniques of the spectacle projected onto angry (and shamed) bodies. In particular, the 'indecent exposure' of certain 'queer' practices in heterosexually-coded public spaces ('wink-ins' and 'kiss-ins' in shopping malls and high streets) creates an eruption of body positivity and through this a radical assertion of the right to be noticed – a reclaiming of space which is a reclaiming of elements of a citizenship denied through a multitude of discourses, but most notably in the era of panic sex by the twin discourses of contagion and denial. A refusal to be either constrained or restrained – a refusal of silence and invisibility – is a refusal of death. The enacting of anger thus becomes a tactic of survival, and 'an instrument of cartography. By determining where, with whom, about what and in what circumstances one can get angry and get uptake, one can map others' conceptions of who and what one is' (Frye 1983: 93). To get angry, then, is one way to map the limits of one's spaces of citizenship, and thence, by the sustained action of rage, to transgress the boundaries of those spaces. While John D'Emilio (1993: 220) has recently argued that 'you can't build a movement on anger', the deployment of public rage and civil disobedience in AIDS activist actions is the only way to secure the momentum needed to push through hegemonically-constructed walls of silence that quite literally spell death. The movement of angry bodies across the topography of public space represents the very actions which, to coin another ACT UP slogan, can and do equal life.

The call for action (and for life) is also a call to reject state health discourses which have been constructed on sex-negative principles, offering abstinence as the only sensible choice. In the age of AIDS, as Douglas Crimp (1987) has urged, a response of 'principled promiscuity' offers radical possibilities for a sex-positive strategy of *refusing refusal*. The imperative of AIDS activism thus becomes 'to shift the language game, to speak, demonstrate, and demand in ways that are seen as inappropriate to the game when the game erases them or excludes them from its continual reformulation' (Yingling 1991: 299). Linda Singer's (1993: 121) 'epidemic strategies' for reconfiguring sexual acts in the age of AIDS represent the embodied play of the language of pleasure which is also the language of politics.

The particular corporeal manifestations of what Crimp (1987) describes as the 'AIDS activist aesthetic' – an aesthetic which, as Simon Watney (1990: 190) puts it, 'amounts to nothing less than a guerilla semiotics on all fronts, threatening "normality" with a long, sustained, deliberate derangement of its "common sense"' – intervene most forcefully when they clash with those twin discourses of contagion and denial noted above.

Even more than the vitally important artistic and media strategies which have brought AIDS into the cultural realm, tactics of spatial appropriation and body-terrorism mark the most powerful confrontations of the activist aesthetic, especially since they occur in the context of public spaces governed by an etiquette of 'minding one's own business', of a kind of public privacy of family units and codes of 'acceptable' behaviour. Strategies outlined by Cindy Patton (1990: 131) – 'throwing our blood at insurance companies, setting up illegal AIDS drug counters in front of the Federal Drug Administration, holding die-ins at hospitals and drug companies' – reveal how the activist aesthetic is oriented around a particular configuration of bodies in space; it has its own politics of location which rupture the boundaries of self and other, public and private (Butler 1993). A reclaiming of what Watney (1987) calls 'the spectacle of AIDS', which turns it from a spectacle of tabloid moral panic and conservative self-righteousness[4] into a spectacle of theatrically-articulated radical rage directed *against* the 'moral majority', invokes new political uses of the personal – a new politics of pleasure – where the bedroom comes out onto the street, and desire and anger are enacted through necessarily 'epidemic strategies', including sex-positive safer sex campaigns, 'a reconfiguration of bodies and their pleasures away from an ejaculatory teleology towards a more polymorphous decen-tered exchange' (Singer 1993: 122), and a postmodern theatricality which destabilises accepted notions of identity, community, citizenship and welfare. A deconstructive identity politics fused with a recuperation of the language game away from hegemonic discourse into the hands and mouths of AIDS activists makes the body, as Geltmaker (1992: 609) succinctly puts it, 'a site of public contestation', a 'politically trespassed public space'.

Of course, tactics of embodied anger are not new in the arena of sexual politics; they have always been part of the portfolio deployed by certain segments of sexual minority populations (segments which might currently be seen to be articulating their rage under the banner 'Queer'). In fact, in some respects, the emergence of the particular activist aesthetic of AIDS and the rowdy reclaiming of rights and spaces (and rights *to* spaces) by queers have occurred hand-in-hand. Certainly, to be queer must mean to have an awareness of safer sex and its erotic possibilities; and the much-debated inclusiveness of queerdom is echoed in the refusal of groups like ACT UP to be labelled as 'gay rights' organisations, preferring instead to fight back for all people infected or affected by HIV and AIDS. The 'paradigm of marginality and regulation' (Yingling 1991: 305) haunts both same-sex 'dissidents' and PWAs, meaning that the subjectivities constructed by both AIDS and same-sex desire run more or less parallel. When Teresa de Lauretis (1991) talks of a 'queer subjectivity' she is inevitably talking of one which bears the scars of the AIDS panic, scars which have reopened and deepened many of the old wounds of homophobia and heterosexism but also ripped into new flesh. The famous tattoo across the knuckles of singer Diamanda Galas, 'We are all HIV+', follows the same confrontational route as the rallying cries 'We're here, we're queer, get used to it' and 'Whose fucking streets? Our fucking streets'; the politicisation of pleasure and health – both resonantly *bodily* politics – comes together to forge a sense of the self as a 'political project of care and of hope' (Probyn 1993: 173). The scaling of AIDS, to paraphrase Neil Smith (1993), runs from the global to the local (see also Weeks 1990), from the 'we' to the 'I'.

The public strategies for refusing refusal which the AIDS crisis has prompted – the ways of acting up and the very need *to* act up – thus mark a contestation of the public–private divide which throws into relief the constructedness of these notions as they apply to spaces, acts and identities. Just as the recent interventions of law into the bedrooms of sadomasochists in the UK show the fragility of privacy (and it must be remembered that this focus on the sadomasochist is inevitably underscored by AIDS discourse), marking out, we might say, the limits of what can be private 'in private', so the eruption of the activist aesthetic onto the streets transgresses the constructed (and continually reaffirmed) norms of what can be public 'in public'. By pushing at these limits with the very body itself, AIDS activism articulates paradoxical geographies of the public and private, of selfhood and otherness, of health and illness. Such paradoxical geographies are

> political projects which attempt to challenge the transparent geography created by hegemonic subjectivity from an 'excessive critical position … attained through practices of political and personal displacement across boundaries' … [and which] have created not so much a space of resistance as an entirely different geometry through which we can think power, knowledge, space and identity in critical and, hopefully, liberatory ways.
>
> (Rose 1993a: 158–9)

The AIDS activist aesthetic thus embodies Elspeth Probyn's (1993) notion of 'thinking the self with attitude', an 'in your face' (and, we might add, 'in your *space*') strategy in which the angry body is thrust up against the boundaries of 'hegemonic subjectivity' and its spatial demarcations; articulating a 'different geometry', a geometry of care and of hope.

CONCLUSION: PERFORMING THE SEXED SELF AS A SITE OF RESISTANCE

The three performances of the sexed subject which we have outlined all articulate the relationship between the self and others in very different ways.

However, it is possible to draw together a number of themes which interconnect the spaces, acts and identities that constitute the managed self, the adorned self and the angry self, and which illustrate how performances of the sexed self can become sites of resistance.

All three enunciations of self are embodiments of theatricality, with the self being 'done', being projected, across the contours of the body. There is a recurrent tension between actor and audience in these presentations of self; according to respective viewpoints, the transgressive potential of performative strategies is predicated on the actor's intentions (as Butler, for example, argues) or in terms of audience reception and response. The AIDS activist aesthetic obviously depends for its political effectiveness on shock tactics which are self-consciously provocative; by confronting the audience in straight spaces with an enaction of embodied rage, AIDS activists push the limits of the performance of the self onto the heavily contested terrains of contagion and denial (Butler 1993). In contrast the destabilising potential of the managed self lies in its capacity to infiltrate straight spaces unnoticed (except by those 'in the know'). By operating within a heterosexual index while at the same time being the 'other' *to* heterosexuality, the 'passing' lesbian challenges how spaces and identities become constructed and encoded, by being inside and outside simultaneously. As Butler (1990: 31) famously stated it: 'The replication of heterosexual constructs in non-heterosexual frames brings into relief the utterly constructed status of the so-called heterosexual original.' The 'radical confusion of identities' (Fuss 1991: 6) offered by such a position might have a more sustained resonance because it shakes the very foundations of 'I am' and 'you are'. The 'crucial sense of alterity' necessary to define sexual 'otherness' becomes unstable once the point has been demonstrated that 'borders are notoriously unstable, and sexual identities rarely secure' (Fuss 1991: 7, 3).

In contrast to the measured and managed self, the adorned subject truly plays the 'excessive game of self'. This game, this performance, is however in some ways located between the strategies of publicity and privacy deployed respectively by angry and managed selves. Although not explicitly 'political' in origin, the actions of state and law periodically bring the adorned self into the political arena by outlawing certain acts and identities. At the same time, the actors' power (in some cases[5]) to name their audience, and to choose where, when and to whom the adorned self is revealed or concealed, means that the figure of the 'passing pervert' can occupy a managed position, a radical insiderness.

One thing that all these identities share is an acute awareness of the audience, and of how that audience reads the performance of self; each is, then, as *self-conscious* as the other, although the way this awareness mediates each performance is context-specific. Indeed, it might be said that the pressures of hegemonic sexuality, as it is constantly and repetitively reaffirmed and re-cited, make all sexual dissidents 'close readers' of identities, ever aware of the impact that performative strategies have on the self and on others. In this way, *all* identities become questioned, working on the assumption that things (and particularly selves) are never quite what they seem. Theorising the performativity of the self in this way offers up ways to 'do' the performance differently; as Probyn (1993: 2) says: 'when we speak our selves within theoretical contexts . . . [the] coherency of the self is opened up and its movement into theory creates the possibility of other positions.'

From this perspective, the very notion of stable identities becomes *de*stabilised, opening up new radical spaces for subjectivities freed from rigid binarisms and cultural matrices[6] (although, as Julia Cream's chapter in this collection makes clear, this can be a problematic process). While 'the limits of what we are and can be' may have been 'already mapped by somebody else' (Rose 1993a: 147), rethinking our (sexed) selves in theory and in practice might produce cartographies of these new spaces.

NOTES

1   Curry's note about body modification signalling taking a responsibility for one's body was also used as an argument in the Operation Spanner trial. The men convicted argued that they were aware of the need for care in their SM scenes, and had sterilised all equipment to preclude any infection (notably of HIV/AIDS). The prosecution argued that their very care itself showed that the risks involved in their activities were high, and that by extension the men themselves were 'high (HIV/AIDS) risk' status (Bibbings and Alldridge 1993).

2   For an elaboration of the notion of 'straight space', and the construction of certain (public) spaces as 'straight', see Bell, Binnie, Cream and Valentine (1994) and Rose (1993a).

3   It must be made clear that ACT UP and GMFA are both AIDS activist groups, while Queer Nation and OutRage! have an agenda centred upon sexual politics more generally. This said, in the 'age of epidemic', AIDS activist work is inevitably a significant part of their political project.

4   As Watney (1987: 83) says, 'the spectacle of AIDS operates a public masque in which we witness the corporeal punishment of the "homosexual body", identified as the enigmatic and indecent source of an incomprehensible, voluntary resistance to the unquestionable governance of marriage, parenthood and property.'

5   If that audience happens to be the state or law, of course, then it may be impossible to say what that audience does or does not see. The recent trial and conviction of same-sex sadomasochists in the UK (Operation Spanner) has shown how fragile appeals to privacy can be (Bell 1994).

6   Appeals to deconstruct identities and reveal them to be nothing more than 'necessary fictions', while theoretically seductive, can be difficult to mobilise around issues of politics. The 'strategic' use of essentialism has thus been advocated, with the essentialism of anti-essentialism being noted as a problematic of this deconstructive turn (see Fuss 1989; Rose 1993a).

8

WOMEN ON TRIAL

a private pillory?

Julia Cream

The woman swallowing the oral contraceptive pill makes sense to us in the twentieth century. She has done since 1960. She is 'intelligible' in a way that a man swallowing hormones, or a post-menopausal woman swallowing the pill is not. We accept her body in a way that we do not accept a woman taking steroids for muscle-building, or a transsexual taking the pill. By using the example of the woman on the pill, I want to try to show how bodies are made culturally intelligible. I want to illustrate how the body of the woman swallowing the pill may (or may not) conform to the cultural matrix of what Judith Butler has called intelligible genders: '"Intelligible" genders are those which in some sense institute and maintain relations of coherence and continuity among sex, gender, sexual practice, and desire' (Butler 1990: 17). The sexes/genders/desires (and races) that make sense to us are not natural or inevitable. The heterosexual, fertile woman on the pill, wanting to plan the size of her family, for example, makes sense. Her body is both legitimate and intelligible. Located in another position, such as the post-menopausal single woman, or even as a man, she is less 'intelligible'.

The pill contains a combination of synthetic hormones which modify the female reproductive system, inhibiting ovulation. The pill acts on a body that is sexed female. While it may sound banal, and even nonsensical, to suggest that the subject popping the pill is female, sex is, however, not simply something one has, or is. Despite appearances, sex is an effect, a means by which 'one becomes viable ... that which qualifies a body for life within the domain of cultural intelligibility' (Butler 1993: 2). It is this appearance of sex as prediscursive, prior to culture, that obscures and disavows the constraints that produce a domain of intelligible and unintelligible bodies.

Rather than understanding sex as a biological bedrock upon which the cultural layers of gender are built, gender is the means by which the sexed body is established as natural. I am taking gender to be 'the repeated stylization of the body, a set of repeated acts within a highly rigid regulatory frame that congeal over time to produce the appearance of substance, of a natural sort of being' (Butler 1990: 33). There is, then, no sexed female body awaiting enculturation, or engendering. Instead, gender is understood to be performative, constituting 'the very subject it is said to express' (Butler 1991: 24).

The woman on the pill comes into being through the mark of gender. If gender is performative, constituting identity, then it makes no sense to start with the subject behind the act, the woman prior to the swallowing of the pill. It makes no sense to start with the 'woman', since 'no subject is its own point of departure' (Butler 1992: 9). There is no volitional subject before the act, but the woman on the pill is constituted by the taking of the pill. It is at this location, the doing, the becoming, the repeated making of the woman on the pill that offers an opportunity to intervene in a conventionally ascribed meaning of the sexed subject.

The woman on the pill is 'doing' her gender in a specific way. She does it repeatedly (and never completely) and in interaction with others. She does it in a way that produces an illusion; an illusion of essence. Gender is the 'mundane drama specifically corporeal, constrained by possibilities cultural' (Butler 1989: 261) that produces the illusion that we simply *are* our genders, that they are an expression of our identity. Consequently, the exposure of 'a constructed identity, a performative accomplishment' such as the woman on the pill, reveals the 'performative possibilities for proliferating gender configurations outside the restricting frames of masculinist domination and compulsory heterosexuality'. In this way, the woman on the pill becomes a possible site for 'gender transformation' (Butler 1990: 141).

To understand gender as a performance, and identity as performative, is not to presume that gender is chosen. Gender is not voluntary: 'Performativity is neither free play nor theatrical self-presentation' (Butler 1993: 94). Gender is not a style, or a game that can be played. It is this forced reiteration of norms, the repetition of 'regulatory fictions' that constitute the subject (Butler 1993: 95).

By revealing the constructed nature of the woman on the pill, her body becomes a site for the possible reworking of gender performances. The exposure of the ways in which her sex, gender and sexuality have been produced as inherent and natural, as well as in ways that endorse heterosexuality, racism, colonialism and sexism, provides opportunities for figuration. Figuration, as Haraway (1992: 86) suggests, 'is about resetting the stage for possible pasts and futures'. Indeed, it is precisely because the production of the sexed subject has been so rigidly constrained that Butler (1993: 123) argues that gender (and other regulatory fictions) 'ought to be repeated in directions that reverse and displace their originating aims'.

In the first section, I provide one story of the constitution of the woman on the pill. I identify three 'bodies' of the woman on the pill, the infertile woman, the Puerto Rican woman and the contracepted western woman. In the second section, I try to show how her body was made culturally intelligible: why all the bodies 'make sense'. I end with some thoughts on how the woman on the pill can be made unintelligible: hinting at her different pasts and futures.

THE MAKING OF THE WOMAN ON THE PILL

Infertile woman

The pill was first tested on infertile women. In the early 1950s, endocrino-logical research was highly speculative. Natural hormones were used for treating gynaecological (dis)orders, habitual aborters, cervical cancer and infertility. In 1953, large doses of progesterone were given to eighty childless women in Boston, Massachussets. A second trial with twenty-seven women was regarded as a 'success' when four of the women became pregnant shortly after the cessation of the trial. A state of ovulation-free 'psuedo-pregnancy' was induced using a regime that imitated the 'normal' menstruation cycle.

Naturally occurring progesterone was, however, prohibitive in price. It was not long-lasting when administered orally, and was painful to inject intramuscularly. A second human trial was repeated, using newly synthesised analogues of progesterone, known as the 19-nor-steroids, which had been found to have a powerful progestational effect on animals. Out of the fifty 'unreproductive patients' (Garcia et al. 1958: 82) aged between 22 and 39 years who 'served as subjects' (Pincus 1958: 5), seven women conceived within five months of the last treated cycle. The precise action of the compound on the reproductive tract remained unclear, but it was suggested that a rebound effect occurred. Not only did progesterone, natural or synthetic, appear to be conducive for pregnancy, it also inhibited ovulation. Despite ignorance about the precise action of progesterone, a collaboration of scientists, doctors, feminists, population planners, pharmaceutical com-panies and women 'co-operated' in a large-scale field trial of the oral contraceptive pill.

Puerto Rican woman

Puerto Rican women provided the ' "cage" of ovulating females' who would submit themselves to clinical experimentation (McCormick, in Ramirez de Arrallano and Seipp 1983: 107). A pilot study was carried out in March 1955 with twenty-three female medical students. It lasted only three months, and suffered from a large dropout rate. Each subject (like the Boston volunteers) was 'required to take her temperature every morning, take a daily vaginal smear on a special glass slide, collect a 48-hour urine sample on a monthly basis, and submit to an endometrial biopsy once a month' (Ramirez de Arrallano and Seipp 1983: 110). A second study failed due to the inability to recruit volunteers from either the nursing or the medical staff. Consequently, female prisoners were used, but that failed too due to the difficulty of securing more subjects after prisoners expressed their objections.

Despite these problems, by April 1956 there were 100 women participating in a large-scale trial. The women selected for the trial ranged in age from 16 to 44 years and all were on a low income (Pincus et al. 1958). Each woman was interviewed every time she received another month's supply of pills.

Each 'allegedly faithful user' (Pincus *et al.* 1959: 1056) was questioned about coital frequency, the number of pills missed and breakthrough bleeding, and any other problems that she volunteered were recorded.

Out of the 221 women who had taken the tablet by 31 December 1956 there had been seventeen pregnancies (Pincus *et al.* 1958). The pregnancies were described as 'patient failures', resulting from a failure to take the medication. Blame was laid firmly at the feet of the user, the woman, rather than in anything inherent about the drug itself. It is noted, however, that eight women stopped because of a reaction to the medication. Side-effects, e.g. nausea, headaches, loss of libido, dizziness, lethargy, that could not be verified by physical bodily signs were put down to the 'emotional super-activity of Puerto Rican women' (Rice-Wray, cited in Ramirez de Arrallano and Seipp 1983: 116). Even some side-effects such as weight gain were dismissed as a result of an enhanced sense of well-being, and a placebo test confirmed, for some of the scientists at least, that the side-effects were psychological in origin (Pincus 1958: 23).

Although women's need for contraception was great and they were prepared to suffer pain and discomfort, many women did in fact choose to withdraw from the trial. Of the original 100 women volunteers, 30 had left by June. By August 1957, although there were 141 patients continuing on medication, 123 women had discontinued the 'therapy'. Reasons for discontinuation vary from pregnancy, sterilisation and religion to an interesting group labelled as 'miscellaneous'. This category includes one woman listed as having an 'uncooperative attitude' as well as three women named as prostitutes. It remains unclear why that precluded them from participating in the trial.

The large-scale field trial 'worked' for the majority of the researchers and the pill was declared 'safe'. By 1963, of the 730 women who had participated in the trials, slightly over half had discontinued. There are conflicting accounts about the efficacy of the pill and the health risks incurred by the women involved. Ramirez de Arrallano and Seipp (1983: 119–23) state that there were eight deaths, five of which were due to flooding, and the rest were not autopsied. Enovid, the trade name of the pill launched by the pharmaceutical company Searle, was authorised by the Food and Drug Administration (FDA) largely on the basis of data collected in Puerto Rico. Although this included thousands of treatment cycles, only 123 'Enovid treated women' (Garcia 1963: 50) had taken the pill for twelve cycles or longer. Nevertheless, the 'successful' testing of the Puerto Rican woman on the pill paved the way for further testing.

Contracepted woman

Enovid was approved as a contraceptive in the US by the FDA in 1960, initially for two years' continuous use. Within months millions of women all around the world were swallowing an oral contraceptive pill. In the early 1960s, the pill was available for the married, or those about to be married, and this requirement was racialised and classed. Women's bodies continued to be the sites of experimental testing for new 'generations' of pills, new doses and new regimes. By 1965, an estimated five million women were using the pill in the United States alone.

MAKING SENSE OF THE WOMAN ON THE PILL

Embodied within an anatomical and physiological narrative are cultural values about how the sexes should be ordered, and the roles and spaces that they should inhabit. Just as Butler suggests that we 'do' our gender, I am arguing that we 'do' our bodies. This performance is about constituting sex, gender and race relations. There could be no body of a woman on the pill if there were no discourses to make her body culturally intelligible. This is not to say that there would be no bodies, but that they would mean something very different.

The three stages that I have elucidated above are not discrete. Some women, and some contexts, cross boundaries and blur divisions of race, class and gender. The three different bodies that I have chosen to highlight, the infertile, the over-fertile/black and the contracepted, are all intelligible. They are located at positions that are explicable given the sex/gender/desire regulatory fictions that were operating in the 1950s–1960s. It is to these narratives that I now turn in order to assess how the woman on the pill was constituted.

Infertile woman on the pill

The 1950s in the US was a 'family-focussed era' (Miller *et al.* 1991: 566), experiencing levels of marriage and pregnancy unprecedented in the twentieth century. Women of all ages, ethnic groups and income levels participated in the baby boom; it was a mass phenomenon (van Horn 1988: 85) and 'fewer than one in ten Americans believed that an unmarried person could be happy' (Mintz and Kellogg 1988: 180). There was intense pressure on couples to have children, and within such a context, the body of a woman exhibiting signs of infertility was a legitimate site for the experimentation of fertility agents.

Childlessness was considered deviant, selfish and pitiable (May 1988). Breinnes (1992: 52) notes that 'a study of marriage at the end of the decade found that marriages which [were] childless by choice [were] practically nonexistent'. Being a woman meant being a wife and a mother. The cultural values imbued in having children were very strong: you were not 'fulfilled' until you had children. Women were not only defined by their role as mothers, but if they were unable to 'develop' their maternal role, then, questions were asked and insinuations made. As late as 1968, Campbell wrote in the *Journal of Marriage and Family* that the couples with no children are likely to be considered extremely self-indulgent, and the couples with only one child will probably be regarded as heartlessly denying their unfortunate offspring the companionship of brothers and sisters. Two children is in the lower figure of social acceptability, and four children is widely accepted as ideal.

Investing heavily in the cult of motherhood, women in the 1950s underwent painful and difficult procedures in order to carry a pregnancy to full term.

Doing one's sex/gendered female body correctly in the 1950s meant being heterosexual, married and childbearing. But this was a highly racialised notion of doing one's body. White women, not black women, were defined by their reproductive function. The ideal type of femininity and femaleness was highly racialised and classed. The construction and constitution of the Puerto Rican woman on the pill offers another body, one constituted by sexism, racism, imperialism and colonialism. Her body is rendered culturally intelligible in a way that is different from the privileged, yet infertile, white woman of the Boston suburbs.

The Puerto Rican woman on the pill

It is no coincidence that the British considered Jamaica for the testing of the oral contraceptive pill (Contemporary Medical Archives Centre: A5/162/5) or that the Americans used a (former) colony, the island of Puerto Rico, for mass-scale testing of the oral contraceptive pill. Contraception was not only a taboo subject in the early 1950s, but was illegal in many states in the US. The Puerto Rican woman was not merely any member of an overpopulated, underdeveloped country; rather, Puerto Rico occupied a privileged position in the minds of the American people as well as in the literature of population theorists and planners in the 1950s. Puerto Rico has a legacy of population policies which cannot be extricated from its colonial history (see Mass 1976) and the testing of Puerto Rican women on the pill is no exception.

American fear of the breeding-grounds of communism initiated a serious consideration of birth control. The Rockefellers became one of the principal financial supporters of the Population Council as corporate leaders became increasingly concerned that Asia, Africa and South America, impoverished through overpopulation, would fall to the communists. Eugenic preoccupations also re-emerged in the 1950s in studies warning that the white nations could be submerged by the yellow and the black (McLaren 1990). Conservation measures were deemed futile whilst human breeding continued unabated: 'Unless population increases can be stopped, we might as well give up the struggle' (Vogt 1949: 280). A new contraceptive device was again held to be the only way out of the 'ecological trap' and the only means with which to ensure 'national security'. The future of the human race was understood in explicitly environmentally deterministic terms; population planning was once again posited as the inevitable solution to a burgeoning population and finite resources.

Puerto Rico, that 'very black spot in the map of universal hunger' (de Castro 1952: 108) came to figure large in the minds of many American commentators: 'Like white blood cells around an infection, a social crisis like that in Puerto Rico always draws a flock of commissions and committees, but they are a waste of time and effort unless they lead to effective action' (Cook 1951: 27). The 'effective action' was contraception. Overpopulation was framed as the 'problem', and birth control as the 'solution'. Population reduction was seen as a precondition for economic development, and the advocacy of birth control was aimed at preserving the economic status quo and the prevailing social order. Epitomising the worst-case scenario of world-threatening overpopulation, Puerto Rico was established as a site that not only 'needed' population control measures, but one in which the women were already accustomed to their presence. It is therefore surprising to discover that 'by 1955, with a high level of sterilization achieved and a huge loss of population through emigration [to the US] Puerto Rico's population growth rate had become the lowest in all the Caribbean' (Mass 1976: 93).

Racialised notions of sex and sexuality are played out across the body of the Puerto Rican woman. Constructed as feckless, illiterate and overbreeding she was understood to be desperately in need of contraception. The pill was seen to offer a panacea for the women who were already 'thronging to the inadequate clinics' (Cook 1951: 39). It is clear that the defining role of a Puerto Rican woman was not one of wife and mother. Their geographically and historically specific bodies, marked by race and colonialism as well as gender, were constituted as in need of reproductive control. Her body legitimated the contracepted body.

Contracepted western woman on the pill

The birth control pill helped a woman *plan* her family. It was not intended as a means with which a woman could abdicate by choice from the responsibility of parenthood. Motherhood remained a mainstay of religious and sexual doctrine: women were 'naturally' mothers. Family planning redefined sexual morality in the post-war era. It promoted marital harmony and sexual compatibility: 'from an issue that had once seemed to epitomize the question for female autonomy, birth control had become a matter of insuring family stability' (D'Emilio and Freedman 1988: 248).

As female sexual pleasure became increasingly important for happiness within a 'companionate marriage' (Finch and Summerfield 1991), the 'natural' sexual relationship became predicated on enjoyment rather than procreation. As the advertisement for Volidan (another oral contraceptive pill) says, 'Once the Volidan habit becomes routine, you and your husband are free to enjoy a completely natural sexual relationship at any time without having to worry about unwanted pregnancy. In this way Volidan makes a definite contribution to happy married life' (Contemporary Medical Archives Centre: PP/RJH/A1/7).

Paradoxically, although a woman's role and the definition of femininity and femaleness remained allied to motherhood, throughout the baby boom of the 1950s, women had accomplished their procreative duties. They had earned the right to continue their sexual relationship without doubling the size of their families: 'wives who had two, three or four children while still in their twenties could hardly be accused of seeking contraceptive devices in order to avoid their biological destiny, or to escape the confines of the home' (D'Emilio and Freedman 1988: 249).

As the pill became synonymous with family planning it endorsed the ethos of planning that was prevalent in post-war years. A planned society offered hopes and proffered visions of a new world. Cloaked in scientific language, the oral contraceptive pill appeared as a miraculous invention. Contraceptives had been messy and interfering, but, with the advent of the pill, contraception entered the scientific age, offering couples a means of birth control appropriate for 'modern living'. By enabling couples to marry young, postpone childbearing and space their children, contraception fostered the modernisation and professionalisation of domestic roles. Like labour-saving appliances, birth control devices could contribute to enjoyment at home and heighten the standards of domestic conduct without disrupting the ideologies of motherhood.

Birth control not only improved sexual harmony within marriage and enhanced the stability of family life, the woman on the pill also ensured (and was constituted by) a stable world order. The woman on the pill became an actor in the tense scenes circulating around Britain and the US in the post-war years; playing out the fears and concerns of an unstable society.

Overpopulation was understood to be an urgent problem and the woman on the pill was often privileged as the only viable solution. Family planning also played an important role in redefining Cold War fears of atomic threat. Recurring throughout the literature is the comparison of the population 'explosion' with that of Hiroshima. The concerns circulating about over-population at a local and global level appear to converge around the topic of atomic threat: 'Next to the atom bomb, the most ominous force in the world today is uncontrolled fertility. Unbalanced and unchecked fertility is ravaging many lands like a hurricane or a tidal wave' (Cook 1951: 15). '[W]e are all uneasily aware of the mushroom cloud that first sprouted over Hiroshima, we are astonishingly unaware of the mushrooming global population' (Maury 1963: vii). The fears and anxieties generated by both the population and atomic explosions are frequently linked, as is the urgency with which both need to be tackled. In 1963 Bertrand Russell outlined two antithetical dangers facing the world's viability: the use of H-bombs and the increase of the human population. Russell (1963: 1) warns that 'Nothing is more likely to lead to an H-bomb war than the threat of universal destruction through over-population.'

Elaine Tyler May (1988, 1989) has developed an incisive thesis that links the fears arising out of the Cold War and the atomic age with the intimacies of private life in the US. The family becomes located in, and not outside, the larger political culture as domesticity is embraced in the face of unknown threats. She proposes that a house full of children created a feeling of warmth and security against the cold forces of disruption and alienation. Children were also perceived to be a connection to the future and a means of replenishing a world depleted by war deaths. Planning evolved as a central theme around which much of the US domestic policy was structured.

It was not just nuclear energy that had to be contained, but the social and sexual fallout of the atomic age itself. Fears surrounding sexual chaos were omnipresent. Non-marital sexual behaviour, in all its forms, became a national obsession after the war. A persistent link was made between communism and sexual depravity; moral weakness was associated with sexual degeneracy which allegedly led to communism (May 1988, 1989). Faderman (1991: 140) adds that 'If political conformity was essential to national security, sexual conformity came to be considered, by some mystifying twist of logic by those in authority, no less essential.' Cold War policy instituted a system of sexual surveillance in which any deviation from the married, heterosexual couple with children was a threat to national security. In the baby boom years in which the nuclear family was firmly and virtuously upheld, 'the man or the woman choosing to pursue same-sex intimacy was more than ever going against the grain. Labelled as sexual and moral perverts gays and lesbians were not only seen as flawed individuals, but as dangerous and threatening.' Ironically, as D'Emilio (1983: 52) notes, the vicious attacks and scapegoating of gay men actually 'hastened the articulation of homosexual identity and spread knowledge that they existed in large numbers'.

By the late 1950s family planning in general, and the pill in particular, had been redefined as the solution to a range of ills manifest in society: both global and local. The woman swallowing the oral contraceptive pill was now not only maintaining the health and happiness of her family, but also the future of the whole world.

UNINTELLIGIBLE GENDERS?

By tracing the production of the body of the woman on the pill I have indicated that there was not one body, but competing and contemporaneous ways of understanding and interpreting her. Not only did she change geographical location, but she shifted in race, class, status, education and religion. Women's bodies became a battleground on which reproductive rights were fought as well as the constitution of gender and subjectivity.

By assessing how culturally intelligible bodies are constituted, who sets the cultural norms, and how it is possible to begin to think about new gendered bodies, I have drawn on several writers to indicate how the woman on the pill could be refigured. Butler, Stone and Marcus have all employed the notion of performative gender constituting a subject.

Judith Butler (1990) uses the politics of drag, where gender is freed from sex, to indicate the possibilities of refiguring bodies. With masculine and feminine no longer restricted to 'male' and 'female' our notions of what makes a sexed and gendered body intelligible become disrupted and destabilised. Drawing on the deployment of butch/femme identities in a 'female' body, Butler indicates that this may provide a site at which new dissonance can occur, producing new intelligibly sexed/gendered bodies.

In a later discussion of the subversive effects of parody, Butler (1993: 231) is at pains to emphasise that 'drag is not unproblematically subversive'. Noting that 'many readers understood *Gender Trouble* (1990) to be arguing for the proliferation of drag performances as a way of subverting dominant gender norms' Butler (1993: 125) again reiterates the contingency of subversion. As Bordo's (1993: 292) comments on Butler's work indicate: 'This is ingenious and exciting, and it sounds right – in theory.'

Indeed, Bordo is right to stress the importance of context, of location, to give some flesh to 'the body'. Bordo also highlights the possibility of different responses of various readers and the various anxieties that might complicate their readings. Nevertheless, the importance of understanding gender-as -drag should not be undermined. The radicalism of drag is maintained through its potential to call into question the norms of heterosexuality and denaturalise the categories of gender. Whilst Butler (1993: 125) questions whether 'parodying the dominant norms is enough to displace them; indeed, whether the denaturalisation of gender cannot be the very vehicle for a reconsolidation of hegemonic norms', I argue that the possibility of exposing what appears to be natural as an effect, is a lead worth pursuing. The 'doing' of gender differently still retains the potential of formulating a 'project that preserves gender practices as sites of critical agency' (Butler 1993: x).

Also employing the notion of performative genders, and inspired by the work of Donna Haraway, Sandy Stone (1991) begins to suggest that transsexuals can lay the foundation for refiguring gender, sex and the body. A transsexual 'is a person who identifies his or her gender identity with that of the "opposite" gender' (Stone 1991: 281). A transsexual body is culturally intelligible only within a binary heterosexual matrix. Their bodies make sense because they are 'wrong'.

Using autobiographical published accounts of male-to-female trans-sexuals, Stone demonstrates the complicity of some transsexuals in a 'Western white male definition of performative gender' which reinforces a binary, oppositional mode of gender identification (1991: 286). Demonstrat-ing a clear understanding of why some transsexuals deny the mixture in their lives, she notes (289) how 'Each of these adventurers passes directly from one pole of sexual experience to the other. If there is any intervening space in the continuum of sexuality, it is invisible.'

Stone calls for the articulation of a counter-discourse, a discourse which would 'generate new and unpredictable dissonances'. By claiming a position that is nowhere, a location that is currently 'outside the binary oppositions of gendered discourse' transsexuals are able to disrupt conventional gender practices. But, she notes with some irony, 'it is difficult to generate a counterdiscourse if one is programmed to disappear' (1991: 295). A trans-sexual, by definition, wants to 'pass', to erase his/her past history. Stone goes on to argue that the refusal of a transsexual to 'pass' as either a 'man' or a 'woman', a refusal to eradicate one's past life as a member of the 'opposite' sex, is to 'fragment and reconstitute the elements of gender in new and unexpected geometries' (296), thereby reworking our notion of culturally intelligible genders. By redefining what counts as a culturally intelligible body, we can begin to undermine and disturb the myth that states that 'only one body per gendered subject is "right"' and that 'all other bodies are wrong' (297).

Sharon Marcus (1992) draws on the idea of performative gender to reassess the commonsense notion that women are always either already raped or already rapable. By using a concept of rape as 'scripted interaction which takes place in language' enabled by narratives, rape becomes a process which we can disrupt, rework and ultimately eradicate. This is not to say that the violence to, and violation of, women through rape can be reduced to text, but the way that we understand rape and the treatment of rapists and women who are raped, is a process that occurs in and through language.

Marcus' use of the notion of performative gender is more contentious than the other examples. The mark of gender in this case is the act of rape. 'Rape does not happen to preconstituted victims; it momentarily makes victims' (1992: 391). The rape script, for Marcus, thus becomes one of the 'regulatory fictions' that Butler discusses. There is no subject 'waiting' to be raped. The rape script pre-exists instances of rape but neither the script for the rape act results from or creates immutable identities of rapist and raped. Hence, it is at this point, this 'gap' that interventions can be made. The citation of the subject that can be raped, or the subject that is capable of raping an other, can be disrupted and 'done' differently.

She offers ways of beginning to refuse the social script, of refusing to be positioned as sexualised, passive, vulnerable, violable and penetrable. By ceasing to become 'legible as rape targets', Marcus provides another way in which genders can be made or unmade intelligible. Marcus argues that 'the horror of rape is not that it steals something from us but that it makes us into things to be taken' (1992: 399). Her essay is a powerful reminder that our bodies can be constituted differently, and that there is nothing natural about the ability of men to rape or women being raped. Through the redefinition of our bodies so that 'we do not need to defend our "real" bodies from invasion but to rework this elaboration of our bodies altogether', Marcus provides a practical and theoretical analysis of how important the notion of performative genders can be in constituting our social values, institutions and practices. She asserts that 'New cultural productions and reinscriptions of our bodies and our geographies can help us begin to revise the grammar of violence and to represent ourselves in militant new ways' (400).

By indicating where and how the woman on the pill was made intelligible, I hope that I have laid the groundwork for rendering her not only as constructed/performed, but as unintelligible within the oppressive frameworks of gender that exist today – of finding ways to do the woman on the pill differently.

The woman on the pill could be a site at which the stability and unquestioned binary, phallocratic logic is disrupted and destabilised. Can she do this by refusing to pass, as Stone implies? By publicly announcing the use of contraception is a woman challenging the western taboos on sex and sexuality? Or, as Marcus suggests, is it about reworking the body so that her body is not rendered as already reproducing or potentially contraceptable or impregnable? Were the women who refused to take the pill in Puerto Rico, or those who refused to undergo the arduous physical examinations necessary for inclusion in a trial, refusing to render their bodies as culturally intelligible to medicine and science? Were the women who pretended to be married, or those that forgot to take the pill and conceived, challenging patriarchal restrictions on female sexuality?

I am not sure. I have suggested that there is nothing inevitable about millions of women around the world taking a pill twenty-one days out of every twenty-eight. I have suggested that the gendered performance of the woman on the pill offers opportunities for reworking our bodies and our social values. But what I am unsure about is how important the intentions of the 'doer' are. Do I have the right to reinterpret a gender performance? What makes a gender unintelligible? Is it up to the woman swallowing the pill and the people who participate in structuring the context in which such gender performances occur, or is it up to me to provide ways of reinscribing the performance, offering a way of understanding differently? Gender performances, such as the 'doing' of the woman on the pill, are highly dynamic and interactional, offering ways of redoing both gender and the contexts in which the performances are interpreted.

ACKNOWLEDGEMENTS

I would like to thank David Bell, Jon Binnie, Laura Cream, Claire Dwyer, Peter Jackson, Nuala Johnson, Charlotte Pomery and Katie Smith for all their criticisms and encouragement. The support of the Economic and Social Science Research Council (ESRC) is gratefully acknowledged.

9

MEN, HETEROSEXUALITIES AND EMOTIONAL LIFE

Victor Jeleniewski Seidler

HETEROSEXUAL IDENTITIES

With the challenges of feminism and gay liberation men have had to rethink their relationship to heterosexuality. This has been part of an exploration of remapping what it means 'to be a "man"'. In what spaces do boys grow up into men and how is this related to different masculinities – both gay and straight – that are available? If we can no longer assume the 'naturalness' of masculinity, what does it mean to say that masculinities are 'socially and historically constructed'? Often the dualities between what is 'natural' and what is 'constructed' are asserted too readily before we have fully understood what is at issue. There is still a lot of confusion about mapping the terms in which some of these questions are raised, and whether we are asking the most helpful questions about how men learn their sexual identities. Often men who identify themselves as 'heterosexual' have wanted to leave these issues alone, seeing sexual politics as concerning 'others'.

If heterosexuality is talked about as part of mapping the self it is usually as a relationship of power which serves to normalise a particular pattern of sexual relationships that serves to oppress women, gay men and lesbians. This has been crucial to stress along with the compulsory character of heterosexuality which has presented it as a norm. But if it has been crucial to recognise that heterosexuality exists not simply as a sexual preference but as a powerful institution within a patriarchal society, it has also meant that both heterosexually identified men and women, for quite different reasons, have had little to say about it. If men are regarded as an 'enemy', then sexual relationships with men were at best to be silently tolerated, especially if women felt that they were thereby sustaining an institution which served to oppress other women. But whether we are to think of heterosexuality exclusively as an institutional relationship of power is a question worth asking. Obviously this has consequences for the ways we understand the space of intimate and personal relationships and *how* power operates within relationships. It is one thing to challenge a liberal notion of the integrity of the private and personal sphere but quite another to recognise the different sources of power that might be at work within heterosexual relationships.

What is striking still is the relatively small numbers of men who have responded actively to feminism in the different political generations since 1970. Even though issues about men and masculinities have moved from the margins of cultural and political concern to nearer the centre, there is still relatively little discussion about *how* men relate to their heterosexualities. At the same time we have experienced an enormous diversity of different masculine styles and a softening of the boundaries that would traditionally have separated gay and straight masculinities. Younger men who have grown up in the 1970s and 1980s seem far more relaxed about traversing boundaries and exploring different identities for themselves. There is little doubt that significant changes are taking place. But there is also considerable confusion about what it means to be a man. A younger generation has had to live with the threat of AIDS and this has transformed the possibilities of sexual exploration as part of discovering more about yourself.

Part of the attraction of the notion that masculinities are 'socially and culturally constructed' is the space it helps to create for thinking that there is no single pattern that men have to conform to. In different periods there have been a variety of different codes and ways of learning about what it means to be a man.[1] This cannot be legislated in the ways that our parents and teachers might once have supposed. This helps to challenge the status of traditional authorities as young men claim a freedom that others would so easily deny them of, to decide for themselves what kind of men they want to be. But this freedom can also be daunting if it opens up too many options at once and leaves people feeling that they are inevitably adjusting to codes they have had no part in creating for themselves. On what basis are different men supposed to make these decisions for themselves?

MASCULINITY AND MODERNITY

Traditionally there has been a strong identification between a dominant masculinity and modernity. This has been organised around an identification between masculinity and reason which has meant that men have often grown up to take their reason for granted. This has allowed men to legislate for others before they have really learnt to talk more personally for themselves.[2] This also relates to the ways that a dominant masculinity has taught men to relate to their own lives and sexualities. Men have often learnt to use reason to discern what will bring happiness and fulfilment.

At some level this has served to impersonalise men's experience of themselves, often making it harder for men to share what they feel. Often, within a dominant culture of middle-class white masculinity, men learn to do what is expected of them and so become 'externally' defined. As I have tried to think this through in *Unreasonable Men*, this is because men's reason has been defined in a way that sets it categorically apart from nature. This has left a dominant masculinity in an ambivalent relationship with sexual identity which is treated as 'animal' and so as a threat to our existence as rational selves.[3]

As rational selves 'we' are not sexual beings; rather we are threatened by sexual feelings that potentially remind us of an 'animal nature'. If we want to sustain a position of dominant masculinity we are tempted into thinking that we can govern our lives through reason alone. As men we are supposedly independent and self-sufficient. We do not have emotional needs of our own because we learn to think of them as a sign of weakness. It is only 'others' who have needs and who thereby prove themselves to be inferior. There is no space for these aspects of our experience within these dominant Eurocentric discourses of masculinity. We learn to cut out these aspects so that we can prove ourselves in the eyes of other men. For it is a crucial aspect of modernity that if men can take their reason for granted they can never take their masculinity for granted. We always have to hold ourselves ready to prove our manhood whenever it is challenged. We can never feel relaxed and easy about a masculinity that can be put to the test at any moment.

Kant argues for the superiority of men because they alone can take their reason for granted. This means that in the terms of an Enlightenment vision of modernity white men can legislate what it means to be human and so set the terms through which 'others' have to prove themselves if they are to be able to join the 'magic circle' of humanity. At the same time Kant argues for marriage as a contract in which both parties agree to make use of each other's sexual organs. As Carol Pateman describes it, 'Kant claims that a "relation of equality as regards the mutual possession of their Persons, as well as of their Goods exists between husband and wife". He rejects the suspicion ... that there is something contradictory about postulating both equality and legal recognition of the husband as master' (Pateman 1988: 172).[4]

For Kant it is clear that women need a relationship with a man so as to be able to follow the light of reason in a way that men supposedly do not need women. It was reason, not sexuality, which initially created the kind of dependency which Kant recognises. It is reason which women supposedly lack and which sustains a notion of male superiority. But this also serves to blind us to the workings of heterosexuality as a relationship of power and subordination, for it is in this realm that the sexes are supposedly more equal for Kant. Sexuality has to do with our 'animal natures' and within the realm of nature there are few gender differences. This prepares the ground for thinking of heterosexuality as a matter of individual sexual preference while at the same time it is normalised. Somehow it comes to exist as beyond the pale of rational theoretical investigation.

MODERNITY AND SEXUALITIES

At some level, sexualities remain a difficulty within much poststructuralist writing, for this often assumes a modernist distinction between 'culture' and 'nature'. Supposedly our identities are established within the realm of culture alone. It is within culture that we can be 'free' and 'autonomous'. But paradoxically this has made it difficult to talk about desire and sexual feelings, for these become 'socially and historically constructed' within the realm of culture alone. We often become trapped into a form of intellectualism, for it is through the categories of mind that we supposedly define our sexualities and come to know ourselves sexually. If this promises a freedom that has all too often been lacking in biological conceptions of sexuality, it tends to make sexuality a matter of individual freedom and choice. We are left with little sense of the joys and anxieties that are often attached to an exploration and discovery of our sexualities.[5]

Rather we inherit a whole set of confusions which tempt us to think that if our sexualities are not 'given' by nature they must be freely created by ourselves. This fostered the notion of sexual preference as a 'political choice' and so was a source of the anger that could so easily be felt at those who seemed unprepared to make a clear choice against heterosexuality. It can seem as if it has to follow that if our sexualities are not 'given by nature' then we should be able to transform them through an act of will. We could supposedly reinvent ourselves according to what we would want our sexualities to be. This paradoxically echoes a secularised Protestant tradition which has done so much to shape our visions of modernity. It becomes an issue of 'mind over matter' and since sexualities have to do with the body which is deemed to be a part of nature, we can supposedly shape our sexualities accordingly.

Here it is important to question the traditional notion of sexuality as an 'irresistible urge' that comes from the body and which has traditionally organised a sense of heterosexual sexual desire. This echoes the notion of sex as an expression of our 'animal natures' as men. The idea seems to be that once men have been sexually aroused then they can no longer be held responsible. So it has been women who have been held responsible and made to carry the blame.[6] As men we have been slow to place responsibility where it belongs, with ourselves, and to learn to map our experience in different terms. This has meant different men challenging a Cartesian dualism between mind and body which has left men feeling separate and estranged from their somatic experience. Within this traditional mapping so long taken for granted with 'modern' forms of philosophy and social theory we are often left feeling that our bodies exist in a separated space.

Learning to think of the body within dominant white masculinities in mechanistic terms as something that needs to be trained and disciplined, men are often left with little inner connection to their bodies.[7] We often give up whatever authority we might have in relation to our bodies, accepting that our bodies have little connection with our identities as rational selves. We learn that the body has to be subordinated to the mind and that we have to exert a rigorous control in relation to it. This helps to shape not only the ways we learn to think of ourselves as men but the relationship we can have to different aspects of our experience. The body as part of a disenchanted nature has no voice of its own. If it has desires of its own they have to be 'animal' and have to be externally regulated and controlled. There is no sense of the possibilities of developing a dialogue with different parts of our bodies, for instance giving the pain in our lower back some kind of voice of its own.

Yet developing dialogues with different parts of our bodies might make us aware of how little we really know about our bodies and sexualities, as heterosexual men. Regarded in mechanical terms it is easy to take the body for granted, thinking as if it only deserves attention if it lets us down in some way. Like the car or even the relationship it is there to be taken for granted, part of the background against which we learn to live out our individual lives as men. Often the focus is upon individual success and achievement because this is the way male identities are often sustained. If we get a backache when we still have to do a crucial piece of work we can often get angry at ourselves, rather than wondering what our backs might be trying to say about the ways we have recently been living our lives.

We go to the doctor, for example, often not to understand more of ourselves but to take these bodily symptoms away. We might feel disappointed that orthodox western medicine seems to have so little to offer when it comes to backs. But the point here is that the doctor is the authority and we learn to accept that the body has been appropriated as an 'object' of medical science. It is the doctor who has objective knowledge while what we can have is at best subjective experience. This does not often tempt us into wondering about how our relationships with our bodies have been built up over time and how little we seem to know about ourselves somatically. The idea of getting to know our bodies more can strike us as fanciful within these traditional mappings of the dominant, white heterosexual male. Often we have little sense of what might be involved in getting to know our bodies more or establishing more of a relationship with ourselves.

HETEROSEXUALITIES AND MEN'S BODIES

Relating to our bodies can impact in crucial ways on how we understand heterosexual male sexualities. Since we often have such little sense of what might be involved in giving time and space to ourselves sexually we automatically think that sexuality comes from a space that is outside of ourselves. This connects to a fear we often carry, related to homophobia, about the revelations of our own bodies. At some level we do not want to know more about ourselves. Often we have grown up to think of sexuality in terms of conquest and performance and as a way of proving ourselves in the eyes of other boys. Desire comes from elsewhere and has a particular 'object' upon which it is focused. This vision is also there in Freud and much psychoanalytic work.

But it is also crucial to recognise other resonances in Freud that appreciate the importance of building up more of a relationship to the self. Freud challenges a Cartesian tradition in a crucial way in that he recognises that emotions have to be mapped into the self. Traditionally it was easy to assume that thoughts were placed in the mind, which was the source of identity as rational selves, while emotions and feelings were located elsewhere in the body. This meant that there was no connection between our thoughts and our emotions and feelings. Rather in Kantian terms it was easy to treat emotions as distractions that took us away from the path of reason. This is why it was so crucial for men to learn 'self-control', which meant the subordination of our emotional lives.

Presumably there was little that our emotions could teach us nor any meaningful distinction to be drawn between emotions and feelings. Rather we could only have an inner relationship to our reason, which was therefore treated as the source of freedom and autonomy within liberal moral and political theory. In contrast we were left with an external relationship with our emotions and feelings, which were sources of unfreedom and determination for Kant.

MEN, REASON AND EMOTIONAL LIFE

Within a Kantian mapping of the self there was no way of recognising our bodies and emotional lives as part of our selves. Rather they were deemed to be a threat to the integrity of the self that had to be protected through silencing emotions, feelings and desires. At some level emotions are not 'ours' in the sense that they are placed beyond the framework of the rational self. Often this is echoed in the ways that men learn to relate to their anger as something that comes from outside of themselves and therefore as not something that they can really be held accountable for. It is an episode that has little bearing upon the ways we are. It is not too dissimilar to the person who refuses to accept that she or he is an alcoholic, claiming to be simply a person who likes their drink everyday. There is a fence of denial that is constructed within this mapping which makes it easy to disown responsibility.

If men think of themselves as not being angry sorts of people then it is easy for them to dismiss their anger as an isolated incident. This is especially easy to do if men have learnt to think of relationships as a set of discrete situations. It is as if life is split into a series of discrete moments. Often this shows itself in the ease that men often feel in leaving their emotions behind when they have gone off to work, while women seem to talk much more of how a row early in the morning can cast a shadow over the rest of the day. Some men might pride themselves in this capacity to 'cut off', enabling them to focus upon the job at hand without being distracted by what was going on at home. This might well be a consequence of the ways men learn to discount emotions and feelings as sources of knowledge. Others might argue that this is just another example of the ways in which men are disconnected from their experience.

But it might be more useful to recognise that the identification between dominant masculinity and reason that plays such a crucial role in sustaining notions of male superiority might at the same time create difficulties for men in their emotional lives. Freud recognises that within a rationalist culture of modernity men have the power to set the terms according to which others have to prove themselves. He was concerned to illustrate the harm that was done to both men and women through the repression of sexuality in the West. He also recognised the ways in which the repression of sexuality was connected to the suppression of emotional life.

For Freud it was important to question the ways in which denial of emotions served to produce a sense of 'unreality' for the self. Both men and women needed to reclaim their emotional histories as part of a process of bringing more 'reality' into their lives. But Freud also serves to support particular assumptions around heterosexual desire and the case of Dora has been argued over to show how Freud seemed to think it was 'unreasonable' for a woman to refuse the sexual advances of a man who was in so many ways 'eligible'. This became something that needed to be explained, for Freud, in terms of the workings of unconscious forces.[8]

MEN, BODIES AND EMOTIONAL LIFE

When we think about how psychoanalysis has largely taken for granted particular conceptions of masculinity we have to think about how it has treated the body. A refusal to think seriously about the arguments that separated Freud from Reich has made it difficult to place the body within psychoanalytic theory. If we have learnt recently to think more about the body, it has often been as a site for cultural meanings. We have failed to open up issues about men's relationships with their bodies and how this might impact upon the ways we understand male sexualities. If people have started talking about how bodies desire and relate to each other it has often been hard to relate such discourses to different aspects of gendered experience, because experience itself has become a suspect category within much poststructuralist writing.

But if we are to escape from notions of sexuality as performance then we have to recognise that the ways we talk about sexuality, as men divided by class, race, ethnicity and sexual orientation, have to be connected to the ways we experience ourselves sexually. A simple example might be that if heterosexual men learn to assume that it is only 'others' who have emotional needs but that 'we' have none ourselves, this is bound to create an imbalance within relationships. Often it will be a way that men can feel good about themselves, knowing that they are there to 'support others' while not really needing anything for themselves. It might be that if men open up a space in which they can begin to explore themselves sexually, simply through getting to know their bodies better, they might find it easier to acknowledge their own needs. It might be that it is only when we have learnt how to love ourselves a little more that we can begin to love others.

When we are mapping heterosexual male sexualities it is crucial not to generalise across class, race and ethnicities. It is also important to appreciate that people might need quite different kinds of relationships at different points in their lives. We need to get away from the moralism that has done such damage to discussions around sexual politics. A more healthy respect for individuality is crucial if we are to create spaces in which people can comfortably and safely explore their own sexualities. But there is an issue here about the way that boys, for instance, are often brought up to prove themselves according to external rules. It becomes difficult to develop more of an inner relationship with self if boys learn that any show of emotions is a sign of weakness. With the threat of AIDS it has become harder for young men to explore their sexualities as a way of getting to know themselves more.

Often men find it hard to acknowledge that they have emotional needs and that they need nourishment. Even an insight like this can be threatening, for it throws into doubts traditional mappings of the self that boys often grow up to take for granted. So, for instance, it can be difficult for men to identify ways they like to be touched or held, for this already assumes that you have built up a particular relationship with self. Within sexual relationships it can often be far safer for men to 'go for sex' because this is much less threatening to a sense of male identity. This can so easily be a way of concealing vulnerability rather than sharing it. For there is so little that teaches us as boys that sexuality has to do with vulnerability and contact. This stresses connections that we might have little experience of, for we have long learnt that vulnerability is a risky business.

What we have often learnt to want as heterosexual men is sex without contact or emotional involvement. Often there is a fear that is attached to making ourselves vulnerable. This is a risk that we learn to avoid for we do *not* want to risk rejection. Sex becomes a way in which we can prove ourselves as men while doing our best to minimise the risk of rejection. This is part of the control which we can insist on having as a way of minimising the risks involved. Often we are so accustomed to wanting to control situations that we are often blind to the ways this is happening, because we are often more in touch with the fear of rejection. Often these forms of control are tied up with the ways we have learnt to think and feel about ourselves as men.

MEN, MODERNITY AND HETEROSEXUALITY

An Enlightenment vision of modernity has been tied to a particular notion of dominant masculinity. The identification of masculinity with reason has allowed men to take for granted that they exist at the centre. This is where heterosexual white men are positioned, for they set the terms according to which 'others' have to be prove themselves to be 'human'. Since reason is set in fundamental opposition to nature and sexuality is taken to be part of an 'animal nature', masculine superiority is constructed against sexuality. Rather it is women who are taken to be identified with their bodies and so with their sexualities, while men are to be identified with their reason. As I have argued this goes some way to explaining the disdain that men can so easily feel for women in the context of sexual relationships. It serves to cast heterosexuality as a relationship of inequality and power in which men so easily learn to blame women for their sexual feelings.

Within the framework of dominant, white heterosexual masculinities men learn to take their superiority for granted. It goes along with a strong sense of entitlement where men can feel that they have a right to be heard and listened to. This can be reinforced in patriarchal family relationships where boys are often treated quite differently from girls. Often mothers will do things for them and they will get used to being served as if this is somehow due to them. This can make us accustomed to women doing things for us in relationships. This is why it is so crucial to disrupt some of these patterns, so that boys and girls both learn to take an equal part in housework from the beginning. When brothers are allowed to go out to play while sisters are expected to stay in to help their mothers get the dinner ready, expectations are clearly set that can be difficult to challenge.

At some level men often absorb the notion that women need them in a way that they do not need women. Traditionally it has been men who have seen it as their role to 'keep women in their place' because they are emotional and irrational and supposedly cannot work things out for themselves. Often this can be used to justify men's violence in relation to women. Adam Jukes in his unsettling but also flawed account, *Why Men Hate Women*, introduces a case history of Alan, who is a middle manager in a large public company who had referred himself for therapy after acting very violently towards his partner.[9] Asked what had happened he said she 'just kept on nagging at me and would not shut up'. As Jukes reports it (1993: 267), he says his attacks were a response to her nagging:

> She became a harridan, and she was 'very bad'. He just had to stop her. 'So you hit her because she's bad and you have to stop her from being bad?' I asked. 'I suppose so,' he said. 'I would do anything to stop her nagging. I suppose I do it to control her so that she won't nag me. She's so unreasonable!'

Often men find it hard to come to terms with their partners' unhappiness or depression, thinking that they are somehow to blame for what is going on. Men can feel betrayed by these feelings for they can feel that they are working hard to provide and that their partners should feel grateful for what they do. Rather than feeling grateful they seem to feel frustrated and unhappy about what is going on in the relationship. Again it can be difficult for men to *listen* to what is going on, for we can so easily feel that we are expected to provide some kind of solution to what is going on that might help to take these negative feelings away. Since this is often the way we have learnt to treat our own feelings, we can find it hard to accept that our partners might want something different.

Men can grow up to take responsibility for their partners in ways that are quite inappropriate. In remapping masculine identities we have to accept different ways of regarding responsibility. For at some level men do not learn to take responsibility for their emotional lives. This is what they expect women to do for them without recognising what is really going on, for men often think they do not have emotional needs of their own. This creates an imbalance in heterosexual relationships since it can seem as if it is always the woman who is needy and emotional while men seem to have learnt to cope with their emotions in quite different ways. This can leave women feeling frustrated and unrecognised for it can seem that to have emotional needs is already a sign of weakness and dependency.

Feminism has been crucial in challenging the possessive character of heterosexual relationships. Men have often grown up to consider women as possessions for this is the way we have also learnt to relate to our own bodies. Within a liberal tradition freedom lies in being able to do whatever we will with what is in our possession. This explains why until very recently there was no sense that there could be rape in the context of marriage, for sex was treated as an obligation that was owed by women to men. Even if the law has been changed we have been slow to recognise the profound shift in attitude towards sexuality that is at issue. Within modernity it was as if women's bodies were regarded as men's possessions, for there seemed to be few other ways of casting women's own sexual desires.

With this bitter herstory it can be difficult not to think of heterosexuality in institutional terms alone, as a relationship of power that has reinforced women's subordination and oppression. But if we think about sexual relationships between men and women simply as an exercise in power there is little sense of *how* such relationships might be transformed. Without minimising the power that operates and the ways this is mediated by larger gendered relationships of power, it is also important to recognise the different forms of power that might be brought into play. Though power is often what is at issue in sexual relationships, as is clear in the literature on rape, it can be misleading to reduce sexual contact to power.

MEN, SEX AND POSSESSION

For heterosexually identified men it is still easy to feel that sex is somehow owed to them, and if their partners do not want to have sexual contact they can find it hard to listen. Men seem more able to separate sexuality from contact and intimacy. Again it is not helpful to generalise and hopefully we shall go on to explore some of the sources of this in more detail. But at the moment we might say that men want the sexual contact while women also seem to feel that their sexual feelings come with contact and intimacy. What is harder to unravel are the powers at work when women are made to feel that they are somehow expected to make love when they do not want to. If they have acted against their deeper feelings there will often be a split or separation that takes place. Something similar might well happen for men, but often they are less aware of it.

This does not mean that the sex might not sometimes be paramount for both partners and that the sex might not be wonderful even though there is very little emotional contact that has been made. It has been crucially important for women to be able to recognise the autonomy of their own sexual desires which were so long denied within patriarchy. For so long there was the double message whereby women were identified with their sex while at the same time denied the autonomy of their own sexual desires. This was part of the control which men traditionally exerted.

But this does not deny the injuries that are done when people feel that they are having sex partly because it is expected of them. Sometimes women will withdraw into themselves, feeling a distance from their partners because there is so little contact emotionally. This is something that men often find it harder to appreciate because their sexual feelings can seem more separate. It is hard to think about such gender differences but it seems important to consider them. If we have become more accustomed to thinking about difference, it is still possible to feel threatened by them because they seem to bring certain egalitarian notions into question. It might well be useful to open up issues of polarity within sexual relationships, which have all too often been theorised within masculinist Jungian terms if they have been talked about at all.

In mapping the diversity of men's sexualities we have to recognise an inner *closing off* that can take place when there is no resonance between inner feeling and outer expression. A structuralist tradition was too ready to treat inner emotional life as an inner representation of an external social reality, allowing little integrity to our emotions and feelings. Cast within rationalist terms it found it hard to appreciate that our emotions might follow a different logic, so that we might feel something about a situation without really being able to explain what we are feeling. For a rationalist it is easy to condescend to this notion, thinking that if we cannot explain our feelings rationally this only goes to show that they are really 'irrational'. But this does not have to be so.

Some of these dynamics are not specific to heterosexual relationships and it might be fruitful to traverse boundaries and recognise how similar dynamics might be at work within gay relationships. The sense of rejection that someone might feel at the end of a relationship might encourage a determination not to allow yourself to be so vulnerable next time. It might be that you have to learn to pick yourself up and start over again. Issues of intimacy and vulnerability, of power and inequality, of desire and experience can come up in different kinds of relationships. Again it is difficult to generalise across sexualities and we can only hope that we are open enough to learn from our experience. But there is no guarantee that we will even want to learn the same lessons.

MEN, INNER LIFE AND RELATIONSHIPS

When we feel tempted by psychoanalytic theory to grant the autonomy of an inner emotional and psychic life, we are left with different problems about how to relate inner feeling with the ways we relate to others. One issue that might turn out to be crucial is what *feeling* we have for what we do. This brings the 'outer' and 'inner' into a different kind of relationship with each other. This opens up issues about the nature of the *contact* we have within different relationships and the way sexual contact is linked to feelings. Again it is difficult to generalise, and people might be looking for quite different kinds of contact, so it is important to question the way in which heterosexuality has often been set up as a norm against which other sexualities are to be evaluated and conceived of as 'deviant' or 'pathological'.

Sexual politics has helped to question the normalisation of heterosexuality and so to create a space for the affirmation of different forms of sexual identities. But often it has been trapped through an identification of ethics with moralism, so finding it difficult to open up questions of sexual ethics. We might have learnt to identify the relationships of power and subordination that operate within heterosexual relationships but often there has been a silence around how people might learn to negotiate more equal terms within such relationships.

This is not to wish away the reality of power relations but to recognise that they are also at play within different kinds of sexual relationships. They might take a particular form within heterosexual relationships in which there is an ongoing struggle to establish the equality, independence and autonomy of women. But it is also important to recognise the different kinds of power that are at work, which we have been slow to do because we have assumed this would threaten an acknowledgement of women's subordination within the larger society.

Often women have challenged men for not being more emotionally present and involved in the relationship. Women have often learnt to exert their own power but often this does not bring happiness and fulfilment if there is little meaningful contact. Often this is difficult for men to appreciate for traditional masculinities are defined within the public realm of work and in competitive relationships with other men. It is because men can never taken their masculinities for granted that they have to be ready to defend them at any moment. Men often think that they are working so hard for the good of their relationships and families that they can feel let down and betrayed when they do not feel appreciated. This is partly because men grow up taking it for granted that they will occupy a central space within the life of the family. But more and more men are left feeling that they are dispensable and that the family has organised itself without them.

MEN, INTIMACY AND RELATIONSHIPS

A dominant heterosexual masculinity is sustained within the public realm of work. An emotional and sexual relationship is often taken for granted once it has been established, for male identities are sustained elsewhere. It is as if men live in a different space so that they constantly have to be reminded of their emotional obligations within the family. As we map these men's lives we realise how the separation of relationship and work operates, for so often men are tied to their work in obsessive ways because this is still what matters to the construction of their masculine identities.

Often men's best energies are used up at work and they come into the relationship exhausted and drained. Of course, we have to be specific about age, class, ethnicity and generation, but often at some level there is a fear of intimacy because men have learnt that they need to keep themselves together for work. It is 'others' who have emotional needs and who need support. Often this creates its own imbalance because it leaves women feeling weak and dependent simply because they have emotional needs of their own. Sometimes they learn to silence their own demands because they do not always want to be the one in the relationship who is making demands.

But feminism has supported women in making their own demands in relationships and in insisting that men have to rethink their masculinities. At one level this has meant a refusal on the part of women to do all the emotional work supporting their partners. They have insisted that men find ways of drawing emotional support from other men and so have shifted the emotional geography of relationships. Where this has been hard to do men have often withdrawn into themselves, at least for awhile. But sometimes they have learnt that men's groups are available in which there is a different, if unfamiliar, space in which they can begin to explore some of their inherited patterns of masculinity.

Since men have often depended upon women to interpret their emotions and feelings for them without appreciating or valuing the effort this takes on the part of women, it has been a shock when women have refused to prioritise their relationships with men. In learning to remap their own lives men have had to learn to identify their emotional needs. Often this has been difficult to do because men have felt bereft of an emotional language in which they could translate their needs. For instance, men can be so used to living without contact that they do not know how to recognise what they are receiving in their relationships and how sustaining this might be to them. Often it is only when relationships have broken down that men recognise what they have lost. Often they are so focused upon working out who was responsible and so who to blame for what happened that they do not begin to identify what part they also had to play.

At some level men in heterosexual relationships often seem to feel that women are somehow responsible for the relationship. This is partly because men often learn to conceive of relationships in mechanistic terms. Once the relationship is in place then it supposedly only needs time, space and energy if something is going wrong. Sometimes men feel resentful that they are being called away from more important spaces to do with work and can sullenly blame their partners for not having been able to cope.

This is especially so when issues concern children, as if it is still at some level somehow women's responsibility to take care of the children. Fathers might be ready to help out but still the responsibility lies elsewhere. Often men have very little sense of what time and energy it takes to sustain an emotional relationship and the distances that are created as resentments begin to build. Rather men are often brought up to assume that there is always something that they can do to make things better.

MEN, FEMINISM AND EQUALITY

The pattern of traditional white middle-class heterosexual relationships has shifted over the last two decades with the growing numbers of women who are at work. This creates the material conditions for a more equal relationship as both partners seem able to share what they put into their living situation, if it is a shared space. There is a widespread sense in the 1980s and 1990s that men and women should be more equal in the context of their heterosexual relationships. It is also more common for both partners to maintain their own friendships outside of the relationship. Work can be equally important and both can have ideas of a career they want to follow. This establishes a different pattern of relationship in which both parties are constantly reflecting upon what they get out of the relationship and whether they want to sustain it. Often it is thought about as a lifestyle choice, rather than as a commitment for life. This goes along with changes in the way people conceive of their sexual relationships. But again there might still be crucial gender, class and ethnic differences that need to be explored. It is often reported that young men seem less interested in long-term commitments.

Notions of liberal equality within the context of sexual relationships often break down when babies are born. Often there is an assumption that women will be quite happy to get back to work after a few weeks and that everything will 'return to normal'. Very little, within contemporary culture, seems to prepare young people for the impact that a small baby can have on a relationship and the kinds of dependencies that it creates. Often pregnancy and birth bring emotions to the surface that link back to childhood experiences that are often quite unresolved. A relationship that has been carefully organised, in space and time, on more rationalist principles is often quite unprepared for the changes that often come. Where men have been involved with the pregnancy and where a close emotional relationship has been allowed to develop, there is sometimes a very positive bonding that takes place through the experience of birth. This has been a powerful transformative experience for many men who often want to be much more involved with their children than their fathers were with them.

But after a few days men often go back to work and women feel abandoned, as it seems as if it is only their lives that have radically changed. This can be a most difficult time and women can be left feeling unsupported. There is little 'natural' in the bonding and often they have to learn how to look after their new baby. If men are not equally involved in the emotional space through this period of learning they will often feel shut out and excluded, for their partners might have little patience in teaching their new skills. Sometimes men feel displaced as they experience that there is a strong bond between mother and baby that they feel excluded from. Often in traditional relationships men had expected to be at the centre of the emotional universe in the family, so now they feel bitterly rejected but often unable to express what is going on for them.

What are men responsible for? If men have been responsible for bringing this new life into the world should they not be equally responsible for the baby's care? These are crucial questions for the mapping of contemporary, heterosexual masculinities, for often men feel very unprepared for the changes that are taking place in their lives. At some level we often unconsciously expect to be treated in the ways our fathers were, so whatever we might say we can also feel resentful that we seem no longer to be at the centre of things. Often men seem to feel dispensable, especially in the months after a new baby is born when they can feel as if all the attention and love that used to come their way is now going to the baby. Some men seem to look outside the relationship to affairs as a way of coping with this new situation. But often this lays the seeds for a breakdown in the relationship. At some level men can feel guilty, though they might seek to assuage these feelings by blaming their partners for somehow having forced them away.

If men expect to have things their own way, then it can be difficult when their partner's attention is elsewhere. At some unconscious level there is often a desire to punish or get revenge. But men can also feel guilty for feeling this way so that it is often something they will not talk about. Rather they will suppress their feelings, thinking that they are 'irrational'. This reflects an ongoing difficulty that men have in giving space and relating to their emotional lives, having learnt within a rationalist culture to deny emotions and feelings as sources of knowledge. Often men feel that they have learnt to do without and that it is a sign of their strength and male identity. But this often means that men feel that they can survive without the support and love of others. Often we can take for granted and so devalue what is being offered as support within relationships.

MEN, VULNERABILITY AND EMOTIONAL LIFE

Within the competitive world, men often learn to survive on their own. It is hard to trust other men or make ourselves vulnerable with them for we so often feel that others will take advantage of our 'weakness'. It takes quite a different mapping of masculinities to appreciate that showing vulnerability does not have to be a sign of weakness, but on the contrary can be a sign of strength. Possibly gay men have learnt to relate to each other in more open and vulnerable ways, but this is still very much an issue for heterosexual men. But again we have to unsettle these categories for if heterosexuality is a matter of loving people of a different gender, then we have to recognise that there are very different ways of showing love.

Often there is a great deal of confusion about how to think the differences between sex and love.[10] It is one thing to appreciate the ways that love is 'socially and historically constructed' but quite another to think there is no such thing as 'falling in love'. It is partly because we have so little control over the ways we fall in love that it can be threatening for men who are brought up to assume that life is something to be controlled through reason alone. We cannot so easily control the movements of our desire; but this does not meant that we always have to act upon it.

Rather, within a Protestant culture, there is often very little *space* between our emotions and our actions since we are already judged to be 'evil' because of the emotions that we have. To feel sexually attracted towards someone while you are in a stable sexual relationship is so easily regarded as a sign of betrayal or a proof of our 'animal sexualities' that we often suppress these feelings, not really wanting to acknowledge the revelations of our natures. Part of the antagonism that we often feel, both theoretically and practically, to thinking about 'nature' is that it often exists as a realm that is beyond our control. Within an Enlightenment vision of modernity we like to think, especially as heterosexual men, that we are in control of our experience.

As we need to create more space between our emotions and actions, so we also need to recognise that 'nature' does not need to be linked to 'determination' and 'unfreedom'. It is because arguments from nature have traditionally been invoked to argue that it was to go 'against nature' if women rejected a life of domesticity and child-care, that feminist theories have rightly been suspicious about these arguments. But we have to be careful not to fall into Kantian distinctions between necessity and freedom, when we fall back into a distinction between what is 'determined' and what is 'freely chosen'. This has often muddied discussions in relation to 'essentialism', where we have sometimes been too quick to oppose a notion of 'social constructionism'. Often it takes time and experience for people to begin to know themselves sexually and to define their sexual identities. This is partly a matter of how people come to want to express themselves sexually and the ways they find in which they can give and receive love.[11]

Often this is something that people have to explore for themselves. It is not something that can be worked out in advance or simply be conceived of as a matter of political choice. It is also not something that is fixed but is in process of change, as we come to experience ourselves in different ways. But probably it means questioning the notion that 'experience' can be conceived of as an effect of discourse alone. This is to blind us to the tensions that so often exist between what we experience and how we learn to think about ourselves. Often we are at great pains to try and fit our experience into what is expected of us by the dominant authorities in our lives. Often we are taught to swallow our feelings so that we can do what is expected of us. But this is the way that we maintain an ignorance about ourselves, never really sensing how we might learn from our experience.

Men are often haunted by a fear of rejection so often it is much easier to do what is expected, rather than to explore what we want individually for ourselves. This kind of emotional exploration is threatening because it can unsettle and disturb the way we have learnt to think about ourselves. Often it is disdained, especially within an intellectualist culture that does not want to give space to acknowledge the very different ways in which we can come to know ourselves. Some of these possibilities have been opened up within postmodern discussions which recognise different senses and which open up possibilities that have been conventionally denied within a modernist conception of reason radically separated from nature. But this is also threatening to dominant masculinities which have been so closely identified with reason within modernity.

As men learn to acknowledge their fantasies and their attractions, even if this does not fit their conceptions of themselves, they open up a space for exploration and play. Sometimes we hope that there is such a space for play within our intimate and sexual relationships, but often we are so unused to giving time and attention to ourselves that it can be difficult to care for others. Rather at some level we can feel haunted by a sense that we are not capable of love. Our fantasies might be exciting but somehow this excitement seems to get drained away within the everyday routines of relationships. We might wish it to be otherwise but we sense that it can be hard to care for others if we are still to learn how to care for ourselves. In a redefining of masculinities men would learn to be more open to exploring different aspects of their experience, rather than denying emotions and feelings that are deemed 'unacceptable' because they do not accord with the reasoning we have set out for ourselves.

MEN, LANGUAGE AND CONTACT

Often white middle-class men have learnt to relate to language as a means of self-defence or as a way of proving themselves against others. This can open up a split between the way that men feel inside and the way they present themselves to others. There is often little sense that it might be possible to bring out how we feel because we fear that others will ridicule us and put us down. Of course this is mediated through differences of class, race and ethnicities which help form particular masculinities. But whatever differences need to be acknowledged, there is a sense that men often have to prove themselves as men and that this involves showing that you are a 'real man'. Often men learn to use language as a means of defending the image they inherit of themselves.

But this can make it difficult to reconcile the ways that you feel you need to behave with other men and how you might want to be in the context of an intimate relationship with a woman. Here men often experience a split, especially when they feel that to show their vulnerability is to endanger their very sense of male identity. Sometimes it can be difficult, as I have already mentioned, for men to listen to what their partners have to say because they feel called upon to offer 'solutions' which might take their negative feelings away. Since this is often the way we have learnt to treat negative feelings of depression and sadness as men, we think that this is the kind of support we are being asked to give. But sometimes our partners feel frustrated and unheard, for they were not looking for solutions which they could discover for themselves, but just the experience of being listened to.

Sometimes it is because men feel responsible for the 'negative' feelings their partners are experiencing that they can find it so hard to listen. But this is a responsibility which does not fall to men to carry, though, in the mapping of traditional heterosexual relationships; this might be the way men came to understand what was expected of them. As men become more aware of their own needs for contact, then they can begin to discern when this contact is genuine, because it will be nourishing. It is difficult to recognise this if men continue to insist that they can do very well without the love and support of others. As long as modernity insisted upon defining dominant male identities as 'independent' and 'self-sufficient' there was bound to be uncertainty about what it means for men to be in relationships at all.

This can help to foster a split between sex and intimacy whereby sex becomes a goal, a means of proving or affirming masculinities. This can tempt men into treating sex as a kind of possession which is owed to them as some kind of right. Within this possessive conception of the self sex can be seen as a matter of performance. Within such a mapping of the self, which is familiar within liberal theory, the self comes to be identified with the mind and enjoys an external and possessive relationship to the body. This connects, as I have already hinted, to the *externalisation* of sexual feelings as coming from somewhere else and taking over as some kind of irresistible urge that cannot be contained. In this way men can renounce responsibility for their sexual feelings and can displace blame onto women. This also means that men can silence fears they often learn to carry about their 'animal' natures.

This goes some way to explaining why it is crucial for men to learn *how* to take more responsibility for their emotions and feelings. As men share more of their sexual fantasies with each other they can begin to work on what they might mean. Rather than feeling embarrassed and ashamed about what they reveal about the self, we can recognise fantasies for what they are. This helps to create more space between our emotions and our actions, for we recognise that we will not act upon these fantasies. The more willing we are to acknowledge our emotions, even if we would want them to be different, the more emotional space is created. Within a culture in which we are made to feel that our emotions are shameful, we learn to deny the intimations of our natures. As long as we conceive of our natures as 'evil' or 'animal' we will readily deny what they have to teach us.

Within a rationalist tradition we have been slow to recognise the integrity of our emotional lives. While it has been crucial to recognise heterosexuality as an institution of power, we also need to appreciate the different ways in which men and women can learn to love each other. As long as penetrative sex is seen as essentially coercive there are few ways of exploring different patterns of sexual contact. We need to open up the tensions that are built into sexual contact from men having learnt to condense a whole series of diverse needs into sexual contact. If this goes along with male anxieties about sexual performance it can be hard to open up communication between partners. As men learn to identify discrete needs for being held, touched, caressed in particular ways, they will not feel such an internal pressure to go for sex, even when it is not appropriate for them, let alone their partners.

As long as men feel that talking about sex is the surest way of killing off feelings they will be less inclined to communicate their needs. At some level men might feel ashamed and embarrassed about verbalising what they want if they feel that it somehow compromises an inherited sense of male identity. If language is thought of as killing off passion then people will be very reticent about talking about what they need for themselves. As it is, men often grow up feeling that their partners should already know what they want and that if they do not this is a sign that they are not really loved. Often men are trapped into quite romantic conceptions of relationships, since they carry very idealised notions of love.

As boys grow up into manhood recognising their sexual attraction for the opposite sex they often think of girls as quite 'other', aware at some level that they live in quite a different world. Since boys are fearful about losing face in front of their male peer group they can often feel the power that girls have to reject them. This can make them suspicious of feminist discourses that remind boys of the power they have in relation to girls. Often this does not fit with the anxieties and uncertainties boys feel about themselves. Again it is important not to generalise across different masculinities. But in mapping male identities there is often a fear of intimacy and contact. Often this is reflected in an unease about how to talk to girls, who often seem to have interests that set them quite apart from the everyday worlds of football and computers that many boys seem to inhabit.

Often this fear of intimacy is carried into adult heterosexual relationships. Since men often learn to be self-sufficient and independent, as we have described it, there is often little sense of what it means to be in a relationship. At some level men can be haunted by a sense that they are not loveable and do not know how to care. This is reinforced within a Protestant moral culture in which having needs is a sign of weakness. It is so easy to take relationships for granted because male identities are established elsewhere in the public realm. There is little sense of the 'emotional work' that it takes to sustain a relationship and of the way contact needs to be maintained to keep the excitement within a long-term relationship.

Men often go through a crisis when children are born into their families and their partners have become a 'mother'. This can bring all kinds of unresolved early emotions to the surface that men are just not used to dealing with. If there is little experience of how this might be worked through in the context of a relationship, men can look elsewhere for excitement. The family has become quite a different site and not only might men often feel rejected as the attention of their partner goes to the child, but there is a sense in which the 'partner' they knew is 'no more', now that she has become 'a mother'. Men might feel that their sexual feelings have drained away without really understanding what is going on. Again this relates to the ways in which men relate to their power and have learnt to talk to themselves. They might simply present it to themselves as a matter of 'fancying' someone else, as if they have no control over their feelings but simply have to respond to them. It is not unusual for men from quite various backgrounds to somehow present themselves as the victims of their own emotional lives.

ETHICS, CARE AND EQUALITY

If we recognise heterosexuality as an institution that is in flux as men and women begin to redefine what they want from relationships, we have to explore how personal relationships are mediated through the larger gender relationships of power and dominance. The high rate of separation and divorce is indicative of a broader crisis in what it means to have relationships. If we think in terms of liberal choice then we might accept that if there are issues in a relationship then people might choose to move on to different partners. But this market vision of relationships has often been set on masculinist terms, treating sexuality as a commodity that can be exchanged. For some men this seemed a healthier option than what they knew from their parents, who often stayed together when all feeling and love had evaporated and there was only bitterness and regret.

Often there is an aspiration on the part of men to develop something different from what their parents knew, especially when it comes to fathering, but there are few models around of *how* men are supposed to be. There are few co-ordinates that men seem able to trust in. Sometimes this goes along with an aspiration towards greater equality in relationships. But often there is a complex relationship with feminism because men often cannot recognise themselves in some of the feminist portrayals of masculinity. Part of the recent appeal of Robert Bly's *Iron John* has been the space it has helped to create for a recognition of men's own pain, that they carry from their childhoods.

Men can acknowledge the power they hold in relationships without at the same time feeling that they are responsible for everything that is going on. Sometimes it has been easy for women to claim that virtue is always on the side of the powerless. But often issues of power and dominance within intimate relationships are more complex and sometimes it is important for men to learn to take responsibility for their own emotional lives, rather than to feel they are responsible for everything their partner is going through. Traditionally men have taken responsibility for others, while failing to take responsibility for themselves emotionally. Often there is a strategy of avoidance going on when men refuse to share what is going on for them and insist upon offering 'solutions' for their partner's situation. Often this is not what is wanted and it does not help to open up communication within the relationship.

As men learn to care for themselves emotionally they will begin to develop more sense of what it means to care for others. As men begin to give voice to more of their own emotional needs and desires they will recognise more of what their partners are going through. This involves a different vision of respect as we learn to recognise the integrity of emotional life. For too long men have learnt to trivialise and belittle these aspects of experience, especially within dominant intellectual cultures. But often we have been slow to acknowledge the lack of contact we have, both with ourselves and with our partners. It is as if once a relationship is in place it can be taken for granted until it breaks down. But if we learn to think less mechanistically we can recover a sense of a relationship being more like a garden that needs to be constantly tended and cared for.

As men begin to acknowledge their own needs for nourishment they begin to ask difficult questions about where they get their needs met in their lives. Often men have suffered because within a dominant masculinity they are not supposed to have needs at all. This is why it is crucial to begin to remap masculinities so that men can begin to develop different visions for themselves. Rather than taking their masculinities for granted it could sharpen a sense of critique of a patriarchal society that has offered men power at the cost of central aspects of themselves, as Weber grasped how the identification of masculinity with work meant that men automatically subordinated themselves to work, which became an end in itself. As this connection begins to be loosened men can sense different opportunities, as well as the sacrifices they have been expected to pay.

As men begin to recognise that they are supposed to sacrifice their relationships with their partners and children for work that becomes an end in itself, they will more openly question the terms of this contract. If relationships are to be more equal and if men are to have more meaningful everyday relationships with their children then the organisation of work has to be rethought. Often men are blind to the high price they have been asked to pay because they have not learnt to value a deeper contact with themselves emotionally. But as men learn to want more contact with themselves and their partners and children they will be less prepared to sacrifice other parts of their lives to work. Rather they will seek a different kind of balance between the different areas of their lives.

But as men learn to take greater responsibility for themselves emotionally they might begin to appreciate their relationships in different ways. They might feel more committed to making relationships work and recognise that this means giving more space and putting more time and energy into them. This will help to shape new forms of heterosexual relationships and different visions of respect and equality. This is not an easy task but it remains a vital one in the remapping of masculinities. The way we learn to care for others and the struggles we go through to treat them as equal, free and autonomous is an issue in different kinds of sexualities. No doubt different issues are involved and forms of power remain to be faced but it is still possible to learn from each other, as long as we are ready to acknowledge the integrity of different forms of sexual relationship. This is something we are only just beginning to do.

NOTES

1 Some useful work which shows that 'the historicity of "masculinity" is best shown by cross-cultural evidence on the different gender practices of men in different social orders' is provided by Connell (1993: 597). See also Cornwall and Lindisfarne (1993).

2   The identification of a dominant masculinity with a particular notion of reason was a central theme in Seidler (1986b). I tried to show the ways this identification manifested itself culturally and historically in men's initial responses to feminism in the 1970s.

3   An attempt to subvert an easy identification between men and reason as opposed to women and emotions was crucial to the project of Seidler (1994). I tried to show that men are often emotionally attached to a particular notion of reason separated from nature and that this has centrally informed and shaped 'modern' forms of social theory and philosophy.

4   In Seidler (1986a). I was concerned to explore the difficulties that Kant has in sustaining his notion of respect and equality when it comes to relationships of power and dependence. There are links between the ways he thinks about class and ways he thinks about gender that have had a crucial impact on Democratic Theory.

5   Though Butler (1990) does important work in subverting unproblematised appeals to sex/gender identities, so helping us rethink a categorical distinction between 'sex' and 'gender' that has for so long been the mainstay of structuralist work, it can seem that we are left with a voluntarism in relation to sexual identities. In her later work she seems uneasy about this interpretation, but it seems difficult to shift it.

6   The idea of women's sexuality as a threat to male reason was crucial to shaping an Enlightenment vision of modernity. It has diverse sources within the West. See, for instance, the insightful discussion of Rousseau in Okin (1980).

7   Though there has been much interesting discussion on the body in recent social theory this often fails to engage seriously with feminist work or with the gendered nature of bodily lives. Bryan Turner has done important work in showing the challenges and promise of bringing the body into social theory. See for instance Turner (1992).

8   The case of Dora has been crucial in thinking out the relationship between psychoanalysis and feminism. See for instance Bernheimer and Kahane (1985). There is also a useful discussion in Masson (1989: 84–114). Many psychoanalysts have been too dismissive of this work but I think this is partly because they do not want to face some of the difficult issues it raises. I have critically discussed Masson's work in Seidler (1994: 165–83).

9   What is missing in Adam Jukes' account of male violence in *Why Men Hate Women* (1993) is enough sense of the men themselves. His Freudian confidence in the primal nature of men's violence towards women, because of an early separation from the mother, gets in the way of linking men's violence with the social power and experience of diverse masculinities within a patriarchal society. Somewhat paradoxically it serves to illuminate in its confidence and authority a relationship between masculinity and psychoanalysis. It seems as if Freud himself got tired of listening to the pain of others.

10   An insightful and broad-ranging collection that explores this tricky relationship is Cartledge and Ryan (1983).

11   An early exploration of some of these issues that cut across distinctions between 'gay' and 'straight' masculinities is to be found in Seidler (1992), which brings together various writings from *Achilles Heel*.

Part III

THE LIMITS OF IDENTITY

PART III

INTRODUCTION

Throughout Part II of this book, there is a sense that certain people are simply not allowed to be who they want to be, everywhere and always. This constitutes a limit on subjectivity, where subjects are subjected to social norms, regulations, prohibitions and expectations. There are other senses of limit, however. This part of the book takes issue with the idea of identity summed up in the phrase 'the self-same': how is it possible to make sense of our selves, if the boundaries that tell us who 'we' are are incoherent, or fragmented, or fuzzy, or somehow unreal, or fluid or on the move? Thus, identity itself is limited because it does not mark the same place: no one is identical. This does not stop some people placing other people into specific, yet sweeping, stereotypes – such as madness.

Hester Parr and *Chris Philo* are concerned both with the senses of self and with the actions of those who are commonly labelled as 'mad'. They demonstrate that 'mad identities are to some extent constituted through the geographies of mentally distressed people as they move across and between a diversity of sites where their circumstances, lifestyles, experiences and problems become an issue (are acknowledged, discussed, responded to and acted upon)' and that this process takes place within specific mental health care practices and regimes, which also play their part in constituting 'mad' identities. Parr and Philo work through their argument in two case studies, which they term 'specific "local maps" of situated mad identities'. The first study interprets the presentation of a mad self in one nineteenth-century fictional story about a woman called Mabel Etchell, while the second focuses upon how individuals in Nottingham who are mentally distressed, and who are using the mental health care system, manage their selves on an everyday basis. Parr and Philo conclude that some people in a sense are made by the 'mad' places that they find themselves in, partly because these places (and the people in them) set limits on – or provide the resources for – the identification between the 'mad' individual and their place, and partly because places are also constituted by the mad in their own (mad) terms.

Quite a different limit is suggested by *Marcus Doel*. For many, the question has now become what or who comes after the subject (see the introduction to this book) and Doel takes this to be a symptom of a subject in jeopardy. Doel's sense is that the subject is being marked as a site of catastrophe, a place of exhaustion, a terminal identity. His interest is in the figure of the subject as a machine, which allows him to explore the sense of a body of parts and ultimately trace a body without organs that emerges in 'the aftermath of a schizoanalytic and deconstructive experience'. It is argued that the child has no sense of its own body and must learn to organise its body parts into a functioning whole. Using the body as a marker of distinction, socialisation attempts to place the child within a multiplicity of social categories such as gender, race, class. In this way, the machine-like body is produced in the social factory – but, Doel asserts, this process is never completed and subjectivity always remains 'work-in-progress'. If the subject is 'in progress', then there is no subject 'which could be either situated, embodied, fragmented, decentred, deconstructed or destroyed'. Thus, Doel's position challenges both those who believe that there is a universal, unitary and centred subject and those who have nihilistically killed off the exhausted subject. Instead, Doel asserts the sense of becoming of being – that is of being both in motion and in motionlessness. He concludes that, for the subject, this is a place of deconstruction – neither fixed nor stable, but enduring.

Similarly, *Paul Rodaway* believes that the subject is neither fixed nor stable, but a place 'where human meaning emerges and is contested, and therefore a locus of power'. He argues that changes in contemporary society, such as the development of mass media, the increasing importance of consumption and higher technologies, have produced a (long) crisis of subjectivity. So, while people used to determine their place in the world against the sign-referents of class, race, gender, they now have an unlimited world of significances to choose from – and these significant differences have become estranged from the authenticating discourse of experience: i.e. reality has become hyper-real and the subject has gone with it. Rodaway provides evidence for this assertion through the description of 'the experience of being a subject in the context of hyper-reality', for example, in Disney World, heritage museums, shopping malls, theme-parks, virtual reality, Los Angeles and – by extension – the West. In this (our) world, the (hyper-real) subject is seduced by the continual reinvention and re-creation of technologies, images and signs. From this perspective, it can be seen that subjectivity is about 'choosing' rather than 'being', but in a situation where new choices are ever present. It is ideal for a capitalist society, Rodaway sombrely concludes, that 'the commodity gives the individual subject an identity'.

His return to Britain, after living in the USA and a visit to Israel, leads *Nigel Rapport* to reflect on the experience of migration, of feeling out of place, of being between places. What intrigues Rapport is the stereotypical images that he has built up of the USA, Jerusalem and Britain. The question for him is whether these images are to be dismissed out of hand as being (at best) ignorant or insensitive or bigoted or (at worst) racist. In a world where movement and travelling are normal, the tendency to stereotype requires further, critical examination, invoking the issues of 'home' and 'away'. So, for example, 'home comes to be found far more usually in a routine set of practices, in a repetition of habitual interactions, in a regularly used personal name, in a story carried around in one's head' and these discursively-constituted practices rely on a 'heightened emphasis on the stereotyped, on the clichéd and proverbial and sloganish'. Rapport argues that the stereotype is a cognitive resource: 'the individual *personalises* stereotyped discourse as he [she] puts it into practice, interpreting its implications within the context of his [her] own life'. Stereotypes, far from being a set of commonly-held prejudices, are deeply personal; individuals use them to help them decide what to think and do, to find their place in the world and to help them traverse that world – and they are therefore open to continual change in the light of experience. In order to demonstrate this, Rapport provides a refreshingly honest account of his own experience of migration and stereotyping. Through this narrative, Rapport is able to trace the encounter between self and other, the insider and the outsider. He concludes that – however right or wrong, however fixed or unstable, however coherent or incoherent, however knowing or unknowing, however shared or personal – stereotypes express a set of co-ordinates which the subject uses to map their self into the world.

10

MAPPING 'MAD' IDENTITIES

Hester Parr and Chris Philo

> Alone in a city too big for comfort
> too many people
> too much loneliness
> the spirit gets lost under the noise and clatter
> you can become part of the crowd
> and fade into insignificance
> or you can express your craziness
> and get singled out.
>
> (Bangay 1992: 22)

This chapter is concerned with 'mad' identities, the senses of self, of who one is and is not, of what one can and cannot do, possessed by people who are mentally distressed (people with mental health problems, many of whom will have received the designation of 'mentally ill' at some time in their lives). The terminologies involved here are controversial, and we should say at the outset that we adopt the term 'mad' as a strategy of reappropriating a word usually regarded as prejudicial and mocking: in part to distance ourselves from medical-psychiatric accounts which unthinkingly mobilise the concept of 'mental illness' as if it reveals the complete truth of mental distress, and in part as a recognition of the potential use to which people could put 'mad' as a basis for collective campaigning in the political arena.[1] The underlying motivation for the chapter is a concern for the difficult and often dangerous circumstances which confront many mentally distressed people, particularly those who are discharged from hospital into the 'noise and clatter' of the city, and we believe that rather more should be known about the differing responses of different individuals – whether, as the poem above suggests, they shrink into a sad 'insignificance' or assert a pleading 'craziness' – when seeking to cope with a world which seemingly has little time or money to care. Inquiring into these responses necessarily means thinking about how these particular human 'subjects' are constituted, how their identities are influenced from without (by the material twists and turns of a social life over which they may have little control) at the same time as they are negotiated from within (more or less consciously, with greater or lesser elements of resistance, imagination and even delusion).

The prime objective of what follows is to argue that mad identities are to some extent constituted through the geographies of mentally distressed people as they move across and between a diversity of sites where their circumstances, lifestyles, experiences and problems become an issue (are acknowledged, discussed, responded to and acted upon). Often, but not always, these sites are ones where these people interact with mental health care professionals of various kinds, and in so doing their identities are inevitably shaped by – even if only in their wish to oppose – the prevailing 'establishment' views of what their 'conditions' really entail and require. This is immediately to oversimplify the picture, however, and our hope is to flesh out our claims in this respect by demonstrating the considerable variety of ways in which identities and geographies actually weave together in the worlds of 'mad' people. In so doing we hope to add to the corpus of work currently showing just how important 'spatiality' is in the formation of human subjects, with their fractured identities inevitably bound into the many worldly locations encountered in their personal histories and geographies, and our intention is to provide a specific instance of a broader argument (loosely hung around ideas drawn from Foucault and Goffman) about how geographers and other social scientists might conceive of the 'maps' which all human subjects are simultaneously making and being made by.

We are not here developing these claims through sustained theoretical reflection; instead our chapter revolves around two substantive case studies which comprise not synoptic overviews, but quite specific 'local maps' of situated mad identities: one deals with the fictional experience of a nineteenth-century madwoman called Mabel Etchell, and the other deals with the geographies of deinstitutionalised 'users' in Nottingham. The case studies do not stand entirely by themselves, however, and we situate them alongside a series of preliminary theoretical reflections which tackle both the scope and the limitations of the framework for interpretation being offered here.

MABEL ETCHELL'S MADNESS

Our first case study takes us into a minor British novel of the 1860s entitled *Ten Years in a Lunatic Asylum*, which purports to be an autobiographical account of the early life of one Mabel Etchell,[2] whose mental distress and incarceration in two different asylums provide the backdrop for a story full of unsavoury characters and strange coincidences. The improbable story-line is clearly fictional, but the broader outlines of the novel square with the known experiences of a one-time Huddersfield school-teacher and occasional poet called Charlotte Phillips. From several sources – including biographical notes about her husband, a writer of miscellanea called George 'January' Searle (Hall 1891; Lee 1896; Phythian 1926), and also from a contemporary register of asylum admissions[3] – it can be established that Phillips spent about ten years in mental institutions, initially in a charitable lunatic asylum and then in a public asylum, and that upon regaining her liberty she wrote a 'remarkable book in which her experiences were graphically described' (Hall 1891: 41). Also revealing is the fact that Phillips published a book of poetry under her own name which contained a piece called 'The Lunatic Asylum' (Phillips 1871: 54–6), and there can hence be little doubt that the voice of Etchell in *Ten Years* has a certain authenticity to it, rooted in what this individual herself had been through. And that the novel was intended to be more than just a diversionary read is evident from the presence of a preface in which 'Etchell' openly declares that:

> although [my story] is clothed in the garb of fiction, it aims at something higher than the mere gratification of a taste for amusement or morbid curiosity ... The social condition of our lunatic asylums should not be shrouded in uncertainty or mystery. To create a faithful picture of the inner life there presented is the first object of this little volume. The incidents related may prove interesting and useful to those whose profession brings them into daily contact with the insane.
>
> (Etchell 1868: iii)

In the pages that follow Etchell duly sheds light on numerous aspects of the nineteenth-century 'mad-business' in Britain, notably when exposing the threat of 'wrongful confinement' that so unnerved respectable Victorians, and when showing the acute vulnerability of women to being shut away in asylums at the whim of their husbands, fathers and guardians. Alternatively, her words often criticise the practices of doctors, nurses and attendants, and she rails against the dangers inherent in doctors entrusting patients solely to the care of 'mere hirelings' devoid of specialist expertise and (in many cases) simple human compassion.

The story itself unfolds around a limited number of places which are central both to the plot and to Etchell's progression through her mental distress, and it is apparent to us that a 'geographical reading' of her tale can illuminate the ways in which mad identities are intimately bound up with the personal geographies of the people concerned. Her story commences in a 'sweet village in Warwickshire' which she calls Melford, located not far from either Warwick or the 'native place of our great dramatist and poet, William Shakespeare' (Stratford-upon-Avon), and she tells of a happy childhood with her father, the 'poor curate' to the village, despite never having known her mother who had died not long after giving birth. Everything seemed to be going well for her until almost at once her fiancé, Walter, went off to university and her father died of a 'virulent fever' caught off a sick farmer, which meant that she went to live with Squire Moreland, Walter's father and a man who had just lost his own wife. Very quickly Moreland made his intentions known, announcing his 'passionate love' for his young charge and asking her to become his new wife, the new 'mistress of Melford Hall': when Etchell rejected his advances, indicating her engagement to Walter, Moreland was enraged and claimed that ' "I will make you my wife, or lose all in the attempt" '. Etchell was obviously exhausted by the emotional buffeting to which she was suddenly subjected, and when Walter broke off the engagement (her condition being explained to him as being brought on by her 'faithlessness'), she fell into an extreme bout of mental distress which she herself understood at least partially in medical terms:

> Irritation of the brain [ensued]. I became delirious and knew no one ... I worked myself up into a state of fearful excitement, which retarded my recovery, and gave birth to more alarming symptoms – a derangement of the mental faculties, and fearful prostation of the nervous system.
>
> (Etchell 1868: 62, 65)

Although Etchell does not directly address this point, she gives the impression that in the closed but highly respectable rural community of Melford – one soaked in a pervasive patriarchal order, as tied into an older power structure based on deference to the landed gentry – it was impossible for her to find any resources to help her to combat both her distress and the Squire's designs (which inflamed this distress). Her condition was hence closely linked to the place of Melford, to a collection of factors that combined quite distinctively in such a locality, and it is perhaps unsurprising that elsewhere in *Ten Years* she does pursue a determinedly 'contextual' argument about the causes which can combine to push people into phases of 'mental disorder'.[4]

Having slipped into such a state, Etchell became even more at the mercy of Moreland, and it gave him the opportunity to have her committed to a mental institution from which she could only escape by agreeing to the marriage (and he told her that if she continued to resist his demands he would 'play the tyrant', ensuring that she would 'never leave this place while life continues'). In the story Moreland called upon the services of Dr Williams, the proprietor of a private madhouse called Hygeria Lodge, who examined Etchell – administering blisters, leeches and other drugs, making 'tormenting inquiries', and using terminologies which were meaningless to her – and who concluded that she was 'eligible' for admission to his asylum. Only Mrs Dorothy, the Squire's housekeeper, voiced a view contradicting the label of 'mental illness' being pinned to Etchell by the doctor: ' "what an eye! so quick! send *her* to a 'sylum indeed! it's a burning shame, and I wish the whole place may come down upon that doctor and his crew" '. But Etchell's fate was now fixed, and she was soon taken on the eight-hour journey to Hygeria Lodge, the 'old and popular asylum' located deep in the West Country, where she was to spend a substantial part of the next ten years. Considerable significance attached to this place in the subsequent struggle over and for her identity, as we will explain below, but what might initially be noted is that – in contradiction of 'Jarvis's Law', which in its nineteenth-century variant stated that people send their mentally distressed relatives and friends to the nearest facility (Philo 1995) – Etchell was actually sent to an institution considerably further away than either the Warwickshire public asylum or the several Midlands private madhouses open in the 1850s. One implication is that Moreland wished to lose his charge in a place so far from home that she would be hidden from any potentially meddlesome local acquaintances (notably Walter), and the consequence for Etchell was that the yawning gulf of physical distance separating Hygeria Lodge from Melford greatly exacerbated her sense of abandonment: '"I do not know of one true friend in the world"', she complained to an elderly gentleman visiting the madhouse, '"except indeed ... Rebekah, my old nurse ... but she is so far away."'

Hygeria Lodge was a medical space in so far as Williams was a well-known physician specialising in diseases of the mind, and in so far as the treatments received by patients (the administration of blisterings, 'mustard plaisters', various drugs, shower baths, cold baths) were clearly rooted in the conventional medical assumption that mental complaints were ultimately the product of a disordered physical–chemical constitution in need of 'rebalancing'. Inmates could easily be bludgeoned into believing that Williams did know best in his medical judgements on their woes, and Etchell recalls an occasion when, after being swamped by 'learned names' from the doctor and by 'his technical explanations of the different phases of insanity', she too found herself persuaded that 'they were all developed' in her own tortured mind. In this respect Etchell's experience anticipates that of people who in more recent years have been admitted to mental institutions, and who have there been convinced of the medically-identified 'pathologies' affecting them:

> For six months I was in and out of hospital (several times involuntarily), was given large doses of 'tranquillising' drugs, and was generally made into a mental patient. I was told, and I believed, that my feelings of unhappiness were indications of mental illness. At one point, a hospital psychiatrist told me that I would never be able to live outside a mental institution ... When I was defined as 'ill', I felt 'ill', and I remained 'ill' for years, convinced of my own helplessness.
>
> (Chamberlain 1988: 120)

Mental institutions can therefore be key sites in which people 'learn' to be mad in a medical sense, but from Etchell's words it appears as if this was only partially true of what occurred at Hygeria Lodge. Indeed, in practice here – and the odd meeting with the doctor notwithstanding – inmates were allowed scant insight into the discourses that imposed a medicalised mad identity upon them and then prescribed their fates. When the house was visited by the government's Commissioners in Lunacy, for instance, these inspectors 'walked through [the] various apartments, with their memoranda and pencils in hand, accompanied by Dr Williams, with whom they exchanged telegraphic signs touching the mental intelligence of their protégés': at no moment were inmates let into the secret of what these 'telegraphic signs' meant for their own conditions and futures. The identity which the madhouse cultivated in inmates was thus above all else one of being 'nothing', of being an unwanted cast-off with nothing in mind, nothing to offer, nothing to expect, and who could be expected to understand nothing.

We would argue that various aspects of a place like Hygeria Lodge – the building, its immediate environment and its specific location – must have been bound up with the nothingness that inmates like Etchell evidently felt deep inside. Here are two lengthy passages where Etchell graphically conveys this relationship between the 'abandoned' identities of inmates and the place of the institution:

> The garden of Hygeria Lodge might in some respects be called itself a sepulchre, its high walls and dark foliage scarcely concealing the occupants within; who, amid its silence and seclusion, buried the hopes and joys of their former existence in one common grave, half-hidden in the leaves and flowers which grew upon its margin.
>
> (Etchell 1868: 100)

> How few are there, who – in passing those splendid homes of wretchedness, private madhouses – can imagine the agony, the misery, and, still worse, the mute despair that dwell there ... The costly buildings, the enchanted pleasure-grounds, beautiful strains of music, and the well-dressed inmates and dependants with their autocratic lord (even the doctor himself) at the head, are often the only visions presented to the minds of those who ever take the trouble to associate the poor lunatic even for a few moments in their minds.
>
> (Etchell 1868: 3)

Etchell continually references the attractive rural surroundings of Hygeria Lodge, describing the old mansion house wreathed in climbing plants, the gardens full of trees, shrubs and flowers, the tall outside walls covered in 'thick clusters of purple grapes' and the nearby fields of 'golden corn' and 'richly tinted fruits'. In addition, she describes a location which was clearly immersed in the heart of the Somerset countryside,[5] and her choice of setting reflected both the typical kind of environment that 'higher class' private madhouses tended to occupy and (more specifically) the known prevalence of such institutions in nineteenth-century rural Somerset.[6] It is not hard to detect the anger underlying her accounts of the natural beauty that enveloped Hygeria Lodge, however, and she obviously reckons the country retreat to be serving more than the ostensible purpose of providing a therapeutic *milieu*. Indeed, she appreciates how the gardens, fields and hills could be interpreted by many contemporaries as crucial components in the regimes of care and cure being created for mad people,[7] but she indicates that for inmates of the rural asylum these could also be the curtains hiding their very existence from the eyes of the outside world. To Etchell's thinking, the combination of geography as distance (the separation of the madhouse from centres of population) and geography as rural scene (the hiding away of the madhouse behind the 'leaves and flowers') contributed greatly to occupants having a sense of being entombed, of ceasing to be, of losing any meaningful earthly identity.

Much of what Etchell writes about her days at Hygeria Lodge spirals around activities in and around this rural place, and a difference is hinted at between the total despair (the negativity, the passivity) that she felt when in the immediate confines of the house – notably when staff were present – and the hints of a more optimistic (more positive, more active) cast to her self-identity when walking in the gardens, exploring lanes or being taken to neighbouring settlements. And a significant component of her story relates the times that she met up with a travelling band of Gypsies who sometimes camped near the madhouse, since it was through the help of a Gypsy woman called Brownie that a series of events was set in train which led to Etchell's eventual release from the asylum and reunion with both a long-lost brother and Walter. It is interesting to consider the role that is accorded here to the Gypsies, themselves social outcasts in many ways, and a pivot of the novel is arguably this liaison between different 'marginal' peoples – the mad and the Gypsy – which leads to Etchell rediscovering an identity for herself apart from both the labels of the physicians and the nothingness of the madhouse inmate. Such a liaison between the mad and the Gypsy is one that surfaces in other writings, but of particular note is the contribution that 'gipseys' made to the escape from an Essex madhouse in 1841 by John Clare, the 'rural poet' whose life was marked by a mental affliction which led him to spend many years in mental institutions.[8] In his madness Clare kept muddling up his identity – was he Clare, or was he Shakespeare, Nelson or a prize-fighter? – but at other times he possessed a self-identity which was 'intimately involved with his awareness of his birthplace [deep in rural England, near Peterborough] and of all the living things that he remarked within his locality' (Robinson 1983: x), and it was Gypsies who encouraged him to leave the madhouse and to return to the Northamptonshire countryside where he might rebuild his life:

> Journal Jul 18 – 1841 – Sunday – Felt very melancholly – went a walk on the forest in the afternoon – fell in with some gipseys one of whom offered to assist in my escape from the mad house by hideing me in his camp which I almost agreed ... July 20 – Reconnitered the rout the Gipsey pointed out and found it a legible one to make a movement and having only honest courage and myself in my army I led the way and my troops soon followed.
>
> (extracts from *Journey Out of Essex*, in Robinson 1983: 153)

On a day after his escape Clare was assisted by a tall Gypsy woman, who gave him advice as well as directions to a small tower church in the distance (see Figure 10.1), and – although this is perhaps to conjecture too much – there does seem to be a common thread here between the stories of both Clare and Etchell in that encounters with Gypsies (and with Gypsy places or camps) signalled a route towards personal freedom and renewal.

Although the bulk of *Ten Years* is set in Hygeria Lodge, some of the later

Figure 10.1 This illustration is taken from a wood engraving by John Lawrence found in Robinson (1983: 153). It shows the nineteenth-century poet John Clare meeting a Gypsy woman

narrative shifts from this madhouse to another kind of mental institution. In the story Etchell explains that on account of her 'ingratitude', and also given the 'little progress' that she had made at Hygeria Lodge, Moreland had decided to remove her 'forthwith to a county lunatic asylum in the north of England'. Since it is known that Phillips (the 'real' Etchell) had been confined in the West Riding Public County Lunatic Asylum at Wakefield, it is probable that the public asylum in the novel is modelled on this existing asylum, even to the point of being located in a similar part of the world (and at one point Etchell recollects how the asylum's matron had 'dispensed her Yorkshire hospitality very graciously'). Etchell represents the public asylum rather more favourably than she does the private madhouse, and she explains that the asylum's 'governor', a Dr Cromer, was a knowledgeable and kindly individual with 'the patience of Job' who listened to patients and sought to deal with 'the wants, ailments and complaints of more than a thousand daily'. There is some indication of Etchell gaining a stronger sense of herself during her time under Cromer, and there is the suggestion that – instead of confusing her with medical science whilst insisting that her predicament was a medical one anchored in a 'weak' body – this doctor was effecting a commonsense talking therapy through which patients could arrive at a 'psychological' window on their mad identity. Like Hygeria Lodge the public asylum occupied a rural location, and once again Etchell comments on the rural character of both the grounds (noting 'the fine large green in front of the asylum') and the wider surroundings ('"[i]t is a pretty place"', she remembers saying to another inmate, '"and very pleasantly situated"'). It was perhaps not as buried in rural seclusion as was Hygeria Lodge, though, and Etchell recalls that from the matron's window it was possible to gaze out 'on the garden, and far beyond it to the neighbouring town of V. [possibly Wakefield]'. It may once more be fanciful, but a parallel can perhaps be detected between Etchell moving physically closer to the everyday world (in the shape of a large urban area) and Etchell gradually regaining a sense of herself as an individual with worth and purpose, a process which in the story culminates in her new-found brother securing her release from the asylum and her reunion with Walter.

MAD IDENTITIES AND MAD GEOGRAPHIES

There are innumerable thought-systems which deal with the vexed issue of what this thing we call 'madness' actually *is*, and in their own ways these thought-systems all offer guidance on how best to understand 'the mad identity', whether as an integral part of a person's make-up and self-apprehension *from within* or as a bundle of pressures, stresses, labels and concepts inscribed upon the person *from without*. There are innumerable shades of medical-psychiatric opinion about madness as 'mental illness', where the tendency is to trace the origins of the personality and behaviour disturbances manifested in the mad person to deeper physical-chemical malfunctionings of the body (brain and nervous system); and these discourses inevitably medicalise the mad identity, fixing it as a scientific object amenable to medical interventions such as drug therapy, and in the process the medical profession makes available conceptual resources through which many sufferers will come to view their own conditions. There are numerous shades of psychological-psychoanalytic opinion, moreover, where the tendency is to look directly at the distressing confusion of madness as experienced, and to regard this confusion as the outworking of normally hidden tensions, forces and desires raging in the individual's psyche (tensions which afflict everybody, but which become intensified and irreconcilable within the mad person); these discourses, particularly in their popularisation of Freud, also make available conceptual resources taken seriously by many sufferers seeking to look deeper into their own psyches, pasts and repressions.[9] And there are certainly many other thought-systems – we might term these 'ideologies' – that furnish explanations for madness which focus chiefly on the causes wrapped up within the individual's own body, psyche, being, self. Most religions give accounts of madness as divine inspiration, punishment for evil possession, for instance, and it is likely that distressed people through history have turned more to these sorts of conceptual resources than to any other for insights into their own madnesses.

Rather different from medical, psychological and religious claims about madness are those intellectual positions which are critical of explanations which are 'internalist' (rooting causes within the mad person him- or herself), and which instead see medicine, psychology and religion themselves as phenomena central to the whole problem of where madness comes from. The argument here, as expressed so crisply in Thomas Szasz's *The Myth of Mental Illness* (Szsaz 1974), is that 'mental illness' simply does not exist in the sense of there being a fundamentally different and pathological state of the mind–body axis which sets certain people (sufferers) apart from everybody else. Instead, constructions such as 'mental illness' and (in earlier times) 'madness' are seen as just that: constructions, inventions, labels which society elects to impose on some of its members under specific circumstances, probably as a tool of 'social control' which can be used to justify excluding, locking up and 'operating upon' people whose only crime is to be a little sad, bad, different, deviant, troublesome. Coupled with the brilliance of Michel Foucault's *Madness and Civilization* (Foucault 1967), with its inversion of the usual celebratory tone of psychiatric history-writing, the 'anti-psychiatic' turn initiated a whole new field of sociological-anthropological inquiry into how different societies in different periods and places have developed apparatuses of power (mixing up discourses, institutions, practices) which have identified and responded to madness. An inflexion of this radical turn of thought, meanwhile, accepts that some people really do become distressed – whether or not they receive the official labels of 'mentally ill' or 'mad' – but adds that it is very much in the socio-economic hardships of life, notably as inflicted on the lower classes by industrial capitalism in western nation-states, that can be found the sources of the pressures and stresses which often 'produce' distress in vulnerable individuals. The overall effect of this radical thinking is to usher in an 'externalist' explanation of madness that moves from within the individual to without, and the implication is that the mad identity has little to do with the complexities of a distressed person's own problems, hopes and fears, but everything to do with the purposes (professional, policing, governmental) of the groups and agencies wielding the labels. These are forceful claims, and they undoubtedly play upon our vision of how we want to study madness and its spaces, but what must also be realised is that here again can be found conceptual resources which are being drawn upon by many sufferers themselves as they struggle against the hegemony of the contemporary mental health care 'experts'.

We have here passed lightly over difficult theoretical materials, of course, but our objective is less to debate the relative merits of competing thought-systems and more to highlight how they might touch our own proposals about researching mad identities and mad geographies. What we must underline is that we are concerned with a quite different order of issues to those commonly addressed by medicine, psychology, religion or anti-psychiatry, since we are not seeking to uncover the *causes* of madness acting in or on an individual, and neither are we seeking (except indirectly) to expose the *constructions* of madness leading from the logics of broader apparatuses of power. This is not to deny the importance of tackling questions about causes and constructions – and on other occasions we would do so ourselves – but in the present chapter we wish to focus more narrowly upon how individual people who are mentally distressed, and who for the most part are in contact with the mental health care machine in a more or less formal manner, negotiate their identities (their senses of self, of occupying a 'subject position') on an everyday basis. We wish to look closely at the people concerned, or at least to wonder about an interpretative framework which might prove helpful in looking closely at substantive situations where mad people are striving to come to terms with both their distress and the discourses, institutions and practices impacting upon them. Seen in this light, the relevance of the thought-systems outlined above lies not in what they can contribute to our interpretative framework, but in the obvious significance of such thought-systems as discourses which distressed people may encounter, consult and believe, in the process of reflecting upon their own lives. Their own personal madness – whether and how they acknowledge it, how they explain it to themselves and to others, how they manage it, how they resist it – is going to be intimately associated with the conceptual resources that are available to them from the people, services, facilities and settings with which their own personal geography brings them into contact.

There is nothing all that novel about what we are suggesting in this connection, then, given that we draw upon the spirit of recent developments within human geography – ones that talk of the intersections between identity, space and place – in the course of considering, as well as illustrating, their utility in offering a new window on the interests of a geography of mental health.[10] In order to begin clarifying our arguments here, we should point out that for many writers any discussion of madness is always also a discussion of identity, given that in most theories of identity the emphasis is upon the need for objects to have stable, coherent, bounded identities, allowing it to be stated with confidence that 'X is not Y', which means that intellectual procedures have to be established (requiring rigour in the realms of observation and logic) which are utterly hostile to the chaos of mental activity attributed to the mad person. Furthermore, these expectations bubble over into judgements made of people who fail to accept or to fit in with the normal rules of identity:

> If identity refers to the whole pattern of sameness within a being, the style of a continuing me that permeated all the changes undergone, then difference remains within the boundary of that which distinguishes one identity from another. This means that at heart X must be X, Y must be Y, and X cannot be Y. Those running around yelling X is not X and X can be Y usually land in a hospital, a rehabilitation centre, a concentration camp, or a reservation. All deviations from the dominant stream of thought ... can easily fit into the categories of the mentally ill or the mentally underdeveloped.

> (Trinh 1988: 71–2)

Gunnar Olsson (1980, 1991) has consistently echoed these sentiments, suggesting that those who dispute Enlightenment rules of identity and difference are always likely to be branded 'mad', if not actually condemned to the asylum or the prison where their heresies can be reduced to silence.

In some recent cultural-theoretical writing this madness that refuses conventional approaches to identity has begun to gain attention, and – to cut sharply through a maze of poststructuralist literature – some authors end up celebrating a 'schizo-subject', the schizophrenic wanderer in the city who thoroughly subverts the orthodoxies of modern consciousness by offering new (flowing, displaced, non-dualistic, non-hierarchical) ways of appropriating the world (Gregory 1994: 152–7, mainly discussing Deleuze and Guattari 1984). We are unsure about the precise status of schizophrenia in this theoretical register, and would agree that 'there is also something cruel – at the very least insensitive – about analogising schizophrenia like this' (Gregory 1994: 156), but we would nonetheless wish to press home the claim that for most people identity is actually a much more messy and indistinct stance on self, other and world than the standard 'atomistic' model of the human subject can ever allow. This is the argument of Trinh T. Minh-ha, who insists that it is not only the mad person who possesses an identity which refuses a simple 'me as X, you as Y' logic, since for many people (notably women) identity actually resides in that confused imaginative space where saying 'I am like you' coexists with saying 'I am different to you': thus 'unsettling every definition of otherness [or, indeed, of sameness] arrived at' (Trinh 1988: 76; see also Katz 1992). This honest recognition of the chaos present in the dynamics of individual identity challenges many academic disciplines – it places serious question marks against simplistic treatments of social groups and their collective identities, for instance – but what excites us here is the possibility of taking such a recognition back into research on mad identities.

This approach to matters of identity is filtering into contemporary human geography, as in Gillian Rose's (1994) use of films produced by local community groups to demonstrate identities circulating through the cultural politics of place in a fashion suggestive of a 'third space' between the dualistic logics of 'us and them'. It is perhaps unsurprising that geographers have been attracted to arguments about the fracturing of individual and group identities, given that over recent years they have partaken in an explosion of theoretical and substantive work alert to the complex 'structuring' of both individual human lives and collective societal processes across and through space (Giddens 1985; Gregory 1994; Soja 1989; Thrift 1983a). If it is appreciated that the spaces of the social world are indeed many and various, being heterogeneous in terms of the incredible diversity of phenomena (material and immaterial) contained within them, then it follows that the identity of any one person cannot avoid being in a constant flux depending upon the myriad differing influences to which they are exposed when travelling around this profusion of spaces. Claims of this sort are hardly new: think of Erving Goffman's study of how a large mental hospital contains both very public 'surveillance spaces' where individual identities are 'mortified' and directed towards an ideal institutional model, and more transient 'free spaces' in which inmates sustain some sense of individuality by breaking rules about conduct, pastimes and self-treatment (Goffman 1961). But what geographers provide is a more systematic reflection upon the timing-and-spacing of everyday life, taking into account not just the time-space settings for social interaction but also how the properties of 'locales' enter into the very substance of interactions occurring, and there is surely potential here for painting a more sophisticated picture than hitherto of how fractured identities (at both individual and group scales) and complex geographies (of people existing, moving and interacting around a diversity of sites) are 'always already' knotted together. Isobel Dyck's study of 'negotiating motherhood' in a Canadian suburb may be instructive here, since she shows that '[w]omen's everyday locales, and the routines of which they are a part, are . . . sites for both reproducing and changing activity and identity' (Dyck 1990: 478); and more specifically she describes women carving out time and space for a 'comparing of notes' which allows 'current social identities and the activities through which they are defined [to be] examined and evaluated' (Dyck 1990: 479).

The destination of our arguments here should now start to become clear, then, in that we want to borrow from the above-mentioned thinking about the fusion of fractured identities and complex geographies when opening up new possibilities for research on the geography of mental health. We are hence advocating heightened sensitivity to mad identities (very definitely in the plural), acknowledging that people with mental health problems may internalise an even more chaotic and contradictory assemblage of identities than do supposedly 'sane' people, and we are signposting the need to consider how most mentally distressed people have little choice but to negotiate their identities in relation to the tangled geography of sites comprising the mental health care machine (see also Rowe and Wolch 1990). It is across and through this mish-mash of sites – the hospital, the day centre, the doctor's surgery, the drop-in clinic, the group home, the night shelter, the soup kitchen – that mentally distressed people encounter a proliferation of discourses (and also concrete practices) which influence their identities in different ways, if only partially and if on occasion only because individuals react *against* what they are hearing and experiencing. We have given some flavour of what such an inquiry into mad identities and mad geographies might look like, in that the case study of Etchell's semi-fictional madness reveals how her identity underwent changes – the original collapse into a distressed state; a grudging but brief acceptance of a medical explanation; an intense feeling of nothingness; a gradual regaining of a sense of purpose and worth – and also laid out how these changes were very much linked into the different worldly sites – Melford; Hygeria Lodge; the Gypsy camp; the northern public asylum – where she found herself. What we will do now is to give a second case study, which dips into Hester's research in Nottingham, drawing out a few materials to illustrate further our claims about the interweaving of mad identities, space and place.

NEGOTIATING MAD IDENTITIES IN NOTTINGHAM

This second case study introduces Nottingham, an East Midlands city in Britain, and also introduces research seeking to engage with and to understand the spatialised nature of deinstitutionalisation explicitly from the viewpoint of those people who find themselves caught up in it. Nottingham itself has a special place in the 'map' of deinstitutionalisation. In the 1950s Nottingham's Mapperley Hospital (the old borough asylum) began a radical experiment in community care under the direction of the hospital's medical superintendent, Dr Duncan Macmillan. Local guest houses were used to place discharged patients in the community, their owners being paid for taking in patients, and thus began a long tradition in Nottingham of both community links and private residential care. Macmillan (1958a, 1958b; see also Baldwin 1971: 25) wrote of these innovations, and certainly made important contributions to the debates of the 1950s which are commonly seen as an era of change in the nature and prospects of the old asylum system. In view of the historical significance of community care here, Nottingham is therefore an interesting place to investigate in terms of the possible success and failures of 1990s deinstitutionalisation.[11]

The outcome of legislative and grounded changes in mental health care has unveiled new geographies of treatment, residence and care. Madness can no longer be placed in the one contained location of the remote asylum, and it is now 'out there' in everyday public spaces. Mental health care is thereby occupying many new places in the city, and so too do the people who access that care. If mental distress is seen as a crisis of the self, then how are identities (in and out of crisis, healthy and ill, 'sane' and 'insane') being constituted, maintained, fragmented or destroyed in the contemporary care system in a city such as Nottingham? And are identities themselves even in crisis? Fundamental questions about the nature, cause and condition of mental distress in Nottingham are perhaps too much to consider here (but see Giggs 1973), and yet the social construction and consequent treatment of those labelled 'mentally ill' by those untarnished by the diagnostic brush is important, and may shape the use and constitution of space by people so labelled. This seems relevant especially in terms of this group, who may be sited not only within Nottingham and its care system but also in a dialogue (be that one-sided) with psychiatric professionals whose job it is to dissect, reassemble and analyse the selves of others. For the person with mental health problems in a contemporary city such as Nottingham, space and place are duly reconfigured in a wide array of sites of meaning in which identity, therapy, 'sanity' and 'insanity' (amongst other things) intersect. These intersections can be processes, fragments that are transitory, briefly placed through the creation of defensive, safe, perhaps even aggressive 'insane space' (Parr 1994), and they can be more firmly anchored in the new institutional spaces of a new care system, a variety of schemes and projects that seek to reconstitute the old identity of 'the patient' in the new identity of a social group made up of numerous individuals (be that with specific needs) who are first and foremost people advancing decisions, choices and opinions. In a sense, then, care in the community is contradictory in its ideals and aims. The process of deinstitutionalisation seeks to allow a physical and social space for the identities of these individual people with mental health problems to flourish *independent* from the confines of an older and institution-ridden system. However, the (statutory) replacement care structures are often deliberately geared to aid 'integration' and 'normalisation', and these are goals which will inevitably be imbued with a specific if varied set of assumptions, and goals still heavily *dependent* upon the existing system (its staff, its ethos).

One possible way to conceptualise the interconnectedness of mad identities and the spaces of deinstitutionalisation is to impose a categorisation of the different sites involved. Community care (in Nottingham but also more generally) has resulted in the fragmentation of centralised places of treatment, and the associated provision of a variety of new sites within the city. These form a complex mix of statutory, voluntary and private provisions that differ in their physical location, internal layout, accessibility and philosophical aims. The interactions that individuals have with these sites will of course provide different bases of experience, and consequently different contexts for the negotiation of identity, but it can still be said that these sites have certain characteristics and aims that provide grounds for comparison. We suggest that it will be useful to distinguish between what we are calling 'institutional', 'semi-institutional' and 'non-institutional' spaces. 'Institutional' is used here as a term to describe a whole range of community facilities, from residential care homes to day-care clinics to mental health centres, all of which are specialist, formal, regulated places of mental health and mental illness. 'Semi-institutional' is used here to describe a less regulated, more transitory, less specialist assemblage of sites which are easier to access from the point of view of a 'user'.[12] 'Non-institutional' space hints at created, imagined and undefined space that is accessed and used by people with mental health problems very much on their own terms. This schema provides one way of partially representing the contexts that exist in the lives of mentally distressed people; it opens a window on the interconnecting places and spaces of Nottingham within and between which mad identities are being exposed, fought, accepted and negotiated.

Organised geographies and institutional spaces

In Nottingham it is possible to see a transition taking place from the old institutional geography of the asylum system to a new institutional geography of reprovision. Mapperley Hospital is the residual element of an age of containment in mental health care, but it is one that here began to change slowly over thirty years ago. The hospital is currently in the process of closing, and negotiations regarding the reprovisioning for its remaining occupants are taking place at the time of writing. Concern is being voiced by users of the hospital that they are losing an assured site of support and identification, and this is acknowledged by the reproviders (Nottingham Health Authority) in their concern about former patients revisiting the site after it has ceased to be a hospital. Mapperley Hospital has been an important site of meaning for psychiatric patients for over one hundred years, and the local community has grown used to the presence of an institution on their doorsteps and is therefore concerned about the closure: '"I don't mind them being locked up in the hospital on the hill, but it's another thing having one next door"' (local resident, in Parr 1991: 50). These concerns tell a story about an institution that has in some ways defined the identities of people who have lived and been treated there, and also reveal how that institution has become a home, a dwelling place, a familiar, safe retreat to many who have perhaps in some ways internalised the identity of 'the patient'. Although the future of Mapperley Hospital is now clear, the fate and future 'place' of the people who permanently or temporarily access it is undoubtedly in question.

The basis of institutional care is certainly changing, and the current and ongoing fragmentation of care in Nottingham has provided or even forced a changed and changing identity of the people who access that care. Older geographies of stigma, reliance and retreat have been reconstituted and remapped in different parts of the city, which in themselves now provide different contexts of and for care. The sites of care have changed, and with them the aesthetics of mental health care: here we use the term 'aesthetics' to highlight the importance of the physical, outward appearance of mental health service buildings. Many are anonymous semi-detached houses in residential areas; some are more obviously 'clinics' in part through being attached to general health centres or hospitals. The politics of appearance and the locating of services are large concerns of both service providers and users in Nottingham (as elsewhere). Views are often divided between the necessity on the one hand of ensuring accessibility and actively combating stigma by having a high community profile, and the concerns on the other hand of users who would sometimes want to access an anonymous service which creates little risk of further labelling. The basis for a social construction (from 'within' and 'without') of both the ill self and the role of the 'mental patient' has shifted. The imposition and internalisation of an identity that is derived from the big old Victorian hospital on the hill is to be redrawn, and indeed *is* now being so redrawn in many different ways through these different, fragmented community networks.

Different people access new institutional spaces of care, and are finding new bases upon which to confront and to negotiate their own identities. On one level, these spaces provide settings in which the specificities of individual identity and psychiatric problems are recognised and acknowledged, but on another level these spaces are still being specifically provided for mental health care, and with this comes a set of common external and internal associations, be these informed by psychiatry as a discipline or by the user's experience of being in a psychiatric treatment process. Here we should pay attention to the changing status of the 'mental patient'. It is perhaps easy to view mental illness and those who suffer from it in polarised terms, as either a homogenised social group (reflected and reinforced by past spatial treatment practices) or as a differentiated individual experience or reality that is distinctly divorced from the everyday, from the allegedly 'normal'. Indeed, many users would agree up to a point that the ill self is not easily understood by others or even by themselves:

> I suffer from panic attacks, and anyone who has ever had panic, it can be very frightening, 'cos you feel like you are dying and you can't explain that to another human being. You come over all in a sweat and you lose all sense of reason, it's frightening for you and it's frightening for the people you are with, so you tend to run away from people. They don't want to know you, you have got this terrible disease, they just want to be normal. I don't class myself as being normal.[13]

When accessing specific and definable mental health space in its institutional form in the community (in sites that are provided by statutory bodies concerned with treatment and care in medical and contractual ways), it is perhaps difficult for the user to negotiate individual identity as anything other than 'patient', 'ill', 'mentally ill', 'insane', 'client', 'dependent' and so on. This may be compounded by the fact that some people may themselves be confused as to how their own identity can be expressed and understood.

In recognition of this situation, it is unsurprising that Nottingham has seen the rise of the first collective user movement (outside of individual hospitals) in what is now a growing phenomena in Britain. The ironic appropriation of the term 'user' and the powerful capital of having received mental health services has enabled a group of people to enter a dialogue with an increasingly consumer-oriented health authority. This dialogue has a number of functions, one of which is to be involved in the planning of mental health services in the future. Whilst at present the dialogue seems more consultative than directly engaged or formalised, the fact that there is any dialogue at all marks a radical change in the status of ex- and present sufferers of mental illness. These users are the only representatives of a wide and varied group of people who receive services in Nottingham, and the individuals involved key into a range of institutional sites in both hospitals and the community, negotiating as they do so not only service provision but also their own status within the health sector. Their individual identities are arguably being submerged in a new discourse, one in which the user is at times a 'commodity', as if to add authenticity to new projects, schemes and ideas. Although there is no doubt that the organisation as a whole has achieved much in promoting patient rights in Nottingham and in persuading services to become more 'user-friendly', the participants continue to be drawn into a jargon-laden relationship with medical and professional structures that easily incorporates the new identity of 'the user' into their associated institutional geographies. So in one sense, whilst fighting for a recognition of the heterogeneity of sufferers of mental distress (and thus challenging the monolithic identity of 'the patient') and despite securing other more practical improvements to the mental health service, this group of co-ordinated users is assuming a contradictory corporate identity ('the user') tied quite closely into the city's medical-psychiatric establishment. Those who are organised and who are part of the official dialogue often see their own roles as integral to the overall city care service:

> We have to make sure that in Nottingham that we get more than tokenism, so that somebody down in Gosport, they can say, 'hang on a minute, they don't have tokenism in Nottingham'. It's not tolerated, and the professionals see the very strong commitment and profession-alism of the users. That is what I think, I know that I am a professional in the best sense of the word when I do voluntary work. In other words I try to see and recognise that as a user I have a responsibility, a loyalty to other users, but I must be sympathetic to an extent with the professionals.

We acknowledge that the people involved in this process may not see their own identities as quite so bound into the institutional establishment as is hinted here, and that there may be a 'strategic essentialism' (a playing up of 'the user' as a single coherent identity) which is consciously deployed by these users in their dialogues with powerful agencies to secure certain ends. These manoeuvres are at once political, social and personal, and are tied up in a local renegotiation of what 'mental illness' is and of how to treat it. The individuals who make up these users also have their own spatial networks in Nottingham that key into specific institutional nodes: the clean, official sites of mental health care and mental illness. These places are defined – an individual here is assumed to have a clear identity as a patient, a user, an ex-user or staff – and they are specific places of treatment that comprise for some people who access them a difficult and challenging arena in which to find and to express individual identities, but which also serve as sites of collective reworking of the status and identity of the users of psychiatric services.

Disorganised geographies and semi-institutional spaces

Apart from official sites of mental health care there are many other kinds of facilities (voluntary, charity, church-based) that provide support and shelter to the psychiatric population. These include drop-in centres and self-help groups, situated in a variety of semi-institutional settings, and we use the term 'semi-institutional' to hint at the complex nature of these sites of support. Although these sites are undoubtedly a grassroots attempt at providing places of interaction and shelter, they sometimes receive grants from statutory authorities, and this circumstance then throws up questions about the philosophies of such projects and about whether these grounded responses can be compromised by criteria of funding applications, as instruments of what Jennifer Wolch (1990) would term the 'shadow state'. We would argue that for the disorganised, poverty-stricken ex-patient, the few drop-in centres that have appeared on the fringes of Nottingham city centre provide a consistent, safe space in which 'otherness' is accepted and even expected in terms of behaviour. So the users, ex-users and potential users that utilise these sites may to some extent be accessing an institutional geography not so different to that already discussed, as their lives are drawn into a network of facilities, schemes and groups which, again, are often sited in inner-city districts.

It is an ambivalent relationship that some of the users have with these centres and the people in them, and some feel angry about the stigmatised identity that might rub off on them from the centres (even though they may only access the sites intermittently):

> The fact that I am mentally ill is incidental, it is an incident in my life, I don't necessarily have to mix with people who have been mentally ill or who still are mentally ill or psycho-geriatric, which a lot of them are. These places aren't full of young people, they are full of old people, I don't want to mix with them.

In Nottingham, despite or even because of these semi-institutional sites being chaotic in terms of their overall aims (and a resistance to the overordering of aims is sometimes a deliberate policy in such sites), they do still provide concrete points of identification for an often drifting ex-patient. In these spaces all sorts of behaviour, physical appearance and habits are tolerated, and violence is the only personal expression that is wholly banned. The internal dynamics of such centres are important in that they are bounded locations in which fundamental issues of identity and labelling appear and are negotiated, often in the way that their interior spaces are used by the people who have an investment in them. Other researchers have perhaps paid more attention to this issue than have geographers. Sue Estroff (1985) is an anthropologist who has written an ethnography about a small psychiatric unit in Madison, for instance, and she argues that clients here negotiated their own identities by appropriating certain areas within the centre specifically for drinking, talking and therapy. The Nottingham case shows similarities, in that – although providing a cheap and safe place where disorganised users can interact with others that have the same basic life experiences – the centres here remain in some ways a reminder of an institutional past, and as such there are often designated volunteer and user spaces. Administrative and communal space can thereby be segregated, and the focal communal space in a centre may be used for 'acting out'[14] in ways reminiscent of the ward. Even if space can be (and is) appropriated in these rather Goffmanesque ways, however, these are places that remain ultimately defined and to an extent controlled by external agencies. To some extent an individual identity is here still defined as 'user' or 'client', and participants are still expected to display a variety of characteristics and behaviours within the defined territory of the semi-institutional space.

To suppose that realms of identity and identification for the person with mental health problems are limited to institutional or even semi-institutional arenas of interaction is to deny the significance of numerous more invisible political and social 'movements' in the city. Identities can often be constituted outside of the confines of definable mental health space, and we believe that there is a substantial 'other' geography of informal, disjointed networks which sustains many mentally distressed people. The networks involved in this respect are inherently political, given that in the formation of these quite different spaces the social actors work with, but in various ways resist, a variety of 'structures' in the community (from the police force to retail outlets to cafés to church outreach projects). Perhaps there is a sense in which this is the crucial pivot of what we ought to be talking about under the heading of 'community care': the success or otherwise of many sufferers in finding or creating everyday spaces to exist and to function within.

Here users and ex-users access *undesignated* areas for shelter, interaction and freedom, and these spaces perhaps serve as transient sites of resistance to the imposed identities of both institutional and semi-institutional sites. This phenomena has been recognised if not discussed at any length by geographers, but Robin Kearns has written as follows in an unpublished piece:

> At the outset for instance I presumed that certain donut shops were just what they appeared to be, places for consuming coffee and donuts. But I rapidly discovered that the primary meaning projected onto such establishments is one of sanctuary and congregation with ex-psychiatric peers. For this population, the inner city offers few places of rest.
>
> (Kearns 1986: 12)

Research in Nottingham has shown how it is generally recognised by service providers and users alike that there are places accessed by users which provide a cheap respite from the stress of inner city life:

> they wander all over, there are one or two specific places where people go, the Arboretum can be quite a high concentration of people, especially in the summer, pleasant gardens y'know, you can lie down and go to sleep or have a cup of tea at the booth ... I used to go to the Arboretum as much as I could, it is quite a pleasant place, I got fairly familiar with it. I don't go so much now, but it is a nice haven against oppression.

These non-institutional spaces consist of cafés, parks, city squares and particular nooks and crannies in the cityscape. The users or potential users essentially carve out their own places that seemingly allow for their individual identities to be revealed, whether this is achieved by choosing to interact with non-users who have no connections with the mental health system (which is obviously difficult in the formalised networks) or by using these places to exhibit the bizarre behaviour which, inspired by observations from Estroff, might be referred to as a form of 'crazy theatre':

> One hotel in particular was viewed by most of the staff and clients as housing a collection of strange outcasts who had little or no place else to go. Ben and Sadie rather enjoyed these surroundings, often entertaining other residents with stories of their hospital adventures or demonstrating their 'craziness' in the lobby by talking nonsense, gesturing and otherwise behaving bizarrely.
>
> (Estroff 1985: 56)

These places can provide a variety of functions, from defensive space to finding somewhere just to sit and to think alone.

For individuals to find and to construct such spaces of self-identification may be a complex symbol of internal conflict, with the personal and the public being tied together, however briefly, and the process may also be embodied in particular actions:

> I thought right I'm going to do and feel and think everything that my subconscious wants to. In other words my subconscious had again become stronger than my conscious and instead of suppressing it, again I thought no! I am going to let it have full reign and I'm going to do exactly as it tells me ... be totally selfish ... if I want to be obsessive, then I will be obsessive ... if I want to go and kick someone in the street then I will kick someone in the street ... I'll do it, I won't be a good girl.

Such behaviour is tolerated at times, even in the public space of 'the streets', and let us mention at this point the main market square in Nottingham. The Square (see Figure 10.2) is a place that is sometimes used as a meeting place for the ex-psychiatric population, 'acting out' and using drugs:

> Darren moved in and out of the crowd with his hands raised above his head. 'In the name of the lord I am calling my children to me. YOU! (he pointed at a couple sitting by the fountains) CAN YOU SEE ME? I AM HERE. YOU ARE ALL BLIND.' He continued, away from the couple who had only briefly glanced up and then turned away. His movements seemed jolting and awkward. He would stop every third step or so and shake his right leg and make the same repetitive gestures with his hands. As I looked at him, he turned and stared back for what seemed like a long time. 'Hello', I said. He turned away and continued around the square shouting to the air and occasionally to someone in particular. He circuited the square four times while doing this. No one seemed to react to him. No one rushed to call the police. They only stared at him from a distance ... Eventually I decided to try and talk to him again, 'are you alright?' He stopped and looked at me:
> 'No, I'm fucking mad.'
> 'Are you?'

Figure 10.2 The market-place in Nottingham

'Yes, I'm mad, do you know what that means?'
'No.'
'It means I am mental, do you know what that means?'
'No, I don't.'
'It means I am at Her Majesty's pleasure, now piss off and mind your own business.'

(Extract from research diary, June 1993)

These sorts of actions are policed in different ways at different times, and there is hence a notion of the spaces concerned being transient and constituted as 'mad' or 'insane' (Parr 1994) at times when the barriers to the use of these seemingly purified spaces are not operational. The use of central city space in this form is interesting, and may reflect a need on the part of individuals such as Darren to gain help (by 'acting out', being arrested and maybe referred to a crisis intervention team for medical help). Alternatively, it perhaps reflects the need felt by some members of the mentally distressed population to forge their identities in new locations, and using undefined space in this manner avoids the risk of using stigmatised space (a fear noted earlier in relation to the ambivalent use made of semi-institutional sites). Spaces such as the Square are ones of relative freedom: there are no 'staff', volunteers or (usually) students[15] to define behaviour and to regulate the uses being made of space. These are personal spaces of identification, and perhaps here in such spaces is where it is easier for certain distressed people to 'belong', even if such belonging can only be transitory. There is still a need for retreat and for 'asylum' (in the true sense of the word) within the city, and these spaces may just be a way of providing such things.

CONCLUDING THOUGHTS: LOOKING INTO 'GLASS WORLDS'

In discussing the identities of people with mental health problems we have attempted to journey around a series of possibilities that may form 'mad identities', and in so doing have considered how fragmented mad selves may be constituted across and through a range of sites within concrete reality. Whilst we have acknowledged the conflicting and interconnecting discourses that impinge upon how such people may conceive of themselves from within, we have purposefully limited our discussion of identity to the question of how it is bound up with 'real spaces' (be these the asylum, the day centre or public space). What we have argued is that identity can be constituted by, and perhaps also play a part in constituting, the properties (be these at times complex and confusing) of such real spaces. At certain points we have hinted at less straightforward intersections of processes of self-identification with everyday geographies, though, and it is in closing that we wish to provide pointers for further discussion in this respect. These pointers will offer a partial view of quite 'other' possible geographies that we have not discussed in detail here, but which may yet prove central to debates about mad identities and mad geographies.

In arguing about the relationship between the constitution of mad identities and their grounding in real spaces, we have not paid enough attention to the internal place – the 'imagined place' – of the mad self buried within an individual's own mental world. Internal locations of the ill self and the healthy self, of the 'insane' and the 'sane', of the confused and the understood, are clearly crucial to this debate. In considering these issues we draw on the writing of James Glass, a political scientist, whose work *Private Terrors/Public Life* (1989) is a fascinating account of the inner worlds of psychiatric patients who relay their internal stories, struggles and victories to the author. Glass couches an exciting interpretation of these stories in terms of the often fraught engagement of an internal self with a public community (the latter lying beyond the individual, outside of his or her internal realm). Although we can only offer a caricature of Glass's complex text in this conclusion, it does allow us to highlight the importance and the potential of his analysis.

When writing about a troubled inner self, Glass considers internal boundaries, borders, places and geographies, and – even more suggestively – he talks about the *lack* of such spatialised conceptions within many mentally distressed individuals. The condition of being-in-the-world sometimes referred to as *placelessness* obviously has implications both for material external spaces with which people have interactions (Relph 1976) and for immaterial internal spaces of self-negotiation. As Glass argues:

> To be borderline is precisely to live on the borders of society; it is to feel so alienated, so extruded from the social world, that consciousness finds it impossible to find and locate a sense of place. Psychological placelessness, a horrifying experience of aloneness and disconnectedness, becomes the norm, close to the pull and nearness of delusional identification.
>
> (Glass 1989: 58)

Glass's vision of the 'place' or the 'placelessness' of the mentally distressed person lies at several intersections of internal and external reality, and he thereby sees the person with mental health problems as someone experiencing acute difficulty in existing within and between private psychological worlds and public sites of personal and political citizenship. In his own theorising of these boundaries, he draws on the voices of the 'internally bound', of people who are heavily locked into their own internal nightmares, to highlight the struggle over the reality of unreality:

> When I'm psychotic, reality disappears and my mind moves away from ordinary experience. I find myself in places no one understands, worlds that bear no relation to this one. But I always manage to return; I find my way back and store up the experience. It sits there until I feel the need to express it. But to even talk about what I see and feel, I need a place, a settledness, making the fear endurable.
>
> (Glass 1989: 32)

In the chaos of psychosis certain delusions ('unrealities') can become the metaphorical or imagined place for an accepted internal identity, be this only a temporary place 'to settle', but at the same time there may be a need to find some more worldly place (an ordinary 'reality') where it will be safe to express something to other people of what happens in the middle of the private delusional geographies. It is difficult to engage with such arguments here, partly because of space and partly because of the lack of research as yet completed in this field of concern (the lack of attention as yet paid to the voices of those in the middle of these dilemmas). But what Glass does claim is that sometimes (although unfortunately not always) the drifting and 'unplaced' internal identities of the troubled self *can* be at least partially anchored, and maybe given assistance, by conducive concrete real space. The deliberate fostering of 'therapeutic' places by various agencies (both formal and informal) can hence serve on occasion to help troubled people in their search for a more grounded and non-delusional identity, one free from the unconstrained terrors so often let loose in their imaginings:

> Annie had no fixed, constant sense of reality, no grasp of a place where she should be; throughout her life, her voices would tell her what to do. The ranch however, gave her a sense of meaning in social context; its very physical activity provided her with measures of her capacity and effectiveness; it also framed and defined her will to live and diminished

her persistent confusion over who she was and how she should conceive of her purposes in life.

(Glass 1989: 168)

It may therefore be less than coincidental that so much of the literature outlining alternative strategies for mentally distressed people – the full range of self-help schemes, support networks, therapy cultures – speaks of trying to secure a new, safe, non-threatening space or place which can provide a trustworthy mooring for the recovery of a less damaged identity (Chamberlain 1988; Women in MIND 1986).

In one sense of course this is what we have attempted to argue throughout the chapter: that space and place in a physical and concrete reality intersect and interact with other personal (we might say personally political) processes of self-identification which are being negotiated by people with mental health problems on a variety of separate but cross-cutting levels. Internal spaces and places of identity and identification, which are often quite fleetingly 'occupied', are both shaped by external spatial practices and yet sometimes quite untouched by them. This contradiction runs across what are conventionally held apart as private and public, political and social, and are hence extremely complex. It is in this complexity that the key themes of this chapter are crystallised, since the messy geographies (real and imagined, external and internal) that we highlight here are 'always already' a series of complicated and contradictory sites of negotiation of the self. Whilst acknowledging that we have skimmed many issues and contentious arenas of debate in the process, we hope that our account provides one avenue into discussing spatialised mad identities, and that (perhaps more importantly) it opens up many more possibilities for further consideration.

ACKNOWLEDGEMENTS

We would like to thank the editors for their advice and patience, and to acknowledge the support of our colleagues in Human Geography at Lampeter. Chris is indebted to Felix Driver for help in relation to the Mabel Etchell material, and Hester is indebted to the many people in Nottingham who have given her interviews and information.

NOTES

1   The potential for people with mental health problems (and others) to reappropriate the term 'mad' as a radical political statement – as 'one desperate expression of a radical need for change' – was noted some time ago by David Cooper (1978), and related claims can be found in other texts which pit 'schizophrenia' against the excesses of both capitalism and modernism (notably Deleuze and Guattari 1984, and see below). An equivalent development might be the reappropiation of the term 'queer' by gay activists and other sexual dissidents. We recognise that great difficulties attach to our use of 'mad' in this chapter, however, and we certainly intend no insult to people with mental health problems and wish to cause no offence to readers. We might add that even the seemingly very innocent term

'mental distress' is not always popular with people experiencing psychological unrest, as Hester has found in her dealings with user groups in Nottingham.

2 This text has been largely ignored by historians of psychiatry. Even a work dedicated to tracing the history of mad people 'from below' (Peterson 1982), and containing an extensive bibliography of relevant writings, mistitles the novel as *Two Years in a Lunatic Asylum*. We will be quoting and paraphrasing from it extensively in this section of the chapter, but will only give page numbers for longer (indented) quotations.

3 From two *Registers of Admissions* the following can be traced: a Charlotte Philips (case no. 41399) – note the spelling with one 'l' – is recorded as a 'private' patient admitted to a Manchester asylum (seemingly the Manchester Royal Lunatic Asylum at Cheadle) in January 1856 and discharged as 'not improved' in January 1860; and a Charlotte Phillips (case no. 65514) is recorded as a 'pauper' patient admitted to the West Yorkshire Public County Asylum (near Wakefield) in February 1860 and discharged 'recovered' in October 1866. Various authorities are satisfied that Etchell was indeed a pseudonym for Phillips (Kennedy, Smith and Johnson 1926: 16; Kirk 1891, Volume I: 562; 1891, Volume II: 1232).

4 In the preface to the book Etchell (1868: vi–vii) remarks on the 'great strains' imposed on 'mental powers' in the course of all social classes going about 'their struggles for daily subsistence', and she then keys into an anti-urbanism that permeated much debate about madness and its treatment in nineteenth-century Britain: '[i]n the town in which the writer resides, life is almost overwhelming, and aptly did a popular minister and author exclaim ... "oh, ye Manchester men who are burning the candle at both ends, be warned in time."'

5 There are various clues in the text suggesting a West Country location, including a reference to 'driv[ing] through one of the long lanes for which Somerset is so famed', and also in a passage where the inmates are taken to walk the 'Abbot's Way' near Glastonbury which boasted scenery 'bounded by the Mendip Hills, whose sides afford a scanty herbage for cattle'.

6 The Somerset landscape – particularly around Taunton and in the 'Bristol–Bath region' – boasted eighteen private madhouses at various times during the nineteenth century, including several well-known private madhouses in country mansions such as Brislington House (Parry-Jones 1972: 36, 112–15).

7 There was great concern during at least the first three-quarters of the nineteenth century about the need to create what might be termed a 'moral geography' for British asylums, one that was able to utilise the supposed natural healing qualities of the rural environment in effecting a successful 'moral treatment' of mad patients (Philo 1987, 1992: esp. chs 3 and 4).

8 Between 1837 and 1841, the date of his 'escape', Clare was committed to one of Dr Matthew Allen's private madhouses at High Beech in Essex (Parry-Jones 1972: 94), but for the last twenty-three years of his life he was confined to the Northampton General Lunatic Asylum, a facility that effectively combined the functions of a charitable lunatic hospital and a public county and borough asylum.

9 It might be noted here that there is currently a stirring of interest in what geographers might be able to learn from psychoanalysis as we extend our understanding of how individual human subjects and broader social structures intersect with one another (Pile 1991, 1993). We might have done more to bring psychoanalytic materials into thinking about the constitution of mad identities in time and space, but such a project must wait for another opportunity.

10 The field of interest delimited as the 'geography of mental health' has almost never addressed questions of identity, although one or two intriguing signposts in this respect can be identified. The field is critically reviewed in at least two papers (Philo 1986: pt 3, 1992, ap. 1), but in outline it has split into studies dealing with the spatial distribution of revealed 'mental illness' (research into 'psychiatric geographies') and studies dealing with the locational dynamics of mental health care facilities (research into 'asylum geographies'). The dominant approach in the former case has been an 'ecological' or 'spatial scientific' one, as codified in John Giggs' (1973) influential paper, and very rarely has attention dipped 'below' the level of aggregate patterns and correlations to consider the experiences – the personal circumstances prompting distress in particular people occupying particular environments – of the individuals coded as dots on the researcher's maps. An exception can be found in the efforts of Ken Dean (1979, 1982, 1984), who in part employs an 'interpretative approach' on individual case notes to tease out 'subjective realities' as lived through by diagnosed patients, and who also develops a framework which takes seriously the 'spatial implications' of different stages in the 'psychiatric career' of a typical sufferer

from mental distress. In the latter case – concerning research into institutional locations and related issues – the picture has been more confused, but has focused most obviously on the ways in which deinstitutionalisation is currently restructuring the spaces of mental health care away from large hospitals towards networks of small facilities spread across urban areas (Harvey 1983; Wolpert 1976; Smith 1983). In so doing geographers have documented the gravitation of facilities into old and decaying inner cities, as bound up with 'community opposition' to such 'noxious facilities' emanating from supposedly respectable neighbourhoods (Burnett and Moon 1983; Dear 1977; Dear and Taylor 1982; Moon 1988; Smith and Hanham 1981; Taylor 1988), and as related to the emergence of 'psychiatric ghettoes' of facilities and their users within the broader entrapment of 'service-dependent populations' occupying many inner-city localities (Dear 1980; Dear and Wolch 1987; Wolch 1979, 1980; Wolpert and Wolpert 1974). The experiences of individuals on the receiving end of these processes of restructuring and entrapment are rarely addressed unless in passing, although Chris Smith (1975, 1977: esp. 5) does ask about the 'spatial and environmental variables' that are likely to make patients discharged to the community 'happy or unhappy'. Glenda Laws and Michael Dear (1988) consider the factors influencing an ex-patient's ability to cope in the community meanwhile, and in a piece seeking to introduce the idea of a 'humanistic medical geography' Andrew Sixsmith (1988: esp. 20) sensitively raises the need to 'focus upon how people use different places in order to preserve mental well-being'. The research of Robin Kearns (1986, 1990; see also Kearns and Taylor 1989) should also be mentioned here, since he has used interviewing techniques in seeking to get close to the everyday experiences of 'chronic patients' discharged into the community, and as such he begins to tackle issues of personal identity and its negotiation in a manner similar to that we are proposing in this chapter.

11   Deinstitutionalisation involves the closure of large mental hospitals and the provision of care in the community for people relocated from these old institutions. This practice grows out of a dialogue between civil libertarians, policy makers and different arms of the psychiatric profession. The 1959 'Mental Health Act' in Britain confirmed government support for care in the community, following the example set by such as Dr Macmillan, but did not make it a statutory obligation (Eyles 1988). Since this time there have been a number of parliamentary papers and reports which outline the future, principles and ideals of contemporary mental health care, notably the 1988 *Griffiths Report* (Bean and Mounser 1988). Increasingly the emphasis has been on the care of the individual (a stark contrast to care practices in the past), with recent legislation outlining the obligation of local health authorities to provide care within a community setting. The individual focus is achieved by the case management approach to care: a user of mental health services should have a key worker, who, although not providing for all of the individual's needs, will facilitate his or her care activities with other interested parties, for example, social workers, self-help groups or perhaps people providing group therapy at a local mental health centre.

12   The term 'user' is employed to refer to people with mental health problems who at present have, or in the past had, contact with some sort of mental health service. This is the primary means of identification used by the people in Nottingham who inform Hester's research project.

13   Quotes here are not attributed to any individual as informants insist on anonymity. All unattributed quotes in the text are taken from interview transcripts.

14   'Acting out' is a term used in a variety of psychiatric and sociological literatures to indicate a state of being of people with mental health problems. It is often argued, particularly in 'anti-psychiatric' writing, that some care practices have encouraged mentally ill 'careerism', in which the patient assumes a sick role and behaves in a way expected of a passive, unproductive person. The use of the term 'acting out' is related to these arguments. Patients or sufferers, it is suggested, may deliberately 'act out' to gain medical attention or to stay in hospital. On the other hand it can be used as a descriptive term to indicate seemingly 'bizarre' behaviour by people at the peak of their condition. In this context it is used to describe repetitive gestures and behaviours that can be viewed as 'institutional', in that what may be seen as commonplace in the ward seems 'out of place' outside of that environment.

15   One of the problems of accessing many of the institutional sites is the concern that these places are over-researched. Psychiatric, sociological and social work students heavily utilise the statutory mental health services as research sites. The lack of privacy of the person with mental health problems is a large issue, and these considerations defined who contributed to the present research and which sites were accessed. When researching in an undesignated space Hester did not always reveal her identity as a postgraduate student.

11

BODIES WITHOUT ORGANS

schizoanalysis and deconstruction

Marcus Doel

The schizophrenic voyage is the only kind there is.
(Deleuze and Guattari 1983: 224)

DAWN OF THE DEAD

The diagnostic of 'liquidation' exposes in general an illusion and an offense, it accuses: they tried to 'liquidate', they thought they could do it, we will not let them do it. The diagnostic implies therefore a promise: we will do justice, we will save or rehabilitate the subject. A slogan therefore: a return to the subject, the return of the subject.
(Derrida 1988a: 113)

The story of the subject needs to be told and its trajectory needs to be mapped. Like any species in jeopardy, the subject should be recorded in terms of its genealogical inscription within different social apparatuses, according to its evolution and mutation within a succession of permeable and shifting contexts. As a point of departure, one could work through the plethora of disciplines and perspectives where there is a growing sense of unease and foreboding over the fate of the subject. Indeed, one can already discern the outline of a dominant motif: the subject as catastrophe site, accompanied by a rapidly ossifying consensus: the dynamism of the subject has finally exhausted itself and is now fated to disappear through a terminal decline. For many, there is a conviction that the catastrophe has already occurred, and that we are living in a dead zone, or waiting period, haunted by the death of the subject. Hence the theoretical, political and ethical urgency of the speculative question: who comes *after* the subject? (*Topoi* 1988). Will it be an-Other subject, a suicidal nihilist, a community, a new form of schizophrenia, a cyborg, a machinic infestation, nothing, something inhuman or non-human? Or perhaps we should be attempting to revive, resuscitate or rejuvenate the subject in order to give it a new lease of life? And yet, in so far as the philosophy of the subject was only ever a pseudo-beginning, a beginning which was always and already in decline, and one which only served to dissimulate, marginalise and repress all of those 'others' from whom it drew its place and its power, many authors have readily accepted and internalised the death, dispersal and liquidation of the subject with glee: the subject, what a horror. Many, however, remain incredulous to such hyperbole. And yet, if the terminal decline of the subject is indeed the case, one can only hope that in the wake of the subject, something more desirable might finally have the chance of happening: cast of the die.

When considering the fate of the subject, the dominant discourse has been one of catastrophe and exhaustion, a discourse which has become associated with the advent of poststructuralism and postmodernism in general, and the writings of Louis Althusser, Jean Baudrillard, Gilles Deleuze, Jacques Derrida, Michel Foucault, Jacques Lacan and Jean-François Lyotard in particular (Dews 1987; Harland 1987; Lawson 1985; Megill 1985). A few have sought to revel in what they perceive to be the apocalyptic consequences of such a virulent form of anti-humanism (Kroker and Cook 1988; Land 1992). Many more have engaged in nostalgia and lamentation for that which has been lost, often giving themselves over to an heroic quest for the restitution of the subject through relocation, rehabilitation and reconstruction (Rosen 1987; Soper 1986). Finally, there have been a number of attempts to literally *embody* the subject, either through the introduction of a series of replacements and substitutes to take over the place of the subject, or else through an encasement of this ethereal term within a variety of body-parts: skin, face, genitals, hands, eyes, feet. In the wake of the subject, it has once again become possible to situate living and breathing human bodies (Nicholson 1990). In short,

> the body is no longer the obstacle that separates thought from itself, that which it has to overcome to reach thinking. It is on the contrary that which it plunges into, in order to reach the unthought, that is life.
>
> (Deleuze 1989: 189)

Consequently, in the wake of the subject there will have been: joy, lamentation, nostalgia, restitution, resurrection, replacement, substitution and embodiment. What unites each of these responses is the fact that they are all predicated upon some negative event befalling the universal and abstract subject. In some versions, this negative event is truly apocalyptic, manifesting itself through motifs such as death, liquidation, dissolution, annihilation and disappearance. And in so far as this negative event is a terminal and irreversible decline, it is both futile and untimely to seek a recovery of such a subject. Hence the inclination to either mourn, laugh or shrug. In other versions, the negative moment is more modest, expressing a relative rather than an absolute decline. In particular, these versions are dominated by the sense of a damaged, defective, dysfunctional or limited form of subjectivity. Specifically, the subject is curtailed through a series of constraints: the machinic arrangements which construct and animate it; the discourses which circulate through it; the languages which occupy it; the desires which propel it; the powers which saturate it; and the material fabric which binds it. In contradistinction to the longing for an immortal, ahistorical, incorporeal, universal and abstract subject, there is an insistence upon the fact that the subject is bound and pinned down within a plethora of social apparatuses. The subject is a machine, to be sure, but a machine which is assembled and articulated in place. Moreover, this machinic production of the contextual subject is only a constraint from the perspective of a *desire* to escape human locatedness and finiteness. From the moment that one sheds the *force* of such a desire, situated singularity becomes life itself. In other words, the subject is the context in which it is produced: a work-in-process; a work-as-process. The subject is articulated twice: the machinic production of a productive machine; producing, a product.

> Everywhere *it* is machines – real ones, not figurative ones: machines driving other machines, machines being driven by other machines, with all the necessary couplings and connections. An organ-machine is plugged into an energy-source-machine: the one produces a flow that the other interrupts.
>
> (Deleuze and Guattari 1983: 1)

Accordingly, whenever one speaks of the absolute or relative decline of the subject, one is indicating that the subject is dispossessed of its self. What is difficult to grasp, however, is that this dispossession occurs through a double movement: once through the re-embedding of the universal 'I' within the singular contexts in which it is expressed; and again through the reinscription of the individuated 'me' within the social apparatuses which animate and sustain it. However, it is important to emphasise that this is not a negative movement in so far as a negation of the subject would necessitate either a negation of the negation (giving rise to a new positivity through sublation: the arrival of an-Other subject), or an extreme form of nihilism which would seek to block and frustrate such a resurrection effect. Consequently, it is important to insist that the ex-appropriation of the abstract and universal subject is affirmative rather than negative in order to avoid becoming trapped within the Möbius spiralling of two lines which appear to pass through the place of the subject. Whilst the first line traces the eternal recurrence of the machinic construction, de-construction and re-construction of the subject (subject there will be), the second traces the movement of a previous construction into an irreversible destruction (subject there won't be). However, although these two lines appear to bifurcate and diverge, with the former progressing through investment and accumulation (a dialectical perfection), and the latter striving for a pure expenditure without return (death pure and simple), they both actually interlace to stake out the limits of a double bind. Whichever line is followed, the *place* of the subject is always made available to an-Other occupant. Hence the fact that every response to the negation of the subject is always accompanied by the speculative question: *who* comes after the subject? Even in death, the subject will subsist through hypertelia: 'I am – dead' (Courtine 1988: 103). The vampiric subject, what a horror! It is precisely in this sense that the decline of the subject in contemporary social theory remains haunted by a resurrection and return of the repressed. In particular, one might note how the de-construction of the subject invariably produces a stream of body-parts which are then gathered up into a host of fragmented bodies and splintered subjectivities: chunks of flesh wrapped within envelopes of skin and stamped with the traits of faciality. Within the duration of this chapter I will endeavour to distinguish this parcelling of body-parts through a succession of arbitrary combinations and permutations from the Bodies without Organs (BwOs) that emerge in the aftermath of a schizoanalytic and deconstructive experience. Specifically, the BwO is not a fragmented body; it is not the fractured and dysfunctional aftermath of a shattered totality.

> Outside the Oedipally organized Symbolic order there is said to exist only an undifferentiated infant body (the OwB: organs without a body) labouring in a prelinguistic state of imaginary confusion between (fusion with) self and mOther ... The so-called fragmentation exhibited by the 'pre-Oedipal' body is in fact the fractality of part-objects ... not the debilitating lack of an old unity but a real capacity for new connection. It is not a negativity in contrast to which a plenitude might be desired. It is a positive *faculty* ... A return *to* the body without organs is actually a return *of* fractality, a resurfacing of the virtual. Not regression: invention.
>
> (Massumi 1992: 85)

However, before moving on to map the schizoanalytic subject in deconstruction, I want briefly to stake out the terrain of the vampiric philosophy of the subject which 'lives on' even in the wake of its own relative and absolute decline. In particular, I want to problematise the fragmentation, liquidation and resurrection of the universal and abstract subject, and underscore the necessity for an affirmation rather than a negation of the fissiparous movements which traverse the place of the subject.

FRAGMENTED BODIES

> Fractured, all. Every step falls in a void. No sooner do we have a unity than it becomes a duality. No sooner do we have a duality than it becomes a multiplicity. No sooner do we have a multiplicity than it becomes a proliferation of fissures converging in a void ... In itself, the event has only extinction. Its accomplishment is its evaporation in the infinite interplay of its seething components ... Being is fractal.
>
> (Massumi 1992: 19–21)

Conventionally, the subject is assumed to be identical to itself; it is the point, the place in the pattern, which endures. It is the centre of identity, stable and unshakeable. Although it is the condition of possibility of identity, presence and difference, the subject precedes all identification, presentation and differentiation. I *am*, before I am some *thing*. The subject is One: universal, indivisible and eternal. The subject is the subject, and thereby accomplishes two distinct functions within the topography of social theory: universalisation and individuation. On the one hand, the subject is a figure of universalisation in so far as it is the degree-zero of humanity, the place to which all human traits indexically refer and defer (I am – subject). In short, re-cognition passes *through* individual bodies and faces to the place of the universal subject. Moreover, this movement from the individual to the universal does not depend upon the actual variation amongst individual bodies and faces: there is universalisation *before* there are individuations. Indeed, the universal is indifferent to all quantification. This is why the proliferation, de-differentiation or fragmentation of faces and bodies will never serve to problematise the universal subject: subject there is. *The subject is the subject. Alone it stands. And in no need of skin, flesh, face or fluid. Body it never is. Bodies are the enemies of the subject.* The subject is what remains when the body is taken away; it is literally *in*human (I am – dead). On the other hand, the subject is also a figure of individuation in so far as it can only express itself through bodies and faces. The subject only exists in its effects, in the subtraction of its effects; without *a* body or *a* face to pass through, the subject could not fulfil its function of universalisation. Hence the complementarity and the paradox: the subject requires individuation in order to express universalisation; but there is always the danger that the gaze and re-cognition will get hooked on the body, encased in flesh, stuck on the face and immersed in fluid. In short, the material fabric of the body may frustrate the passage towards the place of the universal and abstract subject. Hence the fact that flesh and bodies are always sedimented, stratified and traversed by the double movement of universalisation and individuation which envelopes them with skin and stamps them with face – I am wrapped within me; I am unwrapped within you.

Within the double bind or pincer movement of universalisation and individuation, an assemblage of social apparatuses seizes roughly hewn chunks of flesh, encases them within skin, inscribes them with face and encodes them with the striations of race, ethnicity, gender, sexuality, class and so on and so forth. However, the production of human subjects is never complete; it is always a work-in-progress and a site of continuous experimentation. Hence the fact that the human subject is always a full body *to come*; it endures without ever existing *as such*. Being is Becoming. In other words, the subject endures through continually breaking down, but this is not a negative event. As we shall see more fully below, the assumption that there is a universal, unitary and centred subject which could be either situated, embodied, fragmented, decentred, de-constructed or destroyed is precisely what is in question. Indeed, it is the philosophy of the subject which works through identity, resemblance and negation, with its rigid segmentation and despotic territorialisation of molar subjects (I = I = not you). Meanwhile, deconstruction and schizoanalysis affirm the molecular movement in things.

Accordingly, molar identities are not there from the start, like an array of plenitudes or plenipotentiaries which could be selectively actualised within particular contexts, or which could become embroiled in a series of labyrinthine complications, contaminations or confusions. To the contrary, they are appended like so many dendritic prostheses to the swarming mass of fluid multiplicities in order to arrest becomings, regulate movement and impose stability. And like all molar aggregates, the subject assembles itself as an interruption and derivative of the flows which animate, sustain, traverse and discharge it. In short, molar identities endure *and* break down through the stuttering and stammering of an order-word: Freeze!

> Molarity is mode of desire, as is any move away from it .. It is a matter of force: it is a categorical overlay, an overpowering imposition of regularized effects. Because it constricts actions into a limited dynamic range, it is inevitable that it will be experienced by the overcoded body as a physical constraint. Becoming begins as a desire to escape bodily limitation.
>
> (Massumi 1992: 94)

Little wonder, then, that the BwO should so often experience the machinic apparatuses for imposing molar identities on molecular movements as so many instruments of torture. However, it is vital to understand that the desire to escape molarity is a desire to escape limitation rather than locatedness, and sameness rather than singularity. This is why Bordo (1990: 142–4) is mistaken to conflate schizoanalysis and deconstruction with a 'fantasy of escape from human locatedness' through 'a new imagination of dismemberment: a dream of being *everywhere*'. The confusion is a serious one in so far as it diverts attention away from affirmation towards the false problem of quantitative restraint: without some stopping points, endless fragmentation and dispersal would self-destruct and lead to an erasure of the body in a fractal abyss. As Bordo (1990: 145) noted: 'the appreciation of difference requires the acknowledgement of some *limit* to the dance, beyond which the dancer cannot go.' And yet a limit to the fragmentation is precisely what is lacking from the perspective of the vampiric philosophy of the subject: Being either swerves into Nothingness, or else it slides into a becoming-imperceptible; whilst fragmentation either accelerates into a liquefaction, or else it passes over into a fractalisation (Doel 1993). Hence Rose's (1993b: 79) insistence that 'Critique must settle, but settle *contingently*, make *arbitrary* closures, endorse *strategic* essentialism, make *provisional* gestures' in order to address 'the (historical, social) questions: *Whose* truth? *Whose* nature? *Whose* version of reason? *Whose* history? *Whose* tradition?' (Bordo 1990: 137). Nevertheless, one can only feign the ability to locate and identify *who* comes in the wake of the universal and abstract subject, even though such a line of questioning necessarily inaugurates a return of the repressed in so far as the same imperative is always interpolated into the flow of events: Subject there is. Freeze – *who* goes there? All at once, we are back in the double bind of universalisation and individuation and the hypertelia of the vampiric subject.

As we have begun to see, fragmentation, multiplication and embodiment will not suffice to enable an escape from the tyranny of the vampiric philosophy of the subject. The hypertelia of the subject is exemplified and ensured through the stutter and stammer of the order-word *par excellence*: *who* comes after the subject? Rather than demand an eternal return of the subject, what is required is an experience of deconstruction and schizo-analysis in order to sensitise us to the motionless voyaging in place of the full BwO: everything is flux, flow, becoming. In short, we will strive to free *singularity* from the Möbius strip of universalisation equals individuation, *experimentation* from the Möbius strip of negation equals resurrection, and *complication* from the Möbius strip of fragmentation equals totalisation. Moreover, by opening these forced stabilisations to something wholly Other, a crack emerges along which a fractal, crystal or cancer might proliferate, carrying away all of the overcoded flows which have been trapped within the closed circuitry of the molar machines. The full BwO grows in this crack, not as an amorphous and undifferentiated mass, but as a swarm of virtual multiplicities, teeming singularities and experimental complications and inventions. Something would finally have the chance of happening, that's all: cast of the die.

MOTIONLESS VOYAGING IN PLACE

Individual or group, we are traversed by lines, meridians, geodesics, tropics, and zones marching to different beats and differing in nature ... The lines are constantly crossing, intersecting for a moment, following one another ... it should be born in mind that these lines mean nothing. It is an affair of cartography. They compose us, as they compose our map. They transform themselves and may even cross over into one another. Rhizome.

(Deleuze and Guattari 1988: 202–3)

The subject is in decline. It is an assemblage which is continuously breaking down, leaking in all directions. And yet the subject works; it ceaselessly reintegrates everything which would appear to escape its spheres of influence. Everywhere it is a coupling of asymmetrical flows: deterritorialisa-tion and reterritorialisation; decoding and overcoding; de-construction and re-construction; so many double articulations and pincer movements which render (the place of) the subject an inescapable work-in-progress: subject there will be. But it is also a site for endless experimentation, complication and invention; a site that is only ever actualised as the singularity of the context in which it is produced as a recording surface. In relation to these social apparatuses, deconstruction and schizoanalysis seek to accentuate and intensify the processes of deterritorialisation, destratification and decoding so that they detach themselves from the circuitry of the machinic assemblage and become instead a line of flight towards something wholly Other. In other words, deconstruction and schizoanalysis de-limit flows, short-circuit stria-tions and scramble codes through a motionless voyaging which carries us from identity to multiplicity, from position to potential, from Being to Becoming, from arborescence to rhizomes, from constants to variables, from fragments to fractals, from OwBs to BwOs and from subjectification to schizophrenia.

Deconstruction: destabilising the subject

In order to recast, if not rigorously re-found a discourse on the 'subject', on that which will hold the place of the subject (of law, of morality, of politics – so many categories caught up in the same turbulence) one has to go through the experience of a deconstruction ... there is a duty in deconstruction. There has to be, if there is such a thing as duty. The subject, if subject there must be, is to come *after* this.

(Derrida 1988a: 120)

We have already touched upon three of the most important features of deconstruction: affirmation, movement and responsibility. These features contrast starkly with the prevalent and often mischievous mischaracterisation of deconstruction as negative, static and irresponsible (Margolis 1991; Merquior 1986; Rosen 1987). For whilst it is true that deconstruction goes by way of the undecidable (without which there would be neither theory, nor politics, nor ethics, nor responsibility), it is not at all a 'philosophy of hesitation' which remains neutral, impassive and indifferent to the flow of events (Centore 1991; Critchley 1992; Martin 1992). To the contrary, deconstruction *intervenes*, but rather than intervening in an attempt to enforce the molar order, it intervenes in an endeavour to release the potential of the full BwO. Specifically, it intervenes along lines of force, desire and power in order to lever open, dislocate and displace forced stabilisations into an Open multiplicity: 'if the whole is not giveable, it is because it is the Open, and because its nature is to change constantly, or to give rise to something new, in short, to endure' (Deleuze 1986: 9). Moreover, deconstruction is all the less quarantined within the so-called prisonhouse of language, a new onto-theology or rejuvenated idealism of the Text, in so far as it intervenes within the heterogeneous material and immaterial flows of the entire real-history-of-the-world (Derrida 1988b). It is therefore important to rigorously distinguish between *affirmative deconstruction* on the one hand, and *reactive de-construction* on the other (Doel 1994a). Whilst the former affirms the full BwO, the latter endeavours to recapture it through reterritorialisation, restratification, overcoding and subjectification.

Deconstruction has nothing whatsoever to do with catastrophe or apocalypse. It is neither nihilistic nor destructive, nor does it amount to a 'dissolution of the subject' (Derrida 1992: 7). In short, deconstruction does not come *after* the subject has been constructed, stabilised and emplaced. It is neither a speculative *investment* in negativity on the basis of a rational expectation of an accruable return, nor is it an attempt to transact an absolute *expenditure* without return: it is not part of a regime of accumulation or a site of sacrificial consumption. In other words, deconstruction does not find its proper place within either a dialectical series of speculative investments: construction/de-construction/re-construction, or a metaphysical binarisation of absolute expenditure: construction/destruction (Doel 1992). Any endeavour to de-construct, dismantle or destroy can only ever be a *simulated* catastrophe in so far as its only discernible effect is to furnish the necessary resources required for a re-construction. As we have already seen, the question 'Who comes after the subject?' exemplifies this hypertelia through which the philosophy of the subject 'lives on' despite the utter exhaustion of its resources.

In contrast to the feigned risk of reactive de-construction which is always underwritten by a guarantee of dialectical re-construction and resurrection, affirmative deconstruction follows the movements of destabilisation which traverse (the place of) the subject itself; it affirms the iterability, alterability and alterity of the Same. Consequently, deconstruction does less to disturb, dismantle and destroy the subject than to bring into the Open that which is always and already disturbing and menacing its consistency, coherence, stability and pertinence. In short, deconstruction affirms the destabilisation on the move which Opens (the place of) the subject to that which is wholly Other. From the perspective of the molar organism, the social apparatuses of capture and the encoded strata, these movements may appear as a catastrophic collapse and a terminal decline, but from the perspective of the molecular flows they provide expedient lines of disarticulation and escape towards something wholly Other: experimentation, complication, invention and singularity. But who comes after the subject?

> To elaborate this question along topological lines ('What is the place of the subject?'), it would perhaps be necessary to give up the impossible, that is to say the attempt to reconstitute or reconstruct what has already been deconstructed (and which moreover has deconstructed 'itself', an expression which encapsulates the whole difficulty).
>
> (Derrida 1988a: 114–15)

Derrida's insistence on a return to (the place of) the subject and a return of (the place of) the subject will no doubt surprise all those who would wish to charge deconstruction with advocating its death, dispersal and liquidation. To the contrary, the subject in deconstruction is precisely that which eschews all of those moments of negativity, catastrophe and apocalypse which so readily graft themselves onto the misreading of deconstruction as an architectonic de-construction: dismantlement, disassembly, fragmentation, disintegration, disseverance, dismemberment, decomposition, dissolution, etc.

> It is not at all a question of a fragmented, splintered body, of organs without the body (OwB). The BwO is exactly the opposite. There are not organs in the sense of fragments in relation to a lost unity, nor is

there a return to the undifferentiated in relation to a differentiable totality.

(Deleuze and Guattari 1988: 1964–5)

In other words, the destabilisation on the move which traverses (the place of) the subject does not return us to an amorphous, undifferentiated or homogeneous mass (a state of empirical confusion). Rather, it carries us beyond the molar and the molecular towards alterity and singularity. Hence the fact that the BwO must be made; it is always a full body *to come*. This is why the BwO never belongs to any molar aggregate, least of all an individual; it is always a body in ex-appropriation, both nomadic and rhizomatic, short-circuiting, scrambling and carrying away all claims to propriety. In other words, when everything is taken away, there is nothing left but a distribution of haecceities, singularities, events. However, it is vital to understand that the zero intensity of the BwO is not a negative moment in relation to some positive Unity or Totality. For there to be a negative moment, a negative moment which would befall *a* subject or *an* organism, there would already need to be something assembled in place. But the subject and organism are not at all *constants* (for example, the closed equation: I = I = not you). They are neither stabilised in themselves, nor fixed in place. Consequently, the genealogy of the subject cannot be mapped as if it were the trajectory of so many atoms circulating within a four-dimensional space-time, with their speeds and trajectories, attractions and repulsions, fusions and fissions, orbits and quanta. To the contrary, the subject is a *variable* in continuous and Open modification (for example, the open equation: ... $+ y + z + a$ ...). In short, the subject should be understood neither as a universal, nor as an individual, but rather as a virtual multiplicity.

> The universal, in fact, explains nothing; it is the universal which needs to be explained. All the lines are lines of variation which do not even have constant coordinates. The One, the All, the True, the object, the subject are not universals, but singular processes – of unification, totalization, verification, objectivation, subjectification.
>
> (Deleuze 1992: 162)

This is why the subject is always both a work-in-progress and a social apparatus, undergoing the continuous variation of Becoming-Other through a motionless voyaging in place. It is therefore both nomadic (without home or refuge) and rhizomatic (without roots or anchorage). In short, the subject *endures* through the continuous variation of ex-appropriation and Becoming-Other. Schizoanalysis.

Schizoanalysis: bodies without organs

We have as many entangled lines in our lives as there are in the palm of a hand. But we are complicated in a different way ... schizo-analysis, micro-politics, pragmatism, diagrammatics, rhizomatics, cartography –

have no other goal than the study of these lines, in groups and in individuals.

(Deleuze 1983: 71–2)

Destroy, destroy. The task of schizoanalysis goes by way of destruction – a whole scouring of the unconscious, a complete curettage ... Destroying beliefs and representations, theatrical scenes. And when engaged in this task no activity will be too malevolent.

(Deleuze and Guattari 1983: 311, 314)

On the face of it, the emphasis which schizoanalysis places on destruction would appear to align it with reactive rather than affirmative deconstruction, but this inclination would be mistaken (Bogue 1989; Massumi 1992; Perez 1990). For just as affirmative deconstruction must be distinguished from reactive deconstruction, so too must schizoanalytic destruction be differentiated from paranoiac destruction. Once again, we would discover that schizoanalysis is neither negative, nor catastrophic, nor apocalyptic, nor sacrificial. Like deconstruction, schizoanalysis affirms the eternal recurrence of the motionless voyaging in place of destabilisation on the move and the continuous variation of swarming multiplicities – the full BwO. Similarly, schizoanalysis is not neutral, impassive or indifferent to the social apparatuses of capture which enforce varying degrees of stabilisation upon the heterotopic fluidity of singular events; it intervenes in order to release a full BwO. In short, both deconstruction and schizoanalysis activate multifarious lines of perturbation, agitation and turmoil within (the place of) the subject in order to affirm the alterity of the Same. (The place of) the subject is always and already a teeming multiplicity in continuous variation; the site of a full BwO: 'there is a whole geography in people' (Deleuze and Parnet 1988: 10; Deleuze 1988).

There are many types of line which traverse (the place of) the subject. Some of them ravel and converge to form knots, eddies and vortices of relative stabilisation, binding everything which flows into their midst into molar aggregates. These aggregates may then be called upon by the molar order for further experimentation and complication: reconstruction, reproduction and rearticulation. Meanwhile, other lines break free from this ravelling and entanglement, spinning out movements of relative destabilisation which trace lines of flight, disappearance and deterritorialisation. The aggregates break down, molecularise and decompose into a BwO. But what kind of BwO emerges from such a motionless voyaging? In order to address this question, it is necessary to distinguish between three types of line. First, there are *lines of rigid segmentarity* which confine movement within specific cells, molar aggregates and distinct territories. This type of line works through an unending laceration of the BwO, carving out cells, strata, regions and identities through division and bifurcation: home, family, state, factory, community, face, etc. Second, there are *lines of molecular segmentarity* which produce supple segments, molecular fluxion and destabilisations on the move that are distributed in an entirely different manner; they open onto little fractures, dissimulated lines of disorientation and disarticulation, and unrecognisable particles. In short, a cell begins to depart from its usual metabolism, a flow suddenly overspills its channel or a program momentarily loses its code. But the important thing to note is that these deviations and departures remain *relative* so long as the molar order can clamp down on them through reinvestment, reintegration, reconstruction and overcoding; they remain relative so long as the molar order can capture them within a new segment, stratum or code. For example, every now and again, through a cast of the die, an event short-circuits the segments, striations and codes of race, class, gender and sexuality through a becoming-clandestine, imperceptible and acategorical; but this momentary escape of absolute deterritorialisation – *once detected by the molar apparatuses* – will come to be clamped down upon with the full force of the Law and confined within a new identity. Freeze – who goes there? In short, the molar order ensures that the possibility and force of anomie and transgression will be neutralised and contained under the asymptotic curvature of statistical anomaly: everything will have been accounted for as so many standard deviations within the normal distribution of the Same (Baudrillard 1990; Doel 1994b). From the perspective of molarity, there is no longer any outside, merely events and occurrences which have not yet been recognised and integrated within the normal distribution of an economy of the Same. This is why the molar order is irreducibly despotic and paranoiac in so far as it believes that everything falls within its jurisdiction and spheres of influence. 'At every moment, the machine rejects faces that do not conform, or seem suspicious. But only at a given level of choice. For it is necessary to produce successive divergence-types of deviance for everything that eludes biunivocal relationships.' In short, molarity 'never detects the particles of the other; it propagates waves of sameness until those who resist identification have been wiped out' (Deleuze and Guattari 1988: 177–8). Hence the fact that (the place of) the subject is woven and spun out through the ravelling of these two types of line: a molecularisation of the molar and a molarisation of the molecular. Indeed, molarity functions through the double articulation and Möbius-spiralling of deterritorialisation and reterritorialisation; destabilisation and restabilisation; decoding and overcoding; smoothing and striation. What matters to the molar order is that all of these movements of destabilisation remain *relative* through a containment which is enforced by any means necessary. In short, limits and constraints are interpolated onto the full BwO in order to stop, channel, arrest and break up becoming. Whilst the molar lacerations are forever inclined towards slicing (the place of) the subject into a dismembered, fragmented and dispersed pulp, the molecular movements can always be arranged in order to carry the remains back to the molar apparatuses for perpetual recycling.

The potential complicity of molar and molecular segmentation enables us to clarify what is meant by the final type of line: the *lines of flight*. These lines break free from the Möbius-spiralling of molar and molecular segmentarity, disarticulating the strata and scrambling the codes as they carry away singular events into an *absolute* deterritorialisation: fluid in a pure state, streaming over the BwO without limitation or interruption. The full BwO is what remains when *everything* is taken away: intensity = zero (I am Other). It is the plane of consistency upon which motionless voyages are fated to approach asymptotically. To the question: how far can too far go?, schizoanalysis suggests that a body can never go too far with the deterritorialisation, destratification and decoding of flows. The difficulty, however, resides in knowing how best to traverse (the place of) the subject, with its envelope of skin, covering of face and amalgam of flesh. It is relatively easy to produce an empty or botched BwO through a too violent destratification, or a drugged, paranoid and suicidal BwO through a hatred of the organs, or even a totalitarian, cancerous and viral BwO which attacks the organs and proliferates redundant molar and molecular segments everywhere. Dismantling oneself through a schizophrenic process of desubjectification has its dangers: 'Staying stratified – organized, signified, subjected – is not the worst that can happen; the worst that can happen is if you throw the strata into demented or suicidal collapse, which brings them back down on us heavier than ever' (Deleuze and Guattari 1988: 161). Accordingly, the *full* BwO can only be approached through a cautious experimentation and complication within singular contexts. On each occasion, one must ask:

1. What are your rigid segments, your binary and overcoding machines? For even these are not given to you ready-made, we are not simply divided up by binary machines of class, sex, or age: there are others which we constantly shift, invent without realizing it. And what are the dangers if we blow up these segments too quickly? ... 2. What are your supple lines, what are your fluxes and thresholds? Which is your set of relative deterritorializations and correlative reterritorializations? And the distribution of black holes ... where a beast lurks or a microfascism thrives? 3. What are your lines of flight, where the fluxes are combined, where the thresholds reach a point of adjacency and rupture? Are they still tolerable, or are they already caught up in a machine of destruction and self-destruction which would reconstitute a molar fascism?

(Deleuze 1993: 253–4)

In summary, it is important to clarify that schizoanalysis does not dwell on elements, aggregates, organs, subjects, relations, fragments or structures. To the contrary, it pertains only to *lineaments* which traverse the entire molar order, running through individuals as well as groups: a swarming proliferation and infolding of lines; the 'schiz' of schizoanalysis as traced by the 'random walk' of a space-filling fractal of infinite dimension and immeasurable porosity. As a work-in-progress, the place of the subject is one of interminable ravelling: 'the sole unity without identity is that of the flux-schiz or the break-flow. The pure figural element ... which carries us to the gates of schizophrenia as process' (Deleuze and Guattari 1983: 244). It is in this sense that (the place of) the subject is ex-appropriated through a motionless voyage in place, flowing without interruption, and streaming over the surface of a full BwO. Schizoanalysis and deconstruction simply endeavour to destabilise, discharge and short-circuit the forces, desires and powers which strive to capture, stabilise and limit these flows within a plethora of social apparatuses and molar organisations. Little wonder, then, that the machinically aggregated subject is fated to dis-organise, destratify, fragment and shatter: 'The body is the inscribed surface of events, traced by language and dissolved by ideas, the locus of a dissociated self, adopting the illusion of a substantial unity – a volume in disintegration' (Foucault 1977: 138). It is by following this disintegration and decomposition of the human organism – with its striated flesh, envelope of skin and covering of face – along the lines of absolute deterritorialisation that we are carried towards the full BwO. But as we have seen, this Body is not a return or a regression. To the contrary, the full Body is always *to come*; it is what remains when everything is taken away: zero intensity. It is a Becoming in a pure state beyond the double bind and Möbius-spiralling of universalisation and individuation; decoding and overcoding; deterritorialisation and reterritorialisation. In other words, lines of flight cause the machinic production of human subjects to pass from paranoiac fragmentation to schizophrenic fractalisation: nothing but movement, nothing but flux. They carry the ossified flows held within (the place of) the subject into the Open context of the entire real-history-of-the-world, strangling arborescent hierarchies and instituting involuted rhizomes as they go: complication, experimentation, invention, singularity, alterity.

As the fissiparous figure without limit *par excellence*, the fractal is the perfect motif for schizoanalysis, deconstruction and the full BwO. Nevertheless, the desire for organisation and the power to impose arbitrary limits on fissility should not be underestimated. Indeed, when we look into the fractal abyss, most of us intuitively bring out what Deleuze and Guattari (1988: 200) call 'the terrible Ray Telescope' which is 'used not to see with but to cut with'. Its cutting action works upon 'the movements, outbursts, infractions, disturbances and rebellions occurring in the abyss' in order to restore 'the momentarily threatened molar order. The cutting telescope *overcodes* everything; it acts on flesh and blood, but itself is nothing but pure geometry'. Moreover, the strata, segments and codes which it carves out from the BwO force molecular movements to cram into molar aggregates: a veritable Geology of Morals. You will be one *or* an other, *or* an other, *or* . . . 'The strata are judgements of God (but the earth, or the body without organs, constantly eludes that judgement, flees and becomes destratified, decoded, deterritorialized)' on the way to the asubjective, asignifying and acategorical swarm of the full BwO (Deleuze and Guattari 1988: 40).

As the holding capacity of (the place of) the subject approaches absolute zero, with a haemorrhaging of previously stabilised flows in all directions, there has been a tendency to both recoil from empty BwOs and to refrain from producing a full BwO. Rather than risk experimenting with lines of flight, there has been a general attempt to reinvigorate and rejuvenate the molar order: some fear losing the molar aggregates; others seek to impose supple segments on the molecular flux; others demand that the whole terrain be stabilised through overcoding; whilst still others turn the lines of flight into a passion for destruction. In particular, the decomposition of (the place of) the subject has caused many to cling to the face of the Other as a way to nurture 'an ethical subject-in-process' (Kearney 1988: 365; Critchley 1992). But the machinic production of faciality is precisely the molar apparatus *par excellence* which serves to impose waves of sameness upon a plane of haecceities, events and singularities. 'How tempting it is to let yourself get caught, to lull yourself into it, to latch back onto a *face ... The face, what a horror*' (Deleuze and Guattari 1988: 187, 190). In contrast to this yearning for molar identification and recognition, deconstruction and schizoanalysis intervene in order to dismantle the apparatuses of capture which construct and animate the subject, the body and the face by reterritorialising, restatifying and overcoding the molecular flows. They peel back the automata, simulacra and wraiths which haunt (the place of) the subject in order to affirm the full BwO. Wherever you are, one can never go too far along the lines of flight towards absolute deterritorialisation. Indeed (the place of) the subject swarms with these modalities of disappearance which Open onto the motionless voyaging of Becoming-other. Indeed, even the face of the Other is first and foremost a holey surface. However, which line of flight to follow in any particular context of forced stabilisation can only be determined through a cast of the die. Shake. Rattle. Roll.

12

EXPLORING THE SUBJECT IN HYPER-REALITY

Paul Rodaway

For me, concepts have always been something one can't control and manage. There is no political economy of concepts. It's necessary to destroy concepts, to finish them off. Of course, concepts regenerate themselves, they metamorphose themselves into something different.

(Baudrillard, in Gane 1993: 202)

Everywhere one seeks to produce meaning, to make the world signify, to render it visible. We are not, however, in danger of lacking meaning; quite to the contrary, we are gorged with meaning and it is killing us.

(Baudrillard 1987a: 63)

INTRODUCTION

We perhaps take it as axiomatic that 'we' are each 'a subject', yet what does this 'idea' encapsulate or, more accurately, implicate? This paper explores 'the subject' as cultural construct: both theoretically and practically. The subject is perhaps the location where human meaning emerges and is contested, and therefore a locus of power. Through developing an understanding of the subject – its identity and genesis – we begin to understand ourselves and the world within which we are always and already situated. This argument echoes Heidegger's (1983) famous exploration of the nature of being as always and already a being-in-the-world. From the phenomenological perspective, this is the situatedness of the subject (see also Merleau-Ponty 1962a). In the present paper, it will be argued that 'postmodern' reflections, such as those of Baudrillard (1990b), suggest a breakage of this contextual link between the subject and its specific world in the context of hyper-reality and the reassertion of the power of the object. This line of argument is closely linked with the wider poststructural and postmodern debate about the collapse of sign-original referent relationships.

In contemporary experience, there is a growing feeling that our experience of being subjects – and thus our concept of 'the subject' – is undergoing significant change in the context of the mass media, consumer-oriented, technologically dominated societies of the late twentieth century. Awareness of this 'crisis of the subject' – if we can call it that – has a long history and is especially evident in the writings of Nietzsche, Foucault and Baudrillard – to name just three. Each has recognised the important links between the concept of the subject, meaning and power, as expressed through social practices and spatio-temporal structures (e.g. Nietzsche 1969, Foucault 1979, Baudrillard 1990a). Jean Baudrillard adds a particularly interesting dimension to this debate and some of his writings will form the basis of the present attempt to map the subject in hyper-reality.

Jean Baudrillard identifies the limits of the human subject (as we have hitherto known it) in concepts such as the mass, the ecstatic and the obscene (e.g. Baudrillard 1983b, 1987a, 1990b). His analysis is grounded in a long-run fascination with the relationship of objects and signs, and specifically symbolic exchange (Gane 1993). More than many thinkers, Baudrillard has identified the significance of the breakage of the sign–original referent linkage (and the decontextualisation of human experience – to borrow Fjellman's (1992) term) for the construction of the subject and its significance, or power. The subject as ecstatic or as obscene is a radical subjectivity stretched to the limits of its own being as 'the subject'. This is perhaps 'the crisis' of the contemporary subject, the postmodern subject. Hence the importance of an attempt to map, or re-map, the subject. No longer is it possible naïvely to distinguish an object and a subject when the relationship between sign and original referent is broken, or at least fragmented. We need to question our presuppositions about the subject and the role of subjectivity as it is modified – or metamorphosed (see Gane 1993: 102) – in contemporary experience. The identity of the subject (relative to itself, to other subjects and an object world) and the genesis of subjects (or the power of subjectivity) becomes problematic.

Exploring and mapping the terrain of the subject is not an easy task. In effect, we are participant observers, implicated within the phenomenon we study. To map 'the subject' presupposes a recognised phenomenon, or at least a traceable pattern of fragmentation or metamorphosis. It would seem logical to assume that since the subject emerges out of socio-historical practices expressed in and through spaces and places, we can in part decipher or map the subject through careful and systematic observation of the patterns and processes of person–environment encounter within contemporary spaces and specific places. This is the general logic of Foucault's detailed 'archaeological' investigations of the prison and the madhouse (see Philo 1992a). When shifted to the context of contemporary experience, such analyses are thwart with difficulties, both due to the personal nature of such subjective reflection and because of the on-going processes of fragmentation and metamorphosis. One might also ask, what processes are leading to these changes in the construction of the subject? This is, however, a series of major projects in itself, and first it is important to establish the terrain or – in Jameson's terms (1984) – the map. For the present, we must content ourselves with sketching an emerging morphology of the subject, identifying its plurality and dynamism, and noting possible mechanisms for future research.

Therefore, the paper is not a comprehensive study of the postmodern subject, nor an analysis of the work of Jean Baudrillard – both tasks would fill many volumes – but instead some initial reflections on the experience of being a subject in the context of hyper-reality (a term used to represent the contemporary experience) are presented as alternative mappings of the subject. In several thousand words, it is not possible to explore fully the wider debate about objects and subjects, signification and meaning, the analysis of the concept of power and the critique of the notions of commodity, image, consumer and hyper-reality. Nevertheless, each of these wider debates counterpoint the current more limited exploration. 'The subject' is presented as a transitional form, as effectively in quotation marks, and so fragmented but not dead. This might be described as the hyper-subject – a form of the subject correlate with the hyper-real – or trans-subject – suggesting a metamorphosis of 'the subject' into something of which we are at present only partially aware.

COGNITIVE MAPPING

The map has become a popular metaphor for the investigations of the post-modern experience. Fjellman, in a recent study of the experience of Disney World, Florida, for instance, argues that 'culture is made up of maps and sets of maps that serve as legitimations ... culture serves an important ideological function' (1992: 28). In a much quoted paper, Fredric Jameson (1984) argued that we need to develop 'cognitive mappings' of the contemporary experience (what he called 'late capitalism'). In order to re-establish a critique and a political practice of resistance to oppression and alienation, Jameson argued that it was important to regain our sense of place through developing a conceptual language and theoretical structures of the postmodern. He described these as 'cognitive mappings': 'an articulated ensemble which can be retained in the memory and which the individual subject can map and re-map along the moments of mobile, alternative trajectories' (Jameson 1984: 89). Alternatively, this perhaps can be described as 'mapping the subject'. Yet is mapping the appropriate task, whether practical or metaphorical?

The term 'mapping' carries with it a lot of unwanted baggage and perhaps suggests a journey which is quite unnecessary. The map is a representation of the subject, it is an interpretation and an abstraction. It offers a snapshot of the subject, an essentially dead artefact of the subject as it might have appeared to an observer at some point in time. It suggests a stability and coherence of 'the subject' which is not evident in the postmodern condition or is, at least, most denied by postmodernists (e.g. Deleuze and Guattari 1983). How can mapping the subject, in this context, be anything more than imagining, nostalgia and irrelevance? The metaphor of a map suggests that somehow the subject exists on its own, like an object, with location, form and arrangement. Yet the writings of postmodernists suggest that the subject – as it still survives – subsists, continually becomes through relationships with the world (the other) and itself. Perhaps filming is a more appropriate metaphor than mapping, although both abstract the subject from its situatedness, creating an artefact, a history of its passage, rather than engaging with the subject, as a subject, in its flow into and out of existence.

If we are to retain the 'mapping' metaphor, we must perhaps adopt a technology more akin to satellite imaging, continually updating our images and accepting that it is both a remote and an archaeological strategy. Once we have grasped this wider whole of the 'map of the subject', perhaps then we can engage in the political and psychological experience of the subject, as subject. In this sense, it seems, Jameson was perhaps hoping that 'cognitive mapping' could provide a broader but necessary initial framework for subsequent detailed investigations of specific subject manifestations and generative mechanisms.

HYPER-REALITY AND HYPER-SUBJECT

Hyper-reality is a slippery term. Baudrillard offers many different but closely related 'definitions' – or mutations (see Figure 12.1). Hyper-reality is perhaps best read not as a hypothesis, nor as a concept, but treated as a tool. In this sense, one can identify both a 'soft' and a 'hard' employment of the term in Baudrillard's work. In the first, 'hyper-reality' stands for the contemporary experience, and specifically distinguishes a particular mode of experiencing the world and making sense of it and ourselves characteristically (though not exclusively) found within contemporary mass media, high technology consumer societies. This is hyper-reality as a mode of significa-tion. In the second, 'hyper-reality' illustrates a way of knowing which explores the limits of understanding through a process or 'game' of exaggeration. Here, hyper-reality illustrates Baudrillard's epistemological strategy (in a similar way to Derrida's essay on *différance* (Derrida 1982) illustrates his deconstructive strategy). These two interpretations of hyper-reality are closely related, each grounded in Baudrillard's speculations about the evolution of modes of signification (see Baudrillard 1983a) and essays on the contemporary experience (e.g. Baudrillard 1988a, 1988b, 1990a, 1990b, 1990c, 1990d).

We have not entered the era of hyper-reality – despite Baudrillard's hyperbole: 'America is neither dream nor reality. It is hyper-reality ... It is hyper-reality because it is a utopia which has behaved from the very beginning as though it was already achieved' (Baudrillard 1988a: 28). In this sense, Eco's (1986) use of the term is misleading since it seems to locate the hyper-real within specific places and experiences in contemporary America. However, it is not easy to avoid falling into this trap of naive correlation since specific situations, such as the Disney theme-parks, do demonstrate so well many of the features of the hyper-real which Baudrillard explores (Fjellman 1992, Rojek 1993b). It is perhaps with some justification that Forgacs (in Fjellman 1992) described EuroDisney as a 'cultural Chernobyl'. However, in exploring the subject in hyper-reality it is important to not treat the hyper-real as specific places or situations, but as a potential way (or limits?) of experiencing an associated mode of signification found in contemporary spaces. In other words, the hyper-real subsists in the relationship of self and other, where that 'other' is both one's self, other selves and the world. It is not a thing, but a process. As such, Baudrillard offers many characteristics of the hyper-real, but does not prioritise one over the others. Each of his statements concerning the hyper-real – only a selection is offered in Figure 12.1 – are fragments, in part different dimensions of a single process, and in parts seemingly disparate. The value of the term lies in its provisional character and its continuous metamorphosis.

An epistemological strategy is a method (or methods) by which we come to know, or explicate an understanding, of the world and of ourselves. Within the poststructural/postmodern tradition there have been a number of attempts to develop more effective epistemological strategies. Derrida's deconstruction, Foucault's *dispositif*, Bakhtin's dialogical approach – naming just three – each has explicated fresh insights. Baudrillard's strategy seems to be one of (un)controlled exaggeration, the old gambit of logical extremes, pushing the argument to its flip-point. Reading Baudrillard, it is often difficult to get a handle upon what precisely he is committing himself to, when he is effectively pulling our leg and when he is being deadly serious. Rojek argues that 'Baudrillard's style is one of calculated exaggeration. Reading Baudrillard one feels that he does not control his material, but instead allows it to control him' (1990: 4). The contradiction aside, Rojek identifies the essential feature of this writer in his last phrase. Baudrillard's texts are full of an enthusism for ideas, where the aesthetic experience of the idea seems to almost overwhelm the critical analysis of which the analysis is a part. Baudrillard writes like a prophet, convinced of his vision and confident of the reader's acquiesence. He presents speculations, provocations, delights in paradox and admits contradictions in the attempt to break out of the strait-jacket of conventional readings of the human experience. Typically, he writes: 'you get nowhere by doing a critique of something, because this simply reinforces it' (Baudrillard 1990a: 26). More recently, he stated in an interview, 'the only game that amuses me is that of following some new situation to its very limits' (Gane 1993: 131). This is the 'hyper-', to the point where the phenomenon 'disappears' or more accurately transforms or mutates into something radically different. Hence, in his earlier analysis of the social, Baudrillard identified the death of the social in the emergence of the mass, the mass consumer society (1983b).

The hyper-real as a mode of signification transcends the traditional power relations of representation and speculatively would seem to suggest alternative forms of the subject (and the object). Baudrillard (1983a) identifies three orders of simulacra or modes of signification: counterfeit, production and simulation. Each of these would appear to have correspondence with changes in the construction of the subject, that is, to suggest alternative mappings of the subject.

Counterfeit represents the first order of simulacra. Here, the earth is an imperfect copy of the heavenly realm and the Deity is the ultimate

more real than real
'it is reality itself today that is hyper realist.'

(1983a: 147)

an excess of the real
'hyper-realism is not surrealism: it is a vision which immobilises seduction by sheer visibility. It "gives you more". This is already true of colour film and television ... it gives you so much that you have nothing more to add, which is to say give in exchange. It is totally oppressive.'

(1990a: 147)

the limit of the real
'end of the real, and end of art, by total absorption one into the other? No: hyper-realism is the limit of art, and of the real, by respective change, on the level of the simulacrum, of the privileges and pledges which are their basis.'

(1983a: 147)

as a model
'the abolition of the real not by violent destruction, but by its assumption ... [and] ... elevation to the strength of the model.'

(1983b: 84)

loss of original referent
'the generation of models of the real without origin or reality.'

(1983a: 2)

dissolving fact and fantasy
'illusion is no longer possible because the real is no longer possible.'

(1983a: 38)

transcends representation
'The hyper-real transcends representation only because it is entirely simulation.'

(1983a: 147)

pure repetition
'the hyper-real represents a much more advanced phase, in the sense that even this contradiction between the real and the imaginary is effaced. The unreal is no longer that of dream or of fantasy, of a beyond or a within, it is that of a hallucinatory resemblance of the real with itself. To exit from the crisis of representation, you have to lock the real up in pure repetition.'

(1983a: 142)

equivalent reproduction
'The very definition of the real becomes that of which it is possible to give an equivalent reproduction ... that which is always already reproduced.'

(1983a: 146)

self-referential
'whereas representation tries to absorb simulation by interpreting it as false representation, simulation absorbs the whole edifice of representation as itself a simulacrum.'

(1983a: 11)

'the real is no longer reflected, instead it feeds off itself until the point of emaciation.'

(1983a: 144)

system of signs
'the age of simulation thus begins with a liquidation of all referentials – worse; by their artificial resurrection in systems of signs, a more ductile material than meaning, in that it lends itself to all systems of equivalence, all binary opposition and all combinatory algebra. It is no longer a question of imitation, nor reduplication, nor even parody. It is rather a question of substituting signs of the real for the real itself.'

(1983a: 4–5)

with its own momentum
'images ultimately have no finality and proceed by total contiguity, infinitely multiplying themselves according to an epidemic which no one today can control.'

(1988b: 29)

beyond meaning
'Beyond meaning, there is fascination, which results from the neutralisation and implosion of meaning (by simulation).'

(1983b: 104, bracket added)

Figure 12.1 The character of hyper-reality

controlling agent. Here, the subject is 'fallen', a 'sinner', a 'mortal' and the imperfect copy of a heavenly (or saintly) being. This mortal subject is container for the 'soul', and through its devotion to the preordained order and perfection of the Deity, has access to a life beyond. Religious belief plays a key role in the 'reality' which is constructed and mapping of the subject within it. Medieval European beliefs would, in part, resonate with this kind of model. Production represents the second order of simulacra. This model would seem to describe the post-Renaissance age of science and discovery, and the rise of capitalist economic and social systems. Here the subject becomes an independent entity, who seeks 'salvation' through ingenuity and industry 'here on earth'. This subject is a rational and creative agent, individually oriented and critically willing to challenge accepted conventions through the application of reason and science. In this order, certain subjects – capitalists, entrepreneurs – gain dominance over other human subjects – workers, consumers. This order would seem to be associated with the increasing secularisation of reality and the subject, and an increasing pluralisation of constructions of the subject along socio-economic, cultural, gender and racial lines.

Simulation represents the third order of simulacra and is the correlate of the hyper-real. Here, the technologies of representation and reproduction have gained their own momentum and, according to Baudrillard, the traditional categories of producer and consumer are transcended. In typical bold style, Baudrillard argues that

> abstraction today is no longer that of the map, the double, the mirror or the concept. Simulation is no longer that of territory, a referential being or a substance. It is the generation of models of a real without origin or reality: a hyper-real. The territory no longer precedes the map, nor survives it. Henceforth, it is the map that precedes the territory – PRECESSION OF SIMULACRA – it is the map that engenders the territory.
>
> (original emphasis, Baudrillard 1983a: 2)

In this order of simulation, the reference to an original is lost in a continuous play of signs. Here, the subject is further differentiated and transformed, or more accurately fragmented. The subject becomes bombarded by media images, of new identities it should adopt, alternative maps it should trace. Society becomes the mass, and the subject transforms into the consumer.

Each of these speculations are possible mappings of the subject. In the rest of this paper we will explore the last – which might be distinguished by the term hyper-subject. In his writings of the 1970s and 1980s, Baudrillard seems to use 'hyper-' – the drive to logical limits – as a strategy. In more recent work, he has tended to prefer the notion of 'trans-', as in 'trans-political', suggesting a break beyond the confines of those limits to another region, one more translucent and unpredictable perhaps (see 1990b). Whilst the three orders of simulacra can be presented chronologically, they can be more accurately appreciated as coexisting and being to a certain extent interde-pendent. At any one point in time and in specific situations, one or other of the orders appear to be hegemonic (Rodaway 1994). The hyper-real is evident at all times in human history and mythology (see Kroker and Cook 1988), but Baudrillard seems to suggest – especially in his more fatalistic passages – that the order of simulation is hegemonic in contemporary western culture and at a new level of intensity (an exaggeration or a prophetic warning?).

PHENOMENOLOGY OF THE SUBJECT

Before we explore the hyper-subject, we need to back-track a little and consider some of the presuppositions we tend to hold about 'the subject'. It is for this reason that the term phenomenology is used here. A rigorous Husserlian phenomenology (see Johnson 1983) cannot be explicated in the confines of this paper but it is important to stop and consider some of the basic presuppositions which lie behind common usage of the term 'the subject'.

One immediate way in which we can explicate our presuppositions about the subject is to turn to the dictionary and our everyday word-associations with this often used term. In philosophical discourse, the subject correlates with the conscious self and the thinking mind. The subject is an agent, an entity situated within the world and able to make sense of that world through application of its powers of self-awareness. Linguistically, the subject is that which acts upon an object, and thus subsists within a relationship between the active agent and its world. More generally, 'subject' suggests 'under the power or control of another; owing allegiance; subordinate; dependent; liable to; prone; lying open; exposed – one under the power or control of another' (Collins English Dictionary 1977). The subject is often substituted for terms such as person, individual, agent and self, and yet none of these really correlate with it. Equally, terms such as soul, author and sign seem inadequately to describe its character, even though these might echo the orders of simulacra – counterfeit, production and simulation.

Underlying all these word-associations lie a number of closely interrelated presuppositions. First, we presuppose that the subject is a *knowing agent*, an entity aware of itself and of a wider world, which has power to act upon that knowledge and that world. Second, this subject is in part defined by that world, or in relation to that world, which is described as the object (or other subjects). The identity of the subject is grounded in specific histories (social histories, personal biographies) and located in specific places (corporeal and environmental). In other words, we presuppose that the subject is *con-textualised and developing through time and space*. Third, the notion of the subject suggests the *exercise of power* and the subjection to such power, although it seems ambiguous about the source of that power (see Foucault, for instance). A subject may exercise power over other subjects or the object world, or be the sufferer of the exercise of power by others. The subject therefore implies hierarchical relationships, both social and environmental, presupposing a world beyond the subject (the other) and other subjects (society?). Fourth, there is also a long tradition in western philosophy of regarding the subject as *rational*, and as a 'possessive individual' (McPherson 1962). Yet, even cursory reflection on day-to-day experience reveals that subjects can behave in apparently irrational ways and with much emotional involvement with themselves and the world around them.

Above all, in attempting a 'mapping of the subject' we need to be aware of these presuppositions implicated within our language and thought.

Baudrillard's reflections on contemporary experience, or hyper-reality, implicitly and explicitly bring each of these presuppositions into question. In pushing the notion of the subject to its limit, or even bursting that limit, Baudrillard explores a disappearance of the subject (as we have assumed it), a kind of hyper-subject or trans-subject which is *not* dependent on being a knowing agent, contextualised and developing in time and space, a locus of power, or rational. Baudrillard explores a quite different world, one which is most graphically described in recent reflections on such topics as terrorism and hostages, pornography and sex, obesity, contemporary politics and modern warfare (Baudrillard 1990a, 1990b, 1990c, 1990d, 1991).

BAUDRILLARD AND THE SUBJECT

The problematic of the subject implies that reality can still be represented, that things give off signs guaranteeing their existence and significance – in short, there is a reality principle. All of this is now collapsing with the dissolution of the subject. This is the well known 'crisis of representation'. But just because the system of values is coming apart ... that doesn't mean we are being left in a complete void. On the contrary, we are confronted with a more radical situation.

(Baudrillard, in Gane 1993: 100)

Therefore, Baudrillard takes an extreme position in relation to the notion of the subject. Put most succinctly, he challenges us to think of the consequences of a world in which it is the object rather than the subject which holds centre stage.

We have always lived off the splendour of the subject and the poverty of the object. It is the subject that makes history, it is the subject that totalizes the world ... The fate of the object has been claimed by no-one. It is not even intelligible as such: it is only alienated, an accused part of the subject.

(Baudrillard 1990b: 111)

Throughout much of his work, Baudrillard has shown a fascination with the object, its nature and significance (e.g. Baudrillard 1968, 1990a) and underlying his work, there is also an interest in the ideas of Nietzsche (Gane 1993). He reaches this terrain through exploring the nature of symbolic exchange in contemporary society (e.g. 1981, 1987a) and the limits of the subject as experienced in hyper-reality (e.g. 1990a, 1990b). In exploring his notion of the subject, therefore, it is important to remember this wider interest.

Deleuze and Guattari (1983) develop a concept of radical subjectivity which describes a subject in flux. Baudrillard takes this trajectory a radical stage further and abolishes the subject (Baudrillard, in Gane 1993: esp. 102–3), or more accurately focuses upon the disappearing of the subject. In his most recent writings (e.g. Baudrillard 1990a, 1990b) and more implicitly in earlier work, Baudrillard reasserts the significance of the object. He questions the presupposition of the centrality of the subject as knowing agent, contextualised and a locus of power. In explicating the notion of hyper-reality, he rejects dialectics, senses we are entering an era when meaning is meaningless (see second head-quote) and where traditional concepts such as metaphor and values are bankrupt. This arises due to 'the revenge of the crystal', that is the object (Baudrillard 1990a). Like many writers he expresses concern about the power of mass media and computer technologies which now mediate much of day-to-day life and the hegemony of the continuous circuit of mass media images and signs (eg. McLuhan 1962, Winner 1986). Echoing McLuhan, Baudrillard argues that the medium dominates the message (see also Baudrillard 1987a). 'What has changed is that the means of communication, the medium, is becoming the determinant element in exchange' (Baudrillard, in Gane 1993: 145). No longer is this just a language game (as in Derrida's deconstruction), 'the effects of technology do not occur at the level of opinions and concepts, but alter sense ratios or patterns of perception steadily and without resistance' (Baudrillard 1990a: 89). In other words, our increasing reliance upon, even dependence on, advanced technologies of representation and communication ultimately undermines our independence as self-determining, historically and geo-graphically anchored human subjects.

Where Deleuze and Guattari make desire the basis of the genesis of the subject, Baudrillard looks elsewhere for the determination of the subject and its destiny. The concept of desire offers fresh insights into the nature of the subject but it is still anchored in presuppositions about the centrality of the subject as agent and locus of power. Baudrillard decentres the subject and in exploring hyper-reality explicates an object-centred world characterised by the seduction of the subject by the technologies and images (signs) that continually replicate and circulate. This hyper-real object world gains an independence of human subjects and, through replication and simulation, gains a momentum of its own, metamorphosing and subjugating the mass, or consumers (see Figure 12.1). The analysis of desire is still rooted in the second order of simulacra, whilst Baudrillard seeks to explore the more speculative and emerging order of simulation (and hyper-reality). His analysis, therefore, offers a starting point for the mapping of the subject within contemporary spaces, such as shopping malls, heritage museums, theme-parks and virtual reality simulators. In an interactive immersion virtual reality simulator which is programmed to respond to the human participant's reactions to given sensory stimuli and continually generate specific scenarios, who/what really controls who/what (Rheingold 1991)? It is this kind of human subject which Baudrillard's analysis may help us to map.

Baudrillard shifts attention away from the realm of production (and the blind alley of consumption), to the realm of seduction. This shifts attention away from the subject as active agent, to the subject as fascinated, entranced and entangled with the technologies and images of the contemporary objective world. 'Only the subject desires, only the object seduces' (Baudrillard 1990b: 111). In a recent interview Baudrillard explained this:

the other, enchanting aspect, for me, is no longer desire, that is clear. It is seduction. Things make events all by themselves, without mediation, by a sort of instant commutation. There is no longer any metaphor, rather metamorphosis. Metamorphosis abolishes metaphor, which is the mode of language, the possibility of communicating meaning. Metamorphosis is at the radical point of the system, the point where there is no longer any law or symbolic order. It is a process without any subject, without death, beyond any desire, in which only the rules of the game of forms are involved.

(Baudrillard, in Gane 1993: 102)

The subject world of meaning, language, agency, rationality and desire disappear, and are replaced by the brute world of juxtaposition, repetition, momentum, metamorphosis and the seduction of ready-made packages of meaning (or signs). The subject is made passive, hedonistic and loses its biographical and geographical grounding. This is the hyper-subject and parallels the characteristics of the hyper-real (Figure 12.1).

However, Baudrillard does not wish to suggest the death of the subject since the very continuation of a discourse (of any kind) on the subject would contradict such a notion. The hyper-real is a moment at which the subject reaches its limits as a subject, what we have termed 'hyper-subject'. Baudrillard seems particularly interested in the disappearing of the subject, not as a momentary event, but as a *reiterating process*. The subject in hyper-reality is in continual crisis, a kind of perpetual disappearing, but it remains and in this sense is a hyper-subject. More fatalistically, Baudrillard seems to suggest that as the subject becomes powerless, opportunities for resistance are reduced to the acts of terrorism (Baudrillard 1990b) or simply a strategy of laziness (Baudrillard, in Gane 1993). Perhaps, also, categories such as passive and active become less relevant, since the subject becomes locked into an interactive relationship of consumption, and specifically image consumption. The power of the subject is reduced to a range of choices offered by – in effect – a computer model of consumer preferences which generates the options based on the characteristics of the consumer (income, age, sex, etc.) and previous patterns of purchases. The hyper-subject is thus a ghost of a subject, locked within a simulation of subjectivity. In a recent interview, Baudrillard developed this observation further and argued that 'the subject becomes an integrated circuit, a sort of convolutional system. It becomes self-referential, that is in effect its success' (Baudrillard, in Gane 1993: 173). The subject does not disappear, but is continually disappearing.

At this point of disappearing – the emergence of hyper-subject – Baudrillard identifies ecstatic forms of the subject. 'Ecstasy is the quality proper of any body that spins until all is lost, and then shines forth in its pure and empty form' (Baudrillard 1990b: 9). Here he explores subject forms at the margins of our experience, such as the hostage and the obese. Baudrillard distinguishes between the society of the spectacle (see Debord 1983) and the obscene (1990b: 67). The spectacle is the show, the obscene is the translucent. Is the hyper-subject really a trans-subject? In exploring examples of the subject in hyper-reality, we will explore less extreme forms of the hyper-subject. Nevertheless, ecstatic tendencies are evident in some of these alternative mappings of the subject. Where Baudrillard bursts the limits of the subject, here we will explore the more general stretching of those limits in more everyday experience.

THE SEDUCTION OF THE HYPER-REAL

Baudrillard argues that 'Disneyland is a perfect model of all the entangled orders of simulation' (1983a: 23). In the light of this statement, it is perhaps appropriate to evaluate the proposition of the hyper-subject in the context of the theme-park (see O'Rourke 1990, Richards 1992) and other contemporary environments, notably shopping malls and heritage areas, film and television depictions of landscapes, and the emerging realm of computer simulation and virtual reality (see Hopkins 1990, Soja 1989, Fladmark 1993, Rojek 1993a, Rodaway 1994). This is not to argue that specific places are hyper-real, but to suggest that such situations may offer clues to the re-construction of the subject and in this sense alternative mappings of the subject.

Figure 12.2 offers a collage of candidates for such an investigation, focusing upon museums and leisure parks. In the confines of a short paper only a few of these situations can be considered in detail. In each case, an attempt is made to map the way in which the subject is constructed, that is how self and other articulate. The choice of situations to explore suggests particular contexts where the subject is challenged in a way which may be similar to Baudrillard's notion of hyper-reality and the disappearing subject. Each example, therefore, offers a kind of testing ground for his speculations and reminds us that even in the simulated world of Disney theme-parks the subject is not totally reduced to the state of hyper-subject. First, we will briefly look at the heritage experience (see Uzzell 1989, Fladmark 1993) in which the subject is given the opportunity to participate in a 'living past'. Second, we will briefly consider the Disney theme-park experience in which the subject escapes to a myriad of fantasy worlds of the past, present and future.

The visitor to Beamish Museum, County Durham, is offered a re-creation of the recent past through the collection and re-construction of authentic buildings arranged on a green-field site to re-create a rather idyllic image of the coal-mining villages and industrial towns of North-east England in the early twentieth century. The shops and cottages are filled with artefacts of the period and 'actors' in period dress are on hand to add authenticity and information on day-to-day life. Visitors can wander around at their leisure, just sit in the sun and enjoy the atmosphere, or ride on an authentic tram across the site. This is a recent past and so many older visitors can 'test' the

Jorvik Centre, York, England
A reconstruction of the experience of tenth-century York. A train travels slowly back through time, taking the visitor to the workshops and houses of Viking York. The visitor both sees, hears and smells the day-to-day life of those times. Archaeological research and theme-park dark-ride technology are combined to simulate an 'authentic' experience of Viking York.

Inverary Jail, Inverary, Scotland
'The living nineteenth century prison' (leaflet, 1994). Situated in the original buildings of Inverary Jail and County Court, the museum seeks to tell the story and recreate something of the experience of prison life in the nineteenth century using exhibits and staff dressed as prisoners and warders. The visitor follows a trail first through the court-room of the 1820s with life-like dummies and court-room dialogue, then through the old prison and on to the new prison, with opportunities to meet 'prisoners' and 'warders' who further explain the prison regime, the crimes and the punishments they experienced.

The Forbidden City, De Efteling Park, Eindhoven, Netherlands
A ride which takes the visitor to the fantasy world of Sultan Pasha, the ruler of the city. The journey passes through twenty-one scenes or simulations which feature computer-controlled robots, sounds and odours.

Beamish Open-air Museum, Co. Durham, England
With the assemblage of actual buildings and artefacts from the recent past, this open-air folk museum reproduces an English town street of the 1920s and pit cottages and colliery life of around 1910. The experience is enhanced by an authentic tram, actors in the street, houses and shops, which demonstrate life at the early part of this century in a northern working-class town.

Plimouth Plantation, New England, USA
An opportunity to meet the passengers and crew of the Mayflower. Actors in authentic dress and speech describe the sixty-six-day voyage from England.

Disney World, Florida, USA
The most famous fantasy landscape that forms the 'classic' by which other theme-parks are measured. A visual spectacle assisted with Disney characters on the 'streets' and audio-animatronic figures in the displays. Even the hotels and restaurants are themed (Birnbaum 1989, Sehlinger 1992). Mainstreet is a reconstruction of American small-town life in the past.

This pretty thoroughfare represents Main Street at the turn of the century – you know, life in the 'good old days'! Instead of the traffic horns and garish commercial buildings found in so many American small-town Main Streets nowadays, Disneyland's vision offers the gentle clip-clop of horses' hooves on the pavement, the melodic ringing of the street car bells, and nostalgic old tunes like 'Bicycle Built for Two'.

(Birnbaum 1989: 67)

Wigan Pier, Wigan, England
'Where history comes alive' in the old mills and canal-side buildings. 'Experience: living history … Experience: the way we were.' The visitor is given the opportunity to 'hear, see and touch life as it was in 1900' (Wigan Pier leaflet, 1993).

Busch Park, Williamsburg, Virginia, USA
Four reconstructions of 'typical' European villages – Banbury Cross, Aquitaine, Rhinefield and San Marco. 'This year you can visit England, France, Germany and Italy – without leaving the U.S.A.' (promotional leaflet).

Ironbridge Museum, near Telford, England
Another outdoor museum recreating life in the industrial past. Historical buildings and machines are combined with staff demonstrating. Paul Laxton suggests it replaces 'real life with Hollywood; it does not represent the real landscape' (quoted in O'Rourke 1990).

Tower Hill Pageant, London, England
'An unforgettable experience which brings to life London's sights and sounds – even its smells – from Roman times to the present day. In a computer controlled time-car, you are taken through Roman, Saxon and Medieval waterfront scenes, past the Plague and the Great Fire of London right up to modern times. Tower Hill Pageant brings London to life in every sense' (press release).

Figure 12.2 Place experiences to reflect upon . . .

authenticity against their own experience. Conducting a simple survey of visitors over several days in 1993, it was interesting to observe how old and young alike identified with the fantasy of the past. Apart from the very young, from young adults to the very old accepted the authenticity of the scene, even when noting its cleanliness and order were 'not really true'. Not one respondent noted that the site was created from the assemblage of buildings and artefacts from elsewhere. Furthermore, when comparing their experience to 'real life', by far the majority 'tested' their experience against grandparents' stories, their own childhood, television images and, most interestingly of all, other museums. The survey was not comprehensive enough, both in duration and in assessment of other museums, to make a meaningful statistical analysis, but it was suggestive of particular concepts of reality, verification (or authentication) and constitution of individual and social identities (mappings of the subject).

Beamish illustrates a particular mapping of the subject in hyper-reality. It is experienced as hyper-real because it is seen as 'realistic', or 'more real than real', even though the participants seem to know and accept that it is a fantasy, a 'cleaned up' version as several respondents described it. Furthermore, visitors 'test' their experience more often than not against other media representations – television and film images of the industrial northeast of England. Interestingly, many compared their Beamish visit positively with other more traditional museums, describing the latter with terms such as 'boring' and 'dead'. When asked to explain this assessment, many reiterated (perhaps unwittingly) the text of the promotional literature for Beamish and similar attractions. They felt that the museum brought history to life, they felt that they had participated in it, and it felt 'real'. In contrast, the traditional Victorian-style museum of cabinets and display boards was seen as informative but cold, and they did not feel that they understood the history that was being presented.

A 'living' museum like Beamish, therefore, teaches history through entertainment and through 'bringing to life' aspects of the past. It is highly selective. It offers opportunities for visitors to get inside or 'participate' in the past. It does not present that past classified and explained in detailed texts, but offers that past as themes or, more specifically, a story full of human interest. The shops have names and are stocked with products whose names many remember or which are still available today in modern form. The houses have a lived-in feel and the actors are 'real' people who talk to them, the visitor. The squeaky cleanness of the place, the masses of brightly coloured 1990s visitors on a summer's day, and the signs and labels, which all contradict the illusion, are discounted by the visitor. The story-line – which each visitor can reconstruct as they wander around the buildings and street – holds together a sense of the past which is exciting and 'real'. For many visitors, who are from County Durham or the North of England, Beamish offers them an opportunity to re-establish their roots with the past, their heritage, and to take pride in that heritage.

Throughout the Beamish experience the subject is seduced by the coherence of the story, its apparent authenticity. Each individual subject confirms the experience by their own knowledge of family histories, television depictions and so forth, and partly by the numbers of other visitors who like them are also enjoying the adventure. Therefore, the subject is also seduced by the sharing with others of this experience and the entertainment it brings. Museums like Beamish are not positioned as teacher and do not position the subject as pupil, but offer the subject the opportunity to draw upon and value their pre-existing knowledge (partial or stereotypical) – often reinforcing it, as well as building upon it – and give them the chance to participate, to become 'part of it' – that is, to travel in time and relive the past. The subject is drawn into a nostalgic fantasy, one which recalls childhood memories, the images grandparents' stories gave them, or the images of television dramas and documentaries. For young and old, the living museum draws the subject out of its present context and places it in another place, a re-created past place, a fantasy. It is a fantasy, the more so because it really is not important whether the history re-created is accurate or not. Despite efforts in the designing of such museums to tell a story of the past which is reasonably accurate and informative, much of the re-creation experienced by the visitor is grounded in merging together preconceptions and already known stereotypical histories (often quite inaccurate and romanticised) of the visitor with a collection of bits and pieces of information which attract attention and interest. In this context, the subject is not encouraged to be critical, nor to be investigative. Rather, the subject is encouraged more to passively absorb the atmosphere, add the experience to their preconceptions (which it probably reinforces) and leave the museum feeling happy about the past. This is a construction of the subject as consumer. This is an impression reinforced by the survey finding that those visitors who most enjoyed their day out at Beamish had visited many other such attractions and intended to collect more such encounters with the past.

The distinction between fact and fantasy seems not to be important to visitors; as subjects they suspend their evaluation and somewhat hedonistically enter into the entertainment value of this heritage experience. When asked if they learnt anything from their visit, only the young seem to feel that they acquired new knowledge. Beamish, therefore, decontextualises the human subject and positions it in a circuit of signs, a circuit which describes a particular story of life in north-east England in the early twentieth century. For the duration of their stay, visitors are perhaps reconstituted as hyper-subjects. Yet, once back at the car and on the way home, each visitor seems to regain consciousness, in effect, and a more usual concept of the subject reasserts itself.

A number of heritage experiences ('live' museums) attempt to recreate more distant pasts and exercise a greater control over the environment experienced by the visitor. The whole experience may be contained inside a building, as in the Jorvik Experience, York, and the visitor may be transported through the experience (or scenes) by some kind of vehicle. Throughout the journey, the visitor will be bombarded by an array of sensory stimuli – tactile, auditory, visual and olfactory, each marshalled to tell a relatively simple and direct story. Detailed attention is paid to the appearance of the scenes to maximise the perceived authenticity of characters and places depicted. In experiences such as Tower Hill Pageant, London and the Jorvik Experience, York, the visitor travels on a 'time-car' at a specific speed through a series of scenes which effectively 'leap out' at the visitor with rich sensory stimuli. The subject is seduced by a total environment, a coherent story, and above all by the sheer volume and speed with which information is presented. Fjellman observes a similar character to the experience of Disney theme-park experiences. 'The visitor's attention is focused on countless co-ordinated details passing by at high velocity to the point that one's powers of discrimination can be overwhelmed' (1992: 23). The time-car heritage experience in Britain is perhaps not so rapid and slick as this, but nevertheless draws much of its effect from the steady pace of the movement through a series of scenes and dialogues. The subject as time traveller is constructed in a similar way to the cinema audience and the structure and design of these space experiences is not dissimilar to film. Even though the 'time-car' does not travel like a roller-coaster, but takes a more sedate pace, it nevertheless is a relentless pace and the visitor has no control over the journey and no time to inspect the exhibits in greater detail or go back and look again on that same visit. The experience is presented like a film, flowing from beginning to end. Many of these time-travel heritage experiences owe much in their design to the theme-park ride as well as the cinema. As in these forms, the human subject is placed somewhat passively within the experience. Rather than exploring the experience, it bombards the subject. In this context, the subject loses something of its control over the world and how it experiences it. On leaving the museum, the visitor has a sense of returning to 'boring' reality for the experience was 'more real than real'. Through the duration of the experience, the subject is transposed into another world and experiences itself as almost someone else. As one visitor to Jorvik said, 'I felt I was a viking mother back there. It was ever so hard for us, back then.' Is this the hyper-subject?

The 'Magic of Disney', in films and theme-parks, has come to represent the hyper-real at its most vibrant and widely experienced (Eco 1986, Wakefield 1990, Fjellman 1992). Disney theme-parks typify the postmodern for many and offer a widely shared example of an alternative mapping of the subject. They have also become the model for theme-parks the world over and for many other forms of contemporary environmental design, such as themed shopping malls (Hopkins 1990) and heritage museums (O'Rourke 1990). Richards (1992) states that 60 per cent of Americans, Canadians and Japanese had visited a theme-park attraction in the last year, whilst in western Europe 17 per cent of the population had. It is perhaps with some justification that Baudrillard (1988a) equates the Disney theme-park and its form of hyper-reality with the contemporary American experience, and by implication, the current mapping of the American subject. However, whilst the Disney experience would seem to bear many of the characteristic features of the hyper-real, there is much debate about how we should interpret this experience (e.g. Moore 1980, Eco 1986, Brockway 1989, Wakefield 1990, Bachman 1990, Rojek 1993a). Has 'Distory' really changed the American historical consciousness (Wallace 1985)? Does Disney change who you really are (Johnson 1981)? How might we justify the phrase 'a Disney generation'? Why does the Disney concept seem to work so well from California to Florida and on to Tokyo, and, perhaps, a little less so at EuroDisney, Paris?

In an enthusiastic study of Disney World, Fjellman (1992) provides detailed descriptions of each and every attraction and thought-provoking theoretical analyses of the nature of the Disney experience, its evolution from the concept of the Magic Kingdom to the distinctive world of the Epcot Centre, and the wider issues of the Disney Corporation and contemporary American culture. It is not intended here to repeat Fjellman's detailed coverage, but instead some of the key features he identifies will be highlighted since they reinforce much of the observations of the mapping of the subject in heritage museums. Disney World is a thoroughly designed environment, organised around a series of fantasy themes – such as Main Street and Frontierland – which are each presented with great care and attention to detail, offering a massive selection of attractions designed to inform and entertain, for sight-seeing, riding and consuming, with shops and restaurants appropriately harmonised to the area-specific theme, and 'cast members' (Disney staff in various character costumes) on hand to assist 'guests' to maximise their enjoyment of what the parks have to offer (see Birnbaum 1989). Great care is taken in ensuring that every detail of the design perfectly re-creates the intended effect, whether a fantasy character taken from fairy tales or Disney cartoon films, or a re-creation of an archetypal period and/or place. Equally, great attention is paid to the day-to-day maintenance of the parks and the flows of 'guests' about the park. On this level, Disney is a most professional show – and show is perhaps the most appropriate word. As Mickey Mouse or Goofey walk into Main Street and meet individual guests, make-believe quite literally becomes a 'reality', and the distinction between fantasy and fact seems to become irrelevant. Every guest in some way seems to respond to the experience and so the subject is seduced by the whole simulated world.

Disney theme-parks present themselves as alternative worlds. Great attention is paid to the entrance into the park. Disney World, Florida, is not America, it is another place; EuroDisney is not France, it is a Disney World. We perhaps underestimate the existential significance of this transition (see Lang 1985). These theme-parks are organised cinematically, as scenes (or themes) and with much attention to the visual dimension (Fjellman 1992). In this sense, the subject is positioned somewhat like a member of the audience at a film (see Denzin 1991). However, a park is not just a screen, even though façadism is a major feature of the presentation of the Disney experience. One literally moves 'inside' the world of Disney in a way that is never possible with film. The film is always 'out there' and so not quite 'real'. It is observed. The theme-park is a place, it has a physical reality in itself, and however fantastical its creations, once the visitor enters the park they are a participant in that fantasy. In this sense, the theme-park experience is far more effective at seducing the subject into the realm of the hyper-real. The extent of this seduction is perhaps displayed in the way in which guests behave in the Disney parks. They eagerly plan their visits to the various attractions of the park, queue patiently for specific rides or attractions, talk to cast members and cartoon characters, consume themed food, buy various Disney themed gifts to take home, and above all visit the park time and time again. The subject is not just fascinated by the spectacle, but is submersed within it and eager to be part of it, and quite literally, for some, the Disney experience becomes addictive. This hyper-subject is hedonistically fixed within the logic of Disney stories (circuit of signs) and seems not to care about its own and the wider social histories and realities. Even the fact that the Disney space is a collage of sometimes radically juxtaposed themed spaces, a 'world fair' of historical scenes, place evocations and fantasy worlds, does not seem to undermine the subject's suspension of disbelief. Those guests who most fully identify with the Disney experience, more genuinely seem to enter a hyper-real realm where the distinction between fact and fake is irrelevant; Disney is 'reality'. To the extent that this equates with the hyper-reality of Baudrillard, these guests are constituted as hyper-subjects.

Fjellman's reflections on Disney World, Florida, would seem to suggest that for many guests this degree of transformation of the subject (at least for the duration of the park experience) does occur. 'The volume and velocity of information are just two of the reasons many visitors implicate themselves in the Disney Project' (1992: 24). Main Street, for instance, is not just a 'mock-up'. Minute attention has been paid to every detail, the whole street, every building, the artefacts and merchandise, and cast members are each honed to perfection. The detail is phenomenal. The regular parades add to the over-all overload of stimulation. This same overloading of the visitor is evident throughout a Disney theme-park. There can never be a dull moment in Disney; in every direction there is something to attract the eye and the ears, every moment something of interest is happening, and even in the quiet places on the site, such as the garden areas, there is a richness of detail which can quite mesmerise the guest. It is this which undermines the subject as subject, and opens up the potential for hyper-subjectivity. The subject is totally enclosed with the Disney experience. Some guests enter a kind of ecstatic trance not dissimilar to the kind which Baudrillard equates with the order of simulation (1990b: 9).

Although less complete than Disney World, Florida, the new complex outside Paris, EuroDisney, re-creates much of this formula. It is an enclosed world, a collection of over thirty attractions organised into five main zones: Main Street, USA; Frontierland; Adventureland; Fantasyland; and Discoveryland; with six themed hotels, a themed camp-ground and an abundant supply of themed restaurants and shops. Cast members pose as 'real-life' versions of the famous Disney characters and there are frequent shows to entertain the guests. Above all, the immediate impression of EuroDisney is the usual Disney combination of efficiency and attention to detail, cleanliness and happiness. The fantasy landscapes quite literally are 'more real than real', referring often to films and other representations we have long known since childhood. EuroDisney is not a fake, in the traditional sense of the term, since it does not place itself in an inferior position to an original referent. 'European visitors will be staggered by its authenticity' (Ross, in Terry 1992: 23). Yet, on talking to visitors, it is interesting to hear many who have also visited Disney World eager to make comparisons. Yet, when we consider the 'originals' which Disney themes are based upon – fairy-tales, stories from history, places around the world – then the visitor seems less worried about any inaccuracy of representation and forgets (and possibly does not know) the original referent. Rather, the themed areas fulfil or realise those originals; more authentic, more real and therefore hyper-real.

EuroDisney may only be twenty miles from the centre of Paris, but it offers a world twenty million miles away in the imagination, although Malcolm Ross admits that 'the Disney Magic isn't going to change the weather' (Terry 1992: 22). The resort offers an escape into a fantasy world – guests generally stay a few hours or up to two-and-a-half days, and Disney hope for many repeat visits. It is perhaps this duration and repetition which helps to stimulate the emergence of hyper-subject characteristics in the visitors to a Disney park. The transformation of the subject, the suspension of disbelief and the enjoyment of the hyper-real as the 'real thing' begins at the entrance to the park. The journey from France to EuroDisney is carefully staged. The core of the park is carefully hidden from the approaching coaches, cars and trains. From the entrance gate, one travels through an arcade beneath Main Street station and, like looking down a tunnel to the light beyond, one gains glimpses of the world beyond. Once through the tunnel, the spectacle is awe-inspiring, even to the most sceptical. Already, the subject appears to be drawn into the fantasy. The cinema seems a poor imitation of these 'real life' versions of the Disney fantasy. To meet Mickey, or one of his cartoon friends, and to communicate with one of them, is to admit their personality and even 'reality'. Children and adults seem to fall readily into the fantasy. This is perhaps the fatal slippage into the hyper-subject. The opportunity to stay in themed hotels, and so prolong the fantasy, further assists the realisation of a hyper-subjectivity. By the end of the stay, all the family seem to be buying Disney souvenirs and planning their return.

Yet, once we have left EuroDisney, we seem to return to our senses. It was another place and in this sense Disney offers the subject 'a trip' rather than transforming the subject. Yet, like a drug, some individuals do seem to become quite addicted to the Disney experience (and similar kinds of attraction). The high proportion of visits by Americans, Canadians and Japanese (around 60 per cent, Richards 1992) perhaps suggests that the impact of such environments is far greater than first appears. Furthermore, when a wider range of hyper-real experiences are identified – from virtual reality simulators to heritage museums – the potential impact on the constitution of the human subject would seem more significant. Baudrillard suggests that Disneyland (California)

is presented imaginary in order to make us believe that the rest is real, when in fact all of Los Angeles and America surrounding it are no longer real, but of the order of the hyper-real and simulation. It is no longer a question of false representations of reality (ideology), but of concealing the fact that the real is no longer real.

(1983a: 25)

Leaving EuroDisney and returning to Paris, one is less ready to accept this claim. Yet perhaps one begins to look at the experience of Paris, especially the tourist image of Paris, in a new light. How 'authentic' is this Paris? Is not our visit to Paris also an escape, a journey into a dream and a fantasy? Paris is also a working city, a 'real world', but as the visitor our experience is submerged within a mythology of half-remembered history and images of romantic Paris. To this extent, hyper-subjectivity is realised beyond the confines of Disney-style worlds.

The process of theming, which is used widely in urban design and leisure today, essentially organises physical space cinematically, as both a series of spectacles *and* as coherent stories. The spectacle attracts attention, fascinates and holds the subject. The story carries the subject along from one spectacle to another, scene by scene, giving coherence to the overall experience and securing a sense of a meaning (on film, see Higson 1984). The themed environment, just like the film, must carefully balance spectacle and story. This articulation of spectacle and story establishes the sense of 'realism' in the themed environment (Rodaway 1994) and plays a crucial role in the reconstruction of the subject as hyper-subject. The subject, when placed in a fully themed environment, is given an already-interpreted landscape; it is a ready-made world. The subject does not need to explore that environment critically and seek to make sense of it. It already has a clear sense which is 'projected' at the subject. Furthermore, themed environments often take as their theme the images and stories of other places – famous images of history, exotic images from different parts of the world, widely held fantasy images – and re-create them on top of any pre-existing landscape. The pre-existing landscape is either greatly modified (as in heritage planning in urban areas) or totally obliterated (as in the building of Disney theme-parks). Fjellman sees this 'de-contextualisation' as of fundamental importance. 'By pulling meanings out of their contexts and repackaging them in bounded informational packets, de-contextualisation makes it difficult for people to maintain their coherent understanding about how things work. Meanings become all jumbled together' (Fjellman 1992: 31). In this context, the subject loses its hold on its power to independently make sense of the world, to establish meaning. Specific 'meanings' are presented ready-made, without negotiation, and the subject becomes a consumer of 'meanings'. In Baudrillard's terms, the subject becomes part of the circuit of signs, the articulation of spectacle and story.

The articulation of self and other in Disney theme-parks is strangely old-fashioned (see also Wallace 1985). Gratton (1992) describes the experience of Disney as a form of 'unskilled consumption'. It is predominantly passive and requires a minimum of creative effort on the part of the individual. Three characteristics can be noted: self and other are related in terms of nostalgic references – primarily reference to childhood images of fairy-tales, children's stories and fantasies (see also Steedman, this collection; Walkerdine, this collection). The adult joins the child in participating in the fantasy. Second, it is a social experience – Disney theme-parks are places of crowds, groups and families. The effectiveness of the suspension of disbelief is perhaps strengthened because one is not alone in this fantasy. Its 'reality' is confirmed by a shared experience and memory. Third, self and other are related non-confrontationally, largely passively and above all in a hedonistic fashion. Disney is about happiness and enjoyment. 'The idea of Disneyland is a simple one: It will be a place for people to find happiness and knowledge' (quoted in Rojek 1993b: 126).

This contrasts quite markedly with other forms of simulated realities, such as the Nintendo Gameboy or the glove-and-helmet virtual reality simulators (Davenport 1992). Like the imaginative geographical experiences of novels and television, the worlds enjoyed in these games and simulators are perhaps an even more enclosed and localised articulation of self and other. First, these simulators focus attention on the individual. An individual reacts to a stream of stimuli generated by the machine and becomes encased in their own world. Yet in a sense this is not their own world but the world of the machine. Perhaps it is a model of the hyper-subject? Second, this is more fully hyper-real in the sense of a play of signs having its own momentum. The glove-and-helmet simulator is interactive, continually exciting the tactile, auditory and visual senses and responding instantly to the reactions of the human subject, and immediately simulating the corresponding effect on its world. In this virtual reality who determines whom, machine or human subject (Rheingold 1991)? This perhaps is a more complete model of the hyper-subject than EuroDisney. Third, these virtual realities tend to be projective rather than nostalgic. Whilst the flight simulator trainer seeks 'realistically' to mimic actual flight conditions, the simulator as entertainment tool seeks to thrill and excite by taking the participant through journeys of the imagination (Leisure Management 1992).

Theme-park experiences such as those created by the Disney Corporation can only offer the possibility for hyper-real experiences, and only for some individuals will this undermine their pre-existing subjectivity more than temporarily. Disney may be powerful, but the emergence of the hyper-subject will, it seems, arise from a much broader range of changes in our day-to-day lives, in work as well as leisure. Heritage museums and theme-parks perhaps offer a testing ground to explore the potential implications of current technologies (which here might be given the generic label, virtual reality technologies) for the construction of the human subject in the twenty-first century (see Rheingold 1991).

CONCLUSIONS

It is tempting to equate this alternative mapping of the subject as the subject as consumer (see Featherstone 1990). Yet does the behaviour of the 'consumer' accurately describe the hyper-subject? As a conclusion, it is perhaps worth reflecting on this thought. Fjellman (1992) clearly sees the Disney experience as very much a consumer experience and after all, the managers of theme-parks and heritage museums have an economic motive

(Fladmark 1993, Rojek 1993a). Further, as Fjellman observes, 'the logic of the consumer has already (by the start of the twentieth century in the USA) eviscerated the individual as a reasonable unit of analysis' (1992: 15), or as Baudrillard conceptualises this: the social is replaced by the mass (1983b). The mass behaves not as a collection of individual subjects, as knowing agents with their own biographies and dreams, but as a kind of ecstatic 'object', a continuous circulation of signs, a replicating and metamorphosing body, driven by hedonistic conformity to the order of simulation – represented in part by the images of the mass media, both advertising and programmes.

In exploring the emergence of the consumer in American culture, Fjellman makes the telling statement: 'the consumer is then seen not as a person but as a bundle of characteristics, each one fair game for competition among purveyors of aspects' (1992: 41). This is an alternative mapping of the subject not as coherent individual, a knowing agent, a biographical entity, but as an object, a collection of characteristics, something which responds to specific stimuli with certain expected behaviours. This is the advertiser's view of human beings. This form of the human subject – if the word 'subject' is still appropriate – is associated with the emergence of the commodity form and mass participation in capitalist social relationships. The consumer is defined not by needs, but by wants. Needs are rooted in biological necessities and changing cultural values. We need a certain level of calories to survive, and culturally we have come to expect to have telephones and televisions. Any food with sufficient calorific value can satisfy a need, and the same telephone or television can satisfy a need whilst it continues to function and is regularly maintained. The consumer, however, is governed more by wants than needs. The consumer wants more than just food, the consumer wants a new telephone or new television long before the old one has ceased to function. The consumer is never satisfied and is always chasing after more. As Eco (1986) observes, 'more' is symptomatic of the hyper-real, where 'more' is assumed to be better. Within the capitalist system, producers have a vested interest in selling more so as to maximise their profits, and through advertising stimulate the demand for more.

Through the evolution of the capitalist system, producer and consumer ceased to be directly linked. In the precapitalist period, all but the richest members of society relied heavily upon the produce of their own labours, living off the land. With the rise of urban-industrial society and the market system, specialisation became the norm and individuals increasingly sold their labour to industrialists who deployed them on highly specific tasks within a production process. The 'peasant', who had hitherto been remark-ably self-reliant, became the 'worker', and dependent on selling labour for wages and using those wages to purchase products (foods, articles) in shops. Retailing in the modern sense of the term was born. This was not the market where farmers sold their own produce, it was a location where professional sellers sold other people's products. The commodity form has continued to evolve to such an extent that most people now rely almost totally on purchasing the means to satisfy their wants and often have quite a vague idea how the products they purchase are grown or made. In this sense, the 'consumer' is 'decontextualised' from the physical world. This experience is further reinforced with the general availability of fruits and vegetables of many different types at any time without reference to season, and the popularity of processed foods and preprepared meals.

Consumer subjects define themselves not merely with reference to their own personal experience, but increasingly with reference to the images presented by advertising and through the consumer products they identify with. The subject is defined in relation to the commodity. The commodity is desired since it is felt it will enhance the self in positive ways. Without a specific desired commodity the individual subject feels incomplete, and sometimes even another brand of the same commodity will not suffice. The consumer does not want just a car, nor just a sports car, but a Porsche. The commodity gives the individual subject an identity. The subject looks for its identity not from within itself ('know thyself') nor from its own biography or immediate social experience, but draws its identity from the ready-made packages offered as commodities to buy. Furthermore, the subject may identify with other consumers of the same product, and so establish a 'social' sense quite distinct from their day-to-day social relationships. The competitiveness of the market system has, it seems, led to an increasing search for characteristics which a commodity can be associated with and the consumer can identify with. Essentially identical products can be sold with different characteristics highlighted and so offer alternative identities for the consumer. Each characteristic is aimed at particular consumers – young teenagers with spots, aspiring company executives, nostalgic romantics and so on. Collections of commodities become associated, either deliberately or accidentally, with particular themes or stories. Certain consumers become hooked to the collective identity – or lifestyle – which such a theme offers and purchase all the relevant commodities. The subject becomes defined by commodities, and specific collections of commodities (or themes). However, the market is prone to fashions and these themes change, and with them the consumer's pattern of purchases. Each time, a new identity is adopted by the consumer. This subject as consumer lacks a cumulative identity; rather the subject is constituted externally by the vagaries of fashion.

Baudrillard writes of consumption, 'this now constitutes a fundamental mutation in the ecology of the human species' (1992: 29) and argues that we have come to live in less proximity to other people and more under 'the silent gaze of deceptive and obedient objects which continuously repeat the same discourse, that of our stupefied (*medusée*) power, of our potential influence, and of our absence from one another' (29). Baudrillard suggests that the consumer world is not a social world, nor a world of the subject, but a world of the mass, and of objects. Consumption of commodities is the point at which we can observe the disappearing of the subject, and its persistence maintains what we have called the hyper-subject. Nevertheless, Baudrillard does not see consumption as a passive mode of assimilation. It is an active mode of relations and the basis of our whole cultural system. Significantly, he notes that material goods themselves are not objects of consumption, but rather 'in order to become an object of consumption, the object must become a sign' (Baudrillard 1992: 22). In other words, it becomes external to the relation it signifies. Lifestyle themes represent subsystems of signs. The subject becomes locked in the repetition and metamorphosis of these sign patterns. 'What is consumed is not the objects, but the relation itself – signified and absent, included and excluded at the same time – it is the idea of the relation that is consumed in the series of objects which manifest it' (22). Consumer products offer a chance to feel fulfilled in some way, to share the lifestyle associations which the product signals.

'The consumer, meanwhile, becomes fractionated. Existential coherence becomes increasingly more difficult. People stop living lives and adopt lifestyles as they attempt to put together particular packages of commodities that seem pleasant' (Fjellman 1992: 41). Subject as lifestyle has a number of characteristics which echo the hyper-subject. The subject is defined not by an accumulation of experience, an individual biography and social history, but what it lacks, that is, possession of the current fashionable accessories or encounter with the latest 'experience' (each commodity forms). The subject is not a creative and knowing agent, in the sense discussed earlier, but a hedonistic, relatively passive entity which seems to be dependent – even addicted – to a continuous supply of ready-made identities inscribed in commodities, products and experiences which can be purchased in the market-place. The subject is not only in a perpetual state of lack, and dependent on the market for its satisfaction, but it is also always prob-lematised. The consumer subject cannot be normal, nor can the consumer subject be content. It always has a problem – dandruff, bad breath, an 'old car', last year's model of hi-fi system. The subject is fractionated since it is just a collection of characteristics, and in itself has no coherence. Its coherence is temporarily provided by the ready-made identities which are offered as commodities. In so far as consumption equates with the experience of the hyper-real, the 'consumer' offers a characterisation of the hyper-subject.

13

MIGRANT SELVES AND STEREOTYPES

personal context in a postmodern world

Nigel Rapport

PREFACE

The walled city of old Jerusalem and the new city burgeoning around it are two different worlds. And the difference is abrupt. Indeed, whenever I cross the threshold into the old city I have the feeling that alien manners and competing mores form so thick a mire, the history of each stone so heavy a memory, that once within the walls I might never emerge again.

There are the castes of Armenian Christian, Muslim and Jew, territorially separate but abutting against one another while affecting social blindness: where one is goodness and light, the others are shades of blackness. There are taxi-drivers in Mercedes limousines, and men selling sesame seed rolls from free-wheeling carts which they brake on the steep paths by pressing their weight on old car tyres roped to the carts which drag along behind. There are butchers displaying goats' heads, and bars advertising video games. There are veiled Bedouin women hawking grapes, and bejewelled boutique owners selling fur stoles. There are gangs of men in kaftans sipping Persian coffee and playing backgammon, and squads of soldiers in khaki patrolling with walkie-talkies and submachine-guns. In the Muslim market, unless you look ahead and walk resolutely, you are constantly accosted by, 'Yes Sir! Welcome! This beautiful, no? Only 30 Dollars'; and by the western wall of the Jewish temple, even if you do walk straight, you are intercepted by a black-garbed chassid with, 'Are you from the States by any chance? Would you like to be fixed up for a sabbath meal?'; while in the secretive residential sectors there is the introversion of scuttling women, staring children and meandering alleyways.

The old city is a maze – a maze filled as if with ooze, so that every movement is an effort, but once made supremely visible and, due to the difficulty, no doubt significant. And yet, after wading through I still feel I have discovered little: part of a mass of metropolitan foreigners who do not begin to meet the society that has sought to milk them of their pocket money. And it's strange: the sigh of relief I breathe on escaping the mêlée is mixed with a feeling of loss. For after the claustrophobia of the old city, the new is empty. Wide streets, large buildings, room to gesticulate – in fact, so much room that only if you exaggerate and shout are you socially noticed at all. It is easy to trade claustrophobia for agoraphobia: after the close, teeming life of the old city the space of the new seems to rob it of societal foundation and hub . . .

Visiting Israel shortly after returning home to Britain following three years in North America, and considering what that migration between New World and Old meant to me, the above dichotomy struck me, suddenly, as a reflection of my position. Feeling at ease neither inside the old city of Jerusalem nor outside it echoed my coming back to Britain but feeling out of place – still rather betwixt and between the Old World and the New.

But this imaging of Jerusalem is black and white, a depiction of incommensurability and contrast, of stark abuttals: of inside versus outside, the confined versus the spacious, the close versus the empty, old versus new. Surely these impressions are reductive and I simply stereotype the two Jerusalem milieux. Perhaps so.

INTRODUCTION: STEREOTYPES AND MIGRATION

According to Berger (1984: 55), migration can more and more be portrayed as 'the quintessential experience' of the age, with market forces, ideological conflicts and environmental disasters uprooting an unprecedented number of people. Furthermore, such migration contributes to what can be characterised, within late twentieth-century humanity, as an increasing global homogeneity (Hannerz 1987). That is, with the dismantling by international capital, information technology and mass media of communication, of the structural bases of local social organisations, of traditional cultural differentiations – people more and more speaking common languages, behaving in comparable ways, praying to commensurate gods, engaging with corresponding institutions, participating in contemporaneous rituals – one world system can be seen linking and levelling erstwhile bounded societies in a single social process. Here is what Riesman foresaw as 'a surfeit of inclusions' inducing 'a massification of man' (1958: 376–7).

In such an age of migration and massification, Berger elaborates, the idea of home must undergo dramatic change. From the traditional concept of one's home as the stable physical centre of one's universe – a safe place to leave and return to – a far more mobile conception is now used which is taken along whenever one decamps. For a world of travellers, of labour migrants, exiles and commuters, home comes to be found far more usually in a routine set of practices, in a repetition of habitual interactions, in a regularly used personal name, in a story carried around in one's head: in 'words, jokes, opinions, gestures, actions, even the way one wears a hat' (1984: 64). Hence, increasingly, it is static, limited, even lapidary discursive idioms which come to provide beacons of constancy and recognition through which familiar social order can hope to be secured and stable collective rhythms maintained (Sherif 1967: 157–60; cf. Rapport 1987). One comes to be at home in interactional routines, and routinely these are as fixed as one's experience is fluxional, as straitened as one's itinerary is loose.

A world in migration eventuates, in short, in a heightened emphasis on the stereotyped, on the clichéd and proverbial and sloganish, in discursive usage. Here is a 'clichégenic condition' wherein the stereotype predominates in individual speech, thought, emotion, volition and action, wherein social interaction becomes predominated by the verbal prefab (Zijderveld 1979: 4–5, 16). Here, as Fillmore sums up (1976: 9), 'an enormously large part of natural language is formulaic, automatic and rehearsed, rather than propositional, creative or freely generated'. The felt home of the migrant, of modern massified man, is with the stereotypical.

This essay is an examination of the particular way the individual migrant maintains a home for himself (herself) in stereotypical imagery; how his use of a discourse of stereotypes becomes a home-from-home, in fact, a home *per se*. My purpose is to focus on the relations between the conventions of stereotyping and the construction of the self, thereby better to appreciate the use of stereotypes as a cognitive resort. And my argument is that the individual *personalises* stereotyped discourse as he puts it into practice, interpreting its implications within the context of his own life. That far from amounting to a shared (routine, repetitive) experience, partaking in a common and conventional, even stereotypical discourse can represent a way for the individual to secure a personal preserve: to establish a context for action, to signify a mapping of the world and a journeying across it, which might be highly innovative and special to him alone (cf. Rapport 1990). And thus, from the quintessential experience of migration he may continue to write an individual story of self.

The individual migrant examined here is first and foremost myself, and the stereotyped notions my own; the ethnographic reportage in the essay concerns the stereotypical way in which I would distinguish between life in Britain and North America. I act as my own informant, then, with the academic 'I' who writes this sentence being joined below by a chauvinistic 'I' who indulges in stereotyped notions of national character; here is an attempt to recount some of my impressions as a non-anthropologist and to juxtapose these against more analytical interpretations. By acting as my own informant, I can hope to achieve two things. First, to theorise about the self from a firm grounding in the subjective: to map the contemporary subject from an initial knowledge of myself. Second, to lay the foundation for an intuitive interpretation of the selves of others. If I am conscious of the way I live through discourses, stereotyped and other, how I switch, combine and juxtapose interpretations, identities and selves in securing a home for myself in the contemporary world, then I can construe how those around me might be doing likewise.

STEREOTYPES AND SOCIAL SCIENCE

Resort to stereotypes has not traditionally been well received within the social sciences. Of the three broad analytic approaches to stereotypical discourse, the sociological, the psychodynamic and the cognitive, all concur in linking stereotypes with pejoration and perverse intergroup relations (Ashmore and Del Boca 1981). Hence, it is claimed that stereotypes are the resort of those lacking cognitive complexity, the penchant of those frightened by ambiguity and unsubtle in how they categorise stimuli; or else those emotionally aroused or distracted and unable to attend fully to cognitive classification; or else those fixated on deindividuating themselves and thereafter visiting the same on others (Wilder 1981: 235–40).

Described as 'relatively rigid and oversimplified conceptions of a group of people in which all individuals in the group are labelled with the group characteristics' (Wrightsman 1977: 672), and functioning as 'chunks of attributed traits [which cause] an individual's evaluations of others to come in packaged Gestalten' (Pettigrew 1981: 313–14), stereotypes are said to derive from hearsay and rumour rather than induction from proven fact, and from a simple projection of one's own values and expectations onto the environing world (Allport 1954). Hence, stereotypes allow simplistic and fantastic claims to be made about a group's manifold membership, claims which are all the more ambiguous and gross the higher the societal level to which the collective label is applied.

In short, stereotypes are seen to form a fortress in which groups can barricade themselves, universally convinced of the safety, rectitude and respectability of their own shared traditions while at the same time aroused into making prejudiced (but self-fulfilling) responses not towards real others but towards masquerades and phantasms (Basow 1980: 3–12; also Glassman 1975: 14–20). Either they are examples of 'autistic thinking' which bear possibly no relation to reality at all and which are insufficiently perspicacious to afford valid generalisations, so that any truth they contain is mere chance (Klineberg 1951: 505–11; also Peabody 1985), or else they are part-and-parcel of a group's 'identity rhetoric' and to be understood as a function of the social construction of group characteristics (McDonald 1993: 228–35); so that however much stereotypes might be dismissed as pathological or untrue, still they can be expected to continue to be used in processes of defining socio-cultural 'otherness'. Due to the cultural mismatch of category systems by which different social groups construct the world, then, stereotypical 'otherness' must be appreciated as an autonomous discourse which might predominate within the worlds of the representers quite independently of any 'truth-value', and irrespective of any connexion to what it purports to depict (McDonald 1993: 222–3).

Stereotypes, in sum, are conventionally treated as overgeneralised, over-determined, second-hand and partial perceptions which confuse description and evaluation, which merely reflect ideological biases, instinctual motiva-tions and cognitive limitations. And thus it is that stereotypes may be decried as sources of social pathology: root cause of misconception, and hence of intractable and oppressive sexism, racism and classism (Elfenbein 1989: viii, 158), of misdirected and xenophobic aggression, warring and pogrom (O'Donnell 1977: 23–4; also Lea 1978).

What such portrayals lack, I feel, is an appreciation of stereotyping as one of a number of types of cognitive construction which might be used in concert by an individual, one complementary to another, whereby the import and effect of stereotypes derives from juxtaposition. And so that in the reductiveness of stereotyping can be seen a means of simultaneously conceptualising great newness, multiplicity and flux. At least this is how I intend to analyse my own usage, autobiographically, below.

The general point I wish to make is that as individuals seek to locate themselves in this migrating world, so an imaging of order and collectivity in terms of social stereotypes is a means of positing a wished-for definitional stability while simultaneously being able to come to terms with the continuity of possibly radical personal change. My impressions of Britain and North America run to stereotypes because here is a shorthand way for me to order a vast array of diverse, possibly incompatible data, people, objects and events. But at the same time as I have recourse to stereotypes (and describe 'Britain' and 'America' in ways which might seem at best second-hand, at worst bigoted), the interpretation of experience which stereotyping affords me is far from constricted; stereotypes punctuate my interpreting, serve as a structure, a syntax, a cement for what I say, but they do not determine what I would say. And if I know myself to be thus conscious (however inconsistent) and cognitively autonomous (even if not monadic), then can I not suppose that others might be too? It is individuals who continue to speak and mean, not their stereotyped discourse.

'I WOULDN'T HAVE SEEN IT IF I HADN'T BELIEVED IT'

Let me elaborate by now switching voices. Here is my everyday self whose commonsensical conceptions ordinarily jostle for cognitive precedence alongside the theoretical ones of my academic self, gaining ascendancy and compiling impressions of contrasting social life in Old World and New. As I mentioned above, this imaging derives from a sojourn in North America followed by a shortlived return to Britain followed by a temporary transferral to Israel. Israel represented neutral territory (inasmuch as indigenous depictions of 'Britain' and 'America' were so unfamiliar to me as to be ignorable); hence, an environment (the ancient city of Jerusalem) which came to reflect my own liminal condition. There, 'typically' discomfited by the migration, I considered Britain and America and imagined what future residence in each might entail.

My impressions come together under five contrastive headings: closed versus open society; discriminatory versus indiscriminatory society; community versus travel; contracting versus expanding society; society of the past versus that of the future. I shall recount each in turn.

Closed versus open society

America: a society of immigrants and melting-pots. However tarnished this image has become, America still feeds on a diet of otherness. It engenders an enormous ethnic diversity, a ('beautiful') cultural plurality, and this, it is intended, will ensure equality. Maintaining pluralism has come to be seen as a foundation-stone of New World democracy, for if there is enough difference, then kotels and coalitions will be temporary and specific, and monopolies of any permanence and importance unlikely. It is what may be called the proportional representation system of forming social balance and consensus, for with a diversity of lobbies the lengthy domination of any one will be less likely. Thus, America incorporates newcomers as a means of avoiding internal consolidation (if you keep mixing concrete and adding ingredients, it barely sets into one stratum). So, 'Come and join the US and be part of us'. America invites migrating outsiders.

Of course, a pluralistic society needs some form of common denomination and this is provided by the English language: a medium of money-making and electioneering. Thus, an English of some variety includes all. It represents a hand of welcome, fair, logical and straightforward, in which spelling is rational, vocabulary simple and interactional rules minimised. Furthermore, as long as a modicum of economic and political communication can be secured, greater sharing can be forborne. Hence, the tongue becomes as much an instrument of ethnic differentiation as other institutions. Each group comes to maintain its own linguistic community, and to the extent that it allows English from the market-place into the home at all, fashions it and its information anew. Even the English of American common denomination changes with each new ethnic ingredient, then, and is seldom left to 'set' for long.

The contrast in Britain is marked. Here, the English language is not a means of ethnic inclusion but of social exclusiveness. There is a standard to be aimed for, appositely known as Received English: learnt from the past in order to be carried into the future. For Britain is a society intent on closure, on hierarchy: on cultural singularity encompassing social plurality. Hence, here is linguistic ascription; also, a stratifying of accented speech and writing which subdivides the society into geographical-cum-social groups and then ranks these on one scale.

Not that everyone in Britain is always competitively inclined. For the social plurality within British cultural singularity produces caste-like groupings whose members are bound (by camaraderie and a pride in distinctive histories of hardship, or eccentricity or speciality) to their own ways of doing things. Nevertheless, because these differences are culturally singular, they still come to be seen as part of one interdependent set of doings; the doings and the differences are all known and expected. Hence, while different social groupings may be looked to for the exhibition of distinctive traits, proclivities and skills, their behaviours can still be ranked against one distinct standard. Moreover, it is missing the point to regard the hierarchy simply economically. Money inevitably comes into it – the impoverished aristocracy must marry among the nouveaux riches *in order to maintain the* noblesse oblige *of the 'U', the* parvenus *may mix with the cultured classes and rub off the taint of lucre – but money simply oils the cogs of a cultural apparatus which has its own logic.*

There is also one British universe of information. Hence, the best-selling tabloids are those where inhabitants of Glasgow can read of a Gloucester grandma who recently wed a schoolboy, and where everyone can read of the latest shenanigans surrounding 'The Royals'. There are national radio programmes where Thelma can thank Frank for a fab time, and Norma can have her 'Worst Fright' revealed over the air. Of course, not everyone will always be interested in the same class of gossip, and so different levels cater to suit: radio channels and newspapers and television programmes geared to different hierarchical positions. But still, nation-wide gossip can be expected, and each level knows of the others' and is able to tune in or else keep a polite distance. Here is information which entertains, represents and re-forms the standards of the nation.

Similarly, each social group has its separate kind of physical space: from noble ancestral homes to middle-class detached villas and gardens, to working-class terraces where intercourse spills out of cramped rooms and slippers are worn in the street. But there is one hierarchy here too. So if upper and lower publicly meet, upper patronises and lower defers – before each resumes its distinctive gossiping and hierarchicalising later. When the English-man enters the Welsh pub it is only the Welsh Nationalist who does not switch into English so that the higher-class coloniser feels at ease; when I use the telephone at the front desk of the family business in Cardiff it is the commissionaire who retires to the side of the room, listening in but hoping I am not disturbed.

In sum, if American openness intends a society of great cultural (ethnic, normative) pluralism and diversification, then British closure constitutes a society intent on one cultural game. You accept your hierarchicalised social group and you play your interdependent role (grateful to be part of the same race which bred the Queen or Mrs Thatcher or Will Carling or Cliff Richard), or else you leave and join the colonial free-for-all – independent but adrift.

Discriminatory versus indiscriminatory society

I remember a particularly garish T-shirt on a lad in Toronto (it was in a diner, where he was eating pancakes and eggs and syrup and bacon for breakfast). Above an image of skulls and guns blazoned the words, 'Kill 'Em All. Let God Sort Them Out': divining the nature of the rest of the world, amid its innuendo and hypocrisy, was a superhuman task, so best leave discriminatory foreign policy up to the supernatural.

While British dealings with foreigners occasionally tend towards the indiscriminate ('all orientals are inscrutable'; 'all negroes look alike'), discrimination (reflecting the caste-like differentiation internal to the society) is made with regard to foreign nationalities met on a frequent basis, even if not subtle: 'stupid Irishmen', 'mean Scotsmen', 'thieving Welshmen', 'haughty French', 'mystical Indians', 'regimented Chinese', 'rough-and-ready Australians' and so on. Some of these are retained by Anglo-Saxon descendants in America (as well as a few new ones – 'dumb Poles' – added), but because the discrimination is not tied to hierarchy (WASPish social elitism belied by the proprieties of cultural pluralism), and because the onslaught is such that everyone is now an ethnic American – a Jewish American princess, an African-American, a Newfie – persistent discrimination tends to cancel itself out. There is a mass American society, and attempts to exploit African Americanism or Newfoundland Canadianism notwithstanding, the American empire of clapboarded suburbia and Chevy trucks and MacDonalds and the NFL and cable TV and Hollywood extends across the continent. That Toronto T-shirt epitomises a way of discerning which is applied not just outside the continent but inside it too. The common denomination of simplified social institutions, such as language, which is instrumental in aggregating together vastly different cultural elements and accommodating them within one nation, also entails a loss of subtlety.

That is, subtle discriminations call for the affluence of detailed knowledge, the knowledge of the insider, and a game of cultural subtlety cannot be played if everyone is a cultural outsider to almost everyone else. So subtlety is sacrificed to clarity and simplicity. America represents an open society of gross cultural difference, and it is this cultural plurality which makes it a black-and-white society, figuratively as well as literally. While fledgling attempts at affluent discrimination exist – WASP ancestry, West Point discipline, Wall Street savoir-faire, Yonge Street brazenness, Newfie homeliness – these are always under threat from the 'naivety' of new immigrants, and the writing afresh of the least discerning of constitutions in the clearest possible alphabet. Hence, inside the society as well as outside there are devotees of America and there are Evil Empires: those who would swear allegiance to a New World flag and those trapped by the involutions of the Old.

British society is very different. It maximises discriminatory evaluations and judgements, not just vis-à-vis social groups and accents, but also codes of work practice and dress, aesthetic tastes, times and types of proper food and the places you can properly be seen to dine – from cordon bleu cuisine in all the Good Food Guides, to steak-house chains, to transport 'caffs' and greasy spoons. Life is cut up into an extravagant number of units: one cultural world but a multiplicity of categories, of divisions, which are set up inside. Time is spent manoeuvring between the units of British life and the categories they embody. Time is spent discriminating: determining which food, which radio station, which gossip, which class. Hence, you shop at distinct shops not in a mall, procuring fruit, fresh-baked bread, frozen meat, fish, all from different outlets. You pay rent by the week not the month so you pay more often and exchange more units with your landlord. And you buy your coach ticket at the office for it is against union regulations for the driver to play cashier too.

In sum, Britain represents a world full of discriminatory diversions, finicky, even fetishistic, so that a lifetime is taken up wending your way through the

*social landscape. Here is a society old enough not only to know its own foibles,
but also to relish the subtleties of charades; hence, play-discriminations (such
as in satire) alongside the real, and extravagant discriminations (in pomp)
alongside the everyday. If the New World has the social skin of the newly
born, without characterising blemish, then that of the Old is a weather-worn
and wrinkled hide.*

Community versus travel

*Since it takes a lifetime negotiating the manifold divisions of the British
cultural landscape – a pursuit, moreover, that you are in no hurry to have
done with since its eccentricities are the spice of life, if not its* raison d'être,
*whether these are the risqué titbits in the gutter press or the comedies of
manners and current affairs on* Radio 4 *– British society appears slow-moving
if not stationary. It is stationary in a literal sense in that travel is hard going.
Distances, by American standards, may be small but people hardly fly, and to
drive or go by train is to negotiate the reworked settlements of centuries,
meandering through cities, suburbs and a patchwork of fields all now elided
together. There are too many people, structures and junctions to travel freely,
too many architectural and behavioural logics for time to pass fast. It is like
moving through ooze. But then unencumbered movement is not really the
point; if Britain is chock-a-block with social discriminations, then moving
piecemeal across its compacted landscape is to be occupied in travelling a
world of societies.*

*But British society is also stationary in the metaphorical sense that
neighbourhood or community is where you prefer to spend your time, not on
the road. It is unimportant that distance is difficult to manage because cities,
towns and villages, roads, avenues and crescents are for living in, not driving
through. It is not significant that strangers cannot travel fluently and find little
familiar when they arrive, because an Englishman's home is his castle, and
those removed from theirs have already thrown caution to the wind; it is not
merely coincidence that transport 'caffs' serve the greasiest spoons. Britons
lead stationary lives then, engrossed in different social networks and distinct
vantage-points on the national one. In a world of distinction you do not need
to move in order to abut against boundaries, and have refracted blatant
evidence of your own complex and colourful existence.*

*By contrast, American settlements are not for living in but for whizzing
through, on wide straight roads and a grid system which have been negotiated
even before you arrive. You can have the experience of every settlement
before reaching there, which is perfectly fine because you do not wish to be
slowed down. To keep moving and never to stop is the aim for as the TV
programme on* Health In Ontario *put it, 'The freedom to move is life itself'.
It is the American's car, not his house, which is his castle, so he has drive-in
churches as well as cafés, banks and cinemas: fast service in short pit-stops
before getting back to the track to race onwards. You may be enticed off the
highway for a while by the giant gesturings of the cities' skylines, by*

skyscrapers which shout their presence miles into the spacy hinterland and flaunt their own speed (motion upward instead of forward), but once arrived you do not tarry. Your network, after all, is on the roads, maintained by C-B radio and telephone, while MacDonald diners appear as regularly as meal times and Holiday Inns as reassuringly as bedtimes. Perhaps it is the arterial approaching and departing roads, with their strips of fast-stop services and gas stations, which represent the real heart of the society: a society of strangers in motion. Hence, roads across time-zones merely add to a sense of disconnection, and your experience of movement. That is why the American I met in Israel was so amazed when the unfamiliar STOP signs on German roads caused him almost to crash: fancy having driving practices which were not universal! Surely borders – national, state, temporal or whatever – were essentially for the gauging of velocity and transition, not their obstruction!

If the Briton has his private house and public transport, then, the American has his transport the dimensions of a house (fully accoutred, plush and private) and his accommodation rented (the prefab) or at least mobile (the trailer).

Contracting versus expanding society

If Britons are communitarians, not travellers, then one of the features of a community is members' mooted equality; within the caste-like groupings in which Britons live presides the pragmatic notion that, 'We are all equal here'. Conditions are too cramped and exterior discriminations too close and numerous for any claims of superiority or privilege to be tolerated inside, so each social group upon the British cultural hierarchy behaves as a community of equals. Moreover, since each group is stationary, most interaction takes place inside and amongst people known always. Together with the deference publicly accorded to cultural superiors, it is this which gives rise to a notable reserve and modesty of demeanour. Gestures are small because neighbours are immediate; messages are easily seen and, after long years of repetition, expectable. Furthermore, there is no room for fanciful boasts if you carry your history on your back. But there is also security in this community and dependability in its permanence and your place within it; hence, a British self-containment and smugness. There is no need to go anywhere or do anything because most is available at home: a community of peers who share your level on the hierarchy and deride others, in mutual agreement that here is best. You horde inside the community, turning inward and avoiding unnecessary contact with outsiders. In short, British closure also involves introspection: a society of taciturn expression in which the most valued social life comes from contraction, and keeping things private to one's own.

By contrast, Americans expand into all available space. The public arena is not one in which to keep silent while gathering information and showing pride or deference, but to shout your selfhood to fellow travellers, throwing out life-lines as you hurtle on. Not only is there so much distance to cover but also space to fill, so you find sprawling cities and cars, also people like titans, clothes like lighthouses and voices like loudhailers. The egalitarianism and mutual modesty of British neighbours who know they have a place for ever is replaced by the ostentation of American strangers who pass on the highway or sidewalk and know that only brashness creates an impression and only superfluity renders this more than a fleeting one; only gestures truly extrovert catch the eye and garner a response of the true speed merchant. To procure social recognition and identity you must soak up the space which divides you.

In sum, while the 'anal' Briton retreats inwards, the 'oral' American advances outwards, furnishing himself with larger cultural artefacts to fill the immense gaps and faster travel to bring the edges closer. If the ooze of British social milieux is always in danger of solidifying, then the gas of the American universe is always on the point of dissipation.

Society of the past versus that of the future

I associate a British love of satire with a corresponding appreciation of the weightiness and permanence of the society's discriminations: distinctions are too numerous and too hardy for there to be the risk of any imminent insult or collapse. But an appreciation of permanence entails more besides: a complacency that things in Britain need not change because everything has already been accomplished. As Michael Palin, one of those arch-satirists of Monty Python's Flying Circus, could explain,

> *It's as if we say: 'Oh my God, we've done it, we've been there, empire and all that business and it doesn't mean a thing.' There's this nice, comfortable we're-just-a-little-island-and-leave-us-be type of thing. We all know Shakespeare was the greatest playwright so why bother with anything else.*

> *(1986: C3)*

Britons revere the traditions of yesteryear and the future can only be worse; this is a society of the past. So, after Shakespeare there can be no more drama, after Wellington no soldiering, after Wren no architecture, after Victoria no royalty, after Gladstone no statesmanship, after Woolf no lyricism and so on. You can select other epochs and heroes but you always hanker after what has gone. Here is a reactionary frame of mind wherein life becomes a battle to keep things as they are or at least resist the spectre of change. Thus, there is no reason to rush for that is only to slide sooner into chaos. Far better to queue patiently, retaining as long as possible the dignified manners of statelier times. Meanwhile you can recall past glories and enjoy a TV diet of Restoration and Regency and Romantic drama, and films about the Battle of Britain and the Blitz, themselves recollections of Nelson at Trafalgar and Drake with the Armada and Henry V at Agincourt – occasions when British level-headedness won through and civility gained the day. In short, it is on the past that British society is focused. There is to be found both justification of present practice and also parameters of the possibilities of achievement.

Edmund Burke, one of Britain's most remembered political essayists, is chiefly respected for his scathing pronouncements on the French Revolution.

By contrast, America embraces endemic popular revolt to its bosom; Andy Warhol welcomes a world where everyone will enjoy fifteen minutes of fame before being overtaken by somebody newer. Yet this is the logical conclusion of a way of thinking in which change holds not ghoulish terror but images of better and more. The past is looked to as a baseline still, but it is not harped after. If there were glorious fights for freedom in the past – against George III, against slavery, against the Soviets – then the final battles against communism and dictatorship will be more glorious again. If the Wild West witnessed the heroism of entrepreneurial frontiersmen, then laissez-faire *will in future tame the frontier to the whole world. The past merely provides the measure to be exceeded in future, while devotion to it, believes Susan Sontag (America's cleverest woman, according to another British satirist, Jonathan Miller), represents a disastrous form of unrequited love which should be shunned. Indeed, there are awful American insecurities about the past. Roots have atrophied or been torn away: the continent is a wayward child. Besides, America will never be as old as Europe but it will always be younger, always able to lay claim to having the newest and best. For newer, after all, must equal better, just as American people and society are better than the European ones left behind. America, in fact, is the society of the young, for it is they who invent the future. Hence youthful tastes in food, music, movies, clothes and sex are lavishly catered for, while adults attempt never to appear old; for to be old is to be outside the vanguard, to be satisfied with less; for what you have and are will never be as good ('with-it', 'groovy', 'hip', 'cool', 'beat', 'wicked') as where newer people are going to get. So even if you can no longer experiment with hairstyles every few months, you must still swop house or spouse, or city or car or lifestyle or job or political party, and experience something, someone, new. When there is a continuous stream of youth, then there is no time for the present, and life becomes a continual revolution in fashion.*

In sum, if Britain is determined by a dignified shortsightedness, a disinclination to look very far ahead because everything worth seeing is within (illumined by the smouldering fires of the past), then American society is inclined to plein air *longsightedness, powered by the white heat of futurism and sci-fi. There is no time for the present, and it is pop culture which gets raised to a national culture for idolisation and export . . .*

RESORT TO STEREOTYPES REVISITED

The bad press accorded to the practice of stereotyping can sometimes miss the mark. A better appreciation of stereotyping as a cognitive resort would be to begin by viewing more sympathetically just what such a discourse might be said to offer. Initially, then, stereotypes afford both opposition and exaggeration. From the former, from comparison and contrast, notions of being are to be gained: by continuously 'playing the vis-à-vis', as Boon phrases it (1981: 231), distinctions between self and other are realised. From the latter, from exaggeration, as Douglas suggests (1966: 4), clarity and definiteness are to be derived. Thus, it is in stereotypes that feelings of identity may be seen to inhere, for through the positing of stereotypical images of difference, individuals and groups can maintain their senses of belonging (cf. Cohen 1985: 113). In stereotypical hyperbole, in short, differences between self and other can become ever more clear-cut. Furthermore, rather than scourges of the alien, stereotypes may be seen as facing primarily inward: into the group and, even more, into the individual, furnishing him with comforting shibboleths of self. Here are 'schemata', to borrow from Neisser, which anticipate and direct an exploration of the unknown and potentially chaotic in terms of the personally orderly and known (1976: 53–4). Hence an attempt in my own usage, above, to capture the spirit of a socio-cultural environment I had recently left in contrast to one to which I might return.

However, to stereotype is also to view from a distance, to situate oneself far away, or outside looking in. This certainly affords a unique view – like that, say, from a mountain-top into the valleys on either side – also a distinctive one, but it is a distinctly distant one notwithstanding, and it is the gross lineaments of a landscape which come to be depicted; inside and from close to, things are to be constructed with far more complexity. In my own stereotypic usage, then, I find that the 'spirits' of British and American society are elusive prey. They seem to belong in the rarefied atmosphere of mountain-tops, so that when the valley bottom is regained it is easy to lose sight of them and disbelieve what was seen. Were I still in North America I know that my lumping together of people and place, instance and event, would appear as grotesque parody, an absurd reduction; and having now returned to the niceties (not to mention nastinesses) of day-to-day Britain I find that such stereotypes soon get complicated and overlaid by the material of more mundane particulars. The occasions when I can think in this stereotypical fashion, and represent my situation to myself in this reductionist mode are outnumbered by occasions calling for cognition more finely-planed and proportioned.

Nevertheless, as a Briton (if not as a Welshman, if not as an anthropologist and a liberal) I can and do believe in these stereotypes still, and the identity they facilitate; stereotypes of America enable me to construe the spirit of British society, and image my security within it (cf. Mason 1990). Seen from 'inside' group boundaries, then, the stereotype can serve a further useful purpose. Nor do I refer here initially to 'societal' purpose, to the necessity of inculcating and maintaining a belief in normative stereotypes for a 'properly functioning' community (e.g. Chapman 1968), but instead to individual purpose: the individual can use stereotypes for cognitively mapping and then anchoring himself within a conventional and secure social landscape. That is, stereotypes are a stable and widespread discursive currency, and they provide significant initial points of reference. They afford bearings from which to anticipate interaction, plot social relations and initiate knowing – and from a safe distance, too – however far removed their biases become from the manifold elaborations of relationship and being which eventuate. If two axes must intersect for the identifying of a point in a plane, then in the stereotype the individual finds one ready-made cognitive axis in relation to which to gauge his position (cf. Price 1992: 58–9). However diversely conceived and unpredictably shifting the social universe, still an individual need never be at a loss as to what to perceive and how to commence to act; indeed, the simpler and more ambiguous the stereotype the more situations in which it can be used. Perhaps the stereotype does derive from typifying the world 'outside' in exaggerated opposition, with others' cultural traits being seen as alien and as butting against one's own, but 'inside' the stereotype still provides the cognitive furniture of a secure belonging.

Moreover, stereotypes are never alone. At least one contrast is entailed and very often an entire set; not just Britons and Americans then, but (haughty) French and (mystical) Indians, etc. And if the stereotype is a cognitive anchor, then a set of them anchors the individual to a social world replete with and ready for all occasions. Each stereotype alone may represent a corruption of an immense variety of occurrence, but as a set they provide an all-inclusive, varied and rich array; however fictitious and remote these public labels may be from other individuals' private attributes and penchants, together they constitute a coherent and expectable wider social world, rich in variety and common to all group members.

In sum, the stereotype represents a shorthand: a source of consistent, expectable, broad and immediate ways of knowing of the social world; a ready means by which to embody and express a multitude of complex emotions; a shortcut to generalities, to future possible regularities and uniformities. Such a foundation is very necessary not only as a bulwark against the expected randomness of future events in our entropic western conceptions (cf. Bateson 1972: 8), but also as an encouragement towards action – that vital movement which, if it were not for the bias of the stereotype and the blind spots of perception it incurs, could be replaced by the self-doubt and paralysis of trying to see a social environment from every point of view (Lippman 1947: 89–90, 114).

STEREOTYPES AND CONTEXT

To stereotype is to partake of a cultural discourse: to know of 'French' and 'Indians', of 'haughty' and 'mystical', and of how the words go together; also of how properly to enunciate the words, and combine them with actions, in conventional social exchange. To stereotype, in short, is to evince enculturation into a set of regularly used and possibly widely shared practices. And it is sociologically conventional (*après* Durkheim), not to say modish (*après* Foucault), to argue that such enculturation is all. Disinter the set of discourses which the individual has learnt to employ and there is nothing more in social exchange to describe: the subject dissolves.

The argument of this essay maintains, nevertheless, that there is a sense in which a discourse of stereotypes remains essentially exterior to the individual: something with which he juggles and enters into relationship. The individual might on occasion locate himself within the discourse, as it were, but he comes with his own agenda, his own things to mean and say, and he leaves on his own itinerary, his own route between discourses and usages, between things he would mean and say in future. In this way, the individual can be seen adopting and yet adapting stereotypes, developing his own routine relations with them, posing one against another, personalising what they purport in his own image. It is not that stereotypes contextualise their individual users, then, but that they serve as a vehicle by which the migrating individual can continue consistently to contextualise himself and others. In an original and personal fashion, he very much speaks through his stereotypes; and the context they permit him to construe, for others and for himself, is as original and as personal.

The portrayal of 'context' in this essay, then, is of something internal. Context concerns the way speakers internalise interactional routines and relate them to others in their heads; context represents something more private to the individual speaker than public and intrinsic to an interactional setting. Hence, the same cognitive context may reappear in any number of externally different situations – and the converse: the same interactional setting can be cognitively contextualised in any number of different ways. In short, context is here understood as something cognitive and as prior to the routines of interaction (cf. Rapport 1993: 80–1).

This is not to say that there is no regularity or consistency between cognitive definition and external setting, but that the decision of this relation is an internal one, not forced upon the individual by immanencies of the situation or other partners in the interaction. It is the individual who decides upon the social identity of a particular setting, its links with other habitual settings, and the appropriate behavioural responses called for from himself and others.

For this reason too, it is not a question of stereotypes' rightness or wrongness; one might decry the cognitive need to stereotype but one cannot approach the content of stereotypes with a sense of right and wrong. For beyond the superficial sense in which one partakes of the discourse correctly or incorrectly (remembers that 'the French' are 'haughty' not 'regimented'), what one takes a stereotype to mean, how one puts it to cognitive use, is a matter of personal interpretation; personal contextualisation of the discourse within a head full of other meanings: and hence often a private matter to boot. The above characterisations of 'Britain' and 'America' might seem to some ludicrous, old-fashioned, new-fangled, whatever, but they cannot be wrong. They are personal to me, they accord with my cognitions: I contextualise 'Britain' and 'America' in accord with my wider sense of myself and the world; and if I had simply kept with 'smug Britons' and 'brash Americans', the elaborations of my interpretation would have remained private to me as well.

Finally, the externality of stereotypes, as a discourse, and the internality of their contextualisation, speaks to a further feature of stereotypes: their inertia. The longevity of stereotypes, their persistency and consistency in the face of 'objective' contradictory claims, is widely acknowledged. In this discursive stability, I have suggested, is to be found security, and an assurance of one's possessing interactional currency. But more than such security, it is beneath the conventional discursive forms that life can most creatively and imaginatively be lived. For the very formulaicism permits the freest flights of fancy to be privately construed with the least of public consequences. Hence, far from the pervasiveness of stereotypes necessarily involving a retreat from subtle usage and significance, the very opposite can be the case, as the success of many a dissident pamphleteer, many a coy lover, many a witty satirist attests. Indeed, one can say that the more stereotypically the social environment is imaged the more dynamic and diverse the cognitive play which each individual user may be making of it; stereotyping and personalising are two sides of the same cognitive coin.

CONCLUSION

However much the social world comes to be represented by a single seamless, stereotypical style of speaking and doing, no such uniformity need be posited upon the interpretations made of this style, upon its contextualisation within individual lives. Furthermore, the retention of such stereotypic imagery can be seen as less remarkable, outrageous, despicable, less a threat to exchange and communication, when the stereotype is seen not primarily as an instrument prejudicially to predominate or pre-empt others, and not as evidence of merely thinking in stale, collective terms, but rather as a means for individuals rapidly to project and establish a secure personal belonging in a constantly shifting, satisfyingly complex, modern world. The stereotype is a blunt instrument which flourishes in edgy environments.

Notwithstanding the experience of migration and social flux, therefore, the individual can still cognitively construct for himself a personal place which is holistic and constant. And notwithstanding social massification and superficiality, the individual can construe a personal place with originality and depth. It is personal context which continues to infuse the stereotypical home.

Part IV

THE POLITICS OF THE SUBJECT

PART IV

INTRODUCTION

While in some accounts 'the subject' may be dead or merely an effect, it is clear that there is a politics of subjectivity that remains hard to kill off or deny. Indeed, power relations and the effects of power have been consistent motifs running throughout this book, so – in this part of the book – authors make space for a politicised subject, dealing explicitly with relations of power, whether organised around 'gender', 'race', 'class' or other kinds of 'otherness'. We have seen that subjectivity can be thought of as being contingent on the power relations within which people are placed, through regulatory practices such as the madhouse, fictions, theme-parks, the family, body adornment, care of body and soul, stereotyping and so on. However, it has also been demonstrated that people resist, change and/or appropriate dominant codes of conduct and meaning and that this happens simultaneously in and through place. This Subject-Outside-the-Law is also political. This is the subject of change.

Stephen Frosh is interested in 'otherness', in the ways in which others are othered. He turns to psychoanalysis to appraise the relation between the same and the other because it is constantly struggling 'with the realisation that this "other" lies as much within as without'. In particular, Frosh examines the otherness between masculinity and femininity as a set of boundaries which are drawn to map the individual into a gendered world. On the surface it would seem that sexual difference is about exclusion, about separating 'subject and object – of who is allowed to speak and who is spoken about, but has no voice of her own' (see also Walkerdine, this collection; Rose, this collection). In many ways, such a view accords with Lacan's (1972–3) assertion that '*The* woman can only be written with *The* crossed through' because the essence of T̶h̶e̶ woman is defined only in terms of the phallus, the master discourse. T̶h̶e̶ woman 'is not at all' (1972–3: 144) and woman can only act out her exclusion from speaking positions. Frosh finds this perspective (and its analogues) unacceptable, arrogant and incoherent. Like Rose (this collection), the search is for places from which the subject can speak in order to change things and this involves a critique of the relationship between power and knowledge. Frosh, however, argues that the idea that 'T̶h̶e̶ woman does not exist' is itself a strategy of power and should not, therefore, be taken for granted. By scrambling object and subject, Frosh shows that ~~The~~ woman and by extension ~~The~~ man are contrasts or limits which can be changed or transgressed. This leads to an analysis of (change over) time and (transgression over) space, partly through a discussion of Freud's account of his analysis of 'Dora' (Freud 1905) and partly through a consideration of psychotherapeutic practice. Frosh argues that women's time is not men's time; that women's time carves out a space, while in men's time things are done; and that this difference can be related to a sense of maternal space and father time. He concludes that desire moves in a space between the one and the other and that

> without this movement, all space collapses – there is no difference. So the conventional distinctions and oppositions – feminine space versus masculine time, holding versus doing, repetition versus narrative, hysteria versus obsessionality – get taken up into something else, some other intersection of masculinity with femininity.

Through movement in these 'third' spaces between the one and the other, it is possible both to desire, to imagine and to create new maps for the subject.

Valerie Walkerdine, like Nigel Rapport (this collection), seeks to draw some more general observations about the politics of subjectivity by placing her personal experiences in relation to wider theoretical and political debates. As she says, as academics 'we all have trajectories which implicitly or explicitly fuel our research', but Walkerdine wishes to recover the role of popular imagination in order to counter the popular academic 'sense of the relation between the masses, the working class, the popular, mass consumption, communication, media, as bad, bad, bad'. Walkerdine argues that where the masses, the working class, the popular and so on, are constituted in this way, the effect is very often to permit the academic to distance his or her self from the masses, the working class, the popular and so on and, thereafter, to see it as other (a strategy of power also described by Seidler, this collection). In the practice of knowledge-production, the other is constituted as an object of inquiry. On the contrary, Walkerdine wants to situate the production of knowledge within these social contexts – in order to politicise them and to recognise their blind spots. The point here is to break down the boundaries between the subject and the object: that is, to put the working classes, the masses, the popular and so on back into the field not just of academic vision but also of radical political practice. Walkerdine demonstrates this in a narrative that fuses an account of her autobiography with her politics with the development of her involvement in theoretical debates – and with her personal anger, with her ambivalence and the contradictions of her position – and an analysis of particular films such as *My Fair Lady* and *Annie*. She concludes that

> it has long been women who have had an injunction to speak about the personal, to tell their secrets, just as it has always been the working class who have been asked to tell of their lives, to explain their pathology, while the fact that it takes two classes to tango appears to have escaped

the notice of those who constantly ask us to tell it like it is.

It is therefore necessary to ask questions about where people are speaking from, if we are to learn to talk together.

Walkerdine's concerns with the construction of a working-class subject of radical politics and with the place from which this politics can be articulated are echoed in Rose's analysis of the construction of a feminist political subjectivity. The problem for *Gillian Rose* is this: how are women to find a place to speak from, if every speaking position has already been occupied by masculinist discourse? The ambivalence of power – described in other contexts by Sibley (this collection) and Walkerdine (this collection) – is once again at work; for Rose, 'femininity is thus at once entirely unimportant to the project of the (hu)man subject and yet also central to its fear of and desire for its Other, the non-subject, the abject'. Rose concentrates on the work of feminists such as Teresa de Lauretis who have argued for a political subjectivity which 'acknowledges both relations of power as they constitute identity, and feminist efforts to elude those relations'. This position, Rose argues, sees the subject as constituted through multiple power relations, whether they are mutually supportive or antagonistic. Thus, the feminist political subject becomes a site of differences. The consequence of this move is that commonalities, political allegiances and the formation of political alliances cannot be assumed in advance or taken for granted once identified and/or constructed. The radical move here is to render Authority contingent and contextual, leaving it open to question – because this politics intervenes at the heart of the power/knowledge nexus. For Rose, it is important that feminists use spatial metaphors to elaborate their politics and she further illuminates 'the spatialities of a female subject of feminism (in the urban West) by looking at the work of Jenny Holzer, Barbara Kruger and Cindy Sherman'. Rose concludes by arguing that it is not possible to identify a particular geography or spatiality that would mark the space of feminist subjectivity. Instead the female subject of feminism has 'to be vigilant about the consequences of different kinds of spatiality, and to keep on dreaming of a space and a subject which we cannot yet imagine'.

Michael Keith uses ethnographic empirical material from his research in London's East End to explore racialised subject positions. He argues that there is a new politics of cultural difference which is grounded in the place from which one speaks – beneath this lies a sense that categories of 'race', 'gender' and 'class' are changeable and strategic. Thus, it is around speaking positions that communities of resistance are closed. Again, the use of spatial metaphors to articulate this sense of a collective politics cannot be taken for granted: for Keith, the politics of spatiality 'cannot be measured within a straightforward metric of correspondent truth'. Keith elucidates the ways in which 'the urban' is understood through particular characterisations by scrutinising two characters: the ethnic entrepreneur and the street rebel. Each character seemingly fits a specifically, though historically-embedded, urban scene: the ethnic entrepreneur as the success story of assimilation, and the street rebel as bourgeois nightmare. Another context for Keith's case study is the Conservative government's recent urban policy initiative: the City Challenge. Weaving these stories together, Keith is able to show both how the racialised subject positions of 'ethnic entrepreneur' and 'street rebel' generate distinct forms of political action and how this political manoeuvring itself engenders quite different practices of governance which in turn make the so-called inner city.

14

TIME, SPACE AND OTHERNESS[1]

Stephen Frosh

INTRODUCTION

One of the fascinations of psychoanalysis, in the dual sense of that which makes it fascinating and that which is its own object of fascination, is its concern with mapping the 'other'. This is, of course, a widely shared interest amongst intellectual disciplines, but where psychoanalysis differs from many other colonising endeavours is in its constant struggle with the realisation that this 'other' lies as much within as without – that it is the co-ordinates of inner space which are being mapped, even when the outside world is what is apparently under scrutiny. Put in the literary terms that offer the most compelling contemporary metaphor for the analytic process, this recognition of the other within the self becomes an instance of intertextuality – reading the other, we reconstruct ourselves. 'The discovery of the unconscious was Freud's discovery, within the discourse of the other, of what was actively reading within himself' (Felman 1987: 60). This 'other reader within' – the unconscious, in its simplest interpretation – is a constantly disruptive force, skewing the maps we make of the outer world all the time. It applies to psychoanalysis as a discipline and to individual workers in the psychoanalytic tradition – beginning, as the quotation from Felman indicates, with Freud himself.

In this chapter, one particularly significant aspect of this encounter with the inner world of otherness will be explored. This is where the boundaries drawn on the map concern gender, or rather 'sexual difference' – the experience of masculinity and femininity as separated by a divide, perhaps with a wilderness between. Psychoanalysis has always taken sexual difference as a focus of interest, often explicitly but sometimes not, from the first moments of its origins in Freud's encounters with hysterical women to the recent debates instituted by feminists criticising or utilising psychoanalytic ideas. In this chapter, the reverberations of some of this recent thinking will be explored, with particular reference to the mapping metaphors of space and time.

SUBJECT AND OBJECT

In a passage in 'Women's Time' (1979: 196), Julia Kristeva makes the following comment on sexual difference. 'Sexual difference ... is translated by and translates a difference in the relationship of subjects to the symbolic contract which *is* the social contract: a difference, then, in the relationship to power, language and meaning.' At first glance, this is a relatively linear, straightforward statement in its presentation of the standard feminist insight that power and gender intersect. It seems to say that sexual difference is *equivalent to* ('is translated by and translates') a difference of position within the social/symbolic world, so generating a difference in experience. But Kristeva's sentence is also full of the codes of Lacanian-influenced psychoanalysis. The signifiers 'translation', 'symbolic', 'language', 'meaning', even 'difference', create an associative flow in which sexual difference becomes linked to a division in language, so governing a difference in the production of meaning.

What exactly is this difference? Simply put, one would have thought it to be one of exclusion, of subject and object – of who is allowed to speak and who is spoken about, but has no voice of her own. In the history of patriarchal culture, this by definition means the exclusion of the woman. Man speaks for woman, about woman, naming and placing her and not allowing her her subjecthood, denying her ownership of her own position and voice. Instead, she is idealised and denigrated, made into an object of representation and investigation. This is in part what Lacan (1972–3: 144) is commenting on in his famous slogan, 'There is no such thing as *The* woman'. Speaking more fully of this absence, Lacan claims that the essentially patriarchal organisation of culture, or properly speaking the phallic structuring of language, means that woman takes up her place as the Other, as something which stands outside the Symbolic as its negative, giving it its presence through her exclusion. Provocatively, Lacan claims that this is also the insight of feminists: that is, that all he is doing is putting into theoretical form the complaint made by women who feel themselves to be placed outside of language, to be left out of the corridors of power. In so doing, Lacan dramatises the process whereby men take over women's positions, speaking for women all the time, even when what is being said is that they are not being allowed to speak.

> There is woman only as excluded by the nature of things, which is the nature of words, and it has to be said that if there is one thing they themselves are complaining about enough at the moment it is well and truly that – only they don't know what they are saying, which is all the difference between them and me.
>
> (Lacan 1972–3: 144)

Lacan is saying that he can speak for women because they have no ability to speak for themselves, because they are excribed from language, excluded, other. Indeed, the definition of 'woman' seems here to be 'she who is outside language' ('the nature of things, which is the nature of words') – because if she was 'inside' language, owning it, she would be man. And Lacan can articulate this knowingly, because he *is* inside language, master of it; women can only act their exclusion out.

This is all a fairly clear manifestation of sexual difference as seen by Kristeva, a 'difference in the relationship to power, language and meaning'. Indeed, Lacan makes the 'difference' even more pronounced, apparently insisting on the impossibility of resistance and empowerment. Being excluded from language, women cannot know what they are saying, even when they complain about their exclusion. Consequently, when Kristeva theorises about how sexual difference 'is translated by and translates a difference in the relationship of subjects to the symbolic contract which is the social contract', she cannot know what she is saying – that her own alienation is what is at stake. Additionally, later, when she writes that 'the social contract ... is based on an essentially sacrificial relationship of separation and articulation of differences' (1979: 199), she herself is being placed in the position of having to make the sacrifice – or, rather, of becoming the sacrificial lamb. Women may talk as much as they like, but in this vision of things they cannot, by definition, ever be in command of their own words. And to those who might say that that is true for everyone, men as well as women, there is Lacan's categorical assertion to testify otherwise: 'only they [i.e. women] don't know what they are saying, which is all the difference between them and me.'

The arrogance of this Lacanian claim is quite obvious, and its theoretical incoherence will be described below. But it should be noted that it is not without historical and psychological truth: women have been excluded, consistently and violently, from the male order; and when they have not been quiet about their exclusion, they have been made to suffer. In the Lacanian movement itself, this tableau has been enacted several times, most notably in the prototypical case of Luce Irigaray. Opposing the arrogance of Lacan's self-appointed mastery of femininity, and picking up the performance element in his provocative style, she writes of him: 'The production of ejaculations of all sorts, often prematurely emitted, makes him miss, in the desire for identification with the lady, what her own pleasure might be all about. And ... his?' (1977: 91). Of course the consequence of such an impertinent question – what is Lacan's pleasure? – was to be Irigaray's exclusion from Lacan's school, her sacrifice of her position in the Lacanian sphere. It should be noted, however, that this is not quite because she does not know what she is saying, or even because she speaks and writes in ignorance of the possible effect of her words; she seems quite confident that she has more access than Lacan to the woman's point of view, and that she knows how to put it into words. Rather, her exclusion derives from a specific masculine strategy of control: too much of women's speech, when it opposes the master, is not to be allowed.

This is the first instance of the obvious tautology of Lacan's position – and that of patriarchy as a whole. Lacan claims that women are by definition excluded from language, that it is impossible in principle for a woman to knowingly express herself in the Symbolic, so becoming a full subject of that order of experience. Then, when a woman does act like that, speaking her mind and using Lacanian rhetoric to puncture his claims (*he* cannot know what *she* wants, but consistently misses her point), Lacan actively excludes her, keeps her at bay. Using a psychoanalytic analogy, it is not that the unconscious has no capacity for expressing itself, quite the contrary; it is the active act of repression that keeps it (relatively) quiet. It is necessary to keep it quiet because, if allowed to speak, the unconscious would have so much to say that it would expose as a sham the claim of consciousness to be all there is to psychic life. The analogy here, between the unconscious and femininity, is a familiar one to which I will return.

The power of Kristeva's logic and rhetoric offers another relevant example of the transparent fraudulence of Lacan's claim. Kristeva takes a position on sexual difference in which what might be called feminine and masculine principles are explored in terms of their intertwining and mutual dependence. The order of language which Lacan calls the 'Symbolic' is given great weight in Kristeva's work, but she also argues that a more 'feminine' form, the 'semiotic', is ever-present, existing in relation to the Symbolic order, with each one demanding recognition if the other is to survive. It will be argued below that this formulation is extremely important for possibilities of movement beyond the pessimistic vision of a sexual difference fixed for all time; the point here is that this woman Kristeva is free enough in her own understanding and use of language to add something significant – perhaps even revolutionary – to the scheme of things developed by Lacan. She seems powerful enough here, inscribed in language, using it with force and as her own; there is nothing to suggest that she knows not what she says.

There is another line of reasoning, this time concerning language itself, that seems to make Lacan's claims incoherent. The woman is excluded, has no voice, is other, knows not what she is saying. Yet, in being the negative of the Symbolic she makes it possible for the Symbolic to exist – in having no voice, she articulates a difference that makes speech possible. According to the Lacanian version of Saussurian linguistics, meaning arises only out of difference, as Laplanche and Leclaire (1966: 154) explain in a famously lucid treble-negative: 'If a signifier refers to a signified, it is only through the mediation of the entire system of signifiers: there is no signifier that does not refer to the absence of others and that is not defined by its position in the system.' At its simplest, this promotes a view of language in which what is articulated has its meaning defined by its boundary-conditions: it is only by means of contrast with what is not said that what is said can be known. This in itself is an important enough point, making each signifier dependent upon the whole system of signification for its production of particular signifieds. But there is also something in the tone of this quotation that reveals the source of its dynamic force. In the space of one sentence, there is one 'no', two 'nots' and an 'absence': the negative is startlingly present, keeps raising her hungry head. It is a psychoanalytic truism, beginning with Freud (1925), that the stronger the negation the more important the truth of what has been negated. So, in the context of this discussion about the negation of femininity, the more absent she is, the more excluded the woman is from language, the more speech seems to depend on her voice. Meaning is produced only by difference; Lacan (1957: 154) says he is 'forced to accept the notion' – that is, he does not particularly want to – 'of an incessant sliding of the signified under the signifier'; without the other the whole system falls apart. From all the parallel lines of allusion and denial, what seems to come across is a rather different relationship of femininity to power than that presented by Lacan. Not just historically, in terms of her reproductive function, but also continually, in terms of her impact on the whole order of things – symbolic as well as imaginary – the woman makes the masculine exist.

This vision of the woman who is no longer excluded naturally, but who is kept at bay by an active process of exclusion – by the man making of her the boundary of what can be tolerated – clarifies many of the difficulties in sexual relations which reappear in the therapeutic process itself. Women are constructed as literally *marginal to* (on the margins of) rational, masculine discourse; femininity marks the difference between what is symbolisable and what is not; consequently, between what can be controlled and what threatens to explode, engulf or subvert. Moi (1985: 167) presents this idea in an exceptionally clear fashion, worth quoting at some length.

> Women seen as the limit of the symbolic order will … share in the disconcerting properties of *all* frontiers: they will be neither inside nor outside, neither known nor unknown. It is this position that has enabled male culture sometimes to vilify women as signifying darkness and chaos, to view them as Lilith or the whore of Babylon, and sometimes to elevate them as the representatives of a higher and purer nature, to venerate them as Virgins and Mothers of God. In the first instance the borderline is seen as part of the chaotic wilderness outside, and in the second it is seen as an inherent part of the inside: the part that protects and shields the symbolic order from imaginary chaos.

What this passage is describing is the way the idea of the marginality of 'woman' is actually a method whereby she is placed as an imaginary frontier between rationality and irrationality – indeed, a frontier marking off the symbolic from what is outside it, the sane from the mad. Sometimes this produces an idealisation: as frontier, she is in direct contact with that which lies outside and can offer salvation; more often, she represents a threat to masculine purity, to balance and control. Whichever tendency dominates, 'woman' here is a product of imagination, literally the imaginary; a fantasy that holds masculinity in place. Moreover, she is a *spatial* fantasy, a kind of boundary around a safe terrain – a theme, as will be seen below, which recurs when the gender politics of therapy are considered. It is already implicit in the imagery employed by Moi in the quotation above: when in idealised mode, the woman is 'an inherent part of the inside: the part that protects and shields the symbolic order from imaginary chaos'. In other words, she offers a boundary of containment, something protective allowing what is inside to survive. This, too, is amongst the commonest of all images of the therapeutic task.

The masculine strategy is to exclude the feminine, marking the boundaries of his own unstable identity by reifying and repudiating the other. This process leaves its mark on the man: born out of a terror of disappearance in the other, it creates a division based on negativity rather than on the construction of a positive identity and engagement with difference. Benjamin (1990: 65) comments that, 'The master's denial of the other's subjectivity leaves him faced with isolation as the only alternative to being engulfed by the dehumanised other.' In this situation, desperate strategies of contact are sometimes employed: 'The underlying theme of sadism is the attempt to break through to the other. The desire to be discovered underlies its counterpart, namely masochism' (Benjamin 1990: 71–2). Lacan's 'There is no such thing as *the* Woman' is more playful than this, more knowing of its consequences and of the ripples its rhetoric will create. But it is of the same order as all masculine denials of the feminine, all appropriations of the woman's distinct and powerful voice that does not in fact want to be *spoken for*, in any sense of those words. It denies the other so as to create a boundary around what is experienced as an incoherent self; it uses the woman as contrast or limit, but always as something which will make the man feel safe.

Kristeva's own approach is more attuned to transgression of sexual difference than to repudiation of the other. Writing about the possible position of the psychoanalyst in therapy, she outlines a relatively conventional distinction between masculine and feminine, paternal and maternal, using (admittedly in brackets) a familiar name from British psychoanalysis.

> The analyst situates himself on a ridge where, on the one hand, the 'maternal' position – gratifying needs, 'holding' (Winnicott) – and on the other the 'paternal' position – the differentiation, distance and prohibition that produces both meaning and absurdity – are intermingled and severed, infinitely and without end.
>
> (Kristeva 1983: 246)

This vision of a 'ridge' where the conventions of femininity and masculinity meet, interferes with the apparent clarity of the idea of a fixed sexual difference that produces meaning and is causal in the determination of people's consciousness and of all symbolic relationships. Kristeva implies that the analyst ('he') can transcend this difference, can be both feminine and masculine, maternal and paternal; in fact, the instance of maternity offered in the quotation is a (bracketed) man, Winnicott.

One might ask what magic this is, that de-sexes the analyst? When is a man not a man? One conventional answer is that with the denial of sexual difference we are in the arena of hysteria; does this make all analysts into hysterics? What the material on woman as fantasised limit of man as well as this idea about the androgyny of the analyst suggests, is that there is no certainty when it comes to questions of masculine and feminine, subject and object, speaker and spoken of. So when it comes to thinking about the analytic encounter, the space and time in which a patient and a therapist talk to and fantasise about each other, sexual difference should start to become something else – something fluid and subversive, questioning whatever it is that the protagonists might bring.

TIME AND SPACE

In a collection of papers entitled *Between Feminism and Psychoanalysis* (Brennan 1989), a debate is initiated about the nature of analytic time. One of the contributors, Rosi Braidotti, suggests that an 'ethical aim' of psychoanalysis is to lead the analysand to an acceptance of the 'great master' time. Time here seems to be linked with death and through that to an acceptance of naturalness, of limitation: the passing of generations. According to this view, psychoanalysis is a process that aims to enable the subject to be reconciled with otherness, to acknowledge the power of what lies outside. This involves acknowledging 'the great master' – a gendered word of course, evoking Lacan – who is not the analyst, but time. Despite its Lacanian gloss, this is a message in line with Freud's own impression about the limited nature of psychoanalysis's therapeutic optimism: conversion of 'hysterical misery' into 'common unhappiness', in its most famous, though admittedly early, formulation (Breuer and Freud 1895: 393). Psychoanalysis, in this view of things, can do no more than enable the patient to understand the boundaries of her or his own existence and to comprehend and accept the decline of omnipotent fantasies; that is, it facilitates a more balanced relationship with reality.

Irigaray (1989), in the same collection as Braidotti, makes a comment which relates to this discourse on time, but which is more imbued with sexuality and also with a recognition of the ambiguity of the notion of reality – the way what appears to be necessary (that which must be accepted) might alternatively be seen as constructed and alienated, an aberration rather than a state of nature. She writes, 'Where once there was birth, growth, natural and plant cycles, is now the construction of artificial cultures with strange gods and heavenly bodies, labyrinthine laws and rules, founded in hidden mania, full of terrors, prohibitions, excessive, pathogenic, confused *jouissance*' (Irigaray 1989: 137). Irigaray contrasts what is natural with what is artificial; what is natural is cyclical, organic; what is artificial is labyrinthine, subject to the law. If we are not yet in the world of Kafka here, we are close to that of Oedipus: the natural is maternal, the prohibition paternal. Yet, the use of the word *'jouissance'*, with its connotation of an eruptive and subversive kind of sexualised pleasure, maintains an ambiguity which is characteristic of postmodernism. Irigaray refers back to a previous state – an imaginary time when there was 'birth, growth, natural and plant cycles', when human subjectivity was at one with nature. Nowadays, in contrast, there is alienation born from artificiality, but these artificial cultures, while 'confused' and 'pathogenic', do give us *'jouissance'*, the thrill of a pleasure which cannot be contained. Moving away from the natural to the artificial does not lead only to loss. Ironically, too, the end word given to characterise this apparently paternal nexus, '*jouissance*', is, in Lacan's work at least, usually applied to feminine rather than masculine sexuality. Already in this material there is subversion of what might seem an obvious polarity, between feminine nature and masculine culture; something 'feminine' lives in the latter as well.

The confusion of masculine and feminine positions has already been referred to as part of the discourse of 'hysteria'. Freud's own encounter with hysteria was in many respects the founding moment of psychoanalysis: facing his patients, their symptoms played out on their bodies, he allowed them to speak, and in so doing made a space for 'the irrational discourse of femininity in the realm of science' (Moi 1989b: 196). 'Psychoanalysis', writes Grosz (1990b: 6), 'is formed out of the "raw material" of women's desire to talk and Freud's desire to listen.' The ambiguity of all this is very obvious. Freud transgresses the boundary between masculine and feminine, subject and object: he allows the hysterical female her voice, her subjectivity, and becomes a receptive object for it. In doing so, he allows for the existence of 'another scene' – perhaps femininity, perhaps the unconscious itself. On the other hand, this man, this clever coloniser of the mind and of the discourse of the hysteric, makes of the newly speaking subject woman another object, a dark continent banged into shape, made 'subject to' the rules and regulations of another's ideas. The interplay in operation here is the common one between power and knowledge: irrationality, identified with the feminine, is allowed its voice so as to be better understood, to be subjected to rational discourse. By naming what is going on, we cease to be ravished by it.

Despite the ambiguity of sexual difference symbolised by hysteria, its most common representation is as an encounter between a female patient and a male analyst, gazing at her and occasionally listening to her voice. Freud's (1905) 'Dora' case study is the classic exemplar of this encounter. This piece was originally written with the conscious intention of offering illustrations of the dream theory in practice, but gradually developed into a source for exploration of a number of key issues in psychoanalysis: transference, countertransference and feminine sexuality (see Bernheimer and Kahane 1985). Here, I want to mention briefly just two related points that concern the rendering of hysteria in 'Dora' and the fixedness or otherwise of sexual difference. The first concerns the subversiveness of the text itself. It is apparent from Freud's own remarks and from the tone of much of the case history, that Freud's writing is driven partly by a desire to come to terms with the 'fragmented' nature of his analysis of Dora and its uncertain outcome, in which she leaves him before he is ready for her to go. To some extent this is a scholarly and therapeutic activity of working out and working through what happened, but to some extent it is a form of revenge. Freud writes Dora into history, case history and the history of ideas. From his own account (the only one available) of the analysis, he seems to own Dora, knowing her better than she knows herself, positioning her with his mastery, using her 'case' as an example of his own ideas. For everything she brings, he has an answer, and often one which she does not like. She can deny his interpretations, but he knows that he knows best – and in the end, even though she leaves, she seems to submit. Historically, it would appear that she has to: after all, it is Freud's version of things which comes down to us, and he has a great deal of authority.

Yet the actual history of this text is much more open than might be expected. Presented as an illustrative case history of a hysteric, it has become a number of different things: a source for feminist inspiration, a document of resistance and recovery, a modernist novella, a problematic of Freud and love. As Freud pins Dora down, so she slips away, eventually leaving him, for better or worse. As he writes the story, so it rewrites itself, revealing Freud's own fascinations and inhibitions – for example, famously, when he claims he always speaks openly about sex, but can only reproduce this openness in the text by lapsing into French (1905: 82). The apparent objectivity of the work slides into a complex expression of what might be called a three-way unconscious: Dora's, Freud's and the text's. And now, in contemporary debates, 'Dora' is perhaps the most famous Freudian text of all: open as it is to everyone, most writers on psychoanalysis and sexual difference have had something to say about it, wrestling to produce new meanings that throw light on femininity, masculinity and desire. Freud's mastery has long gone; this textual unconscious, this irrationality, subverts all attempts to conquer it by reason. In this way, 'Dora' demonstrates not only that texts, once written, have lives of their own – which patently they do – but also that when the text is so obviously about sexual difference, all sorts of unexpected pleasures can be found.

The second point concerns the exchange of women in a culture of men. Dora's father brings her to Freud with a request that she should be helped to see reality in the way that *he* sees it – basically, that she should agree to play her part in his affair with Frau K. Freud is both too astute and too honest to work according to another man's agenda, but he does nevertheless find himself caught up in a network of liaisons and identifications from which he cannot easily extricate himself. At its most straightforward, Freud recognises the manner in which Dora is oppressively positioned between two men (her father and Herr K.) working in some kind of collusion with one other, but he also identifies with both of them, particularly Herr K. Moreover, however much Freud creates a setting in which Dora's positive desire might be acknowledged – a space for feminine sexuality – he continues to see her as a term in a masculine economy. In the first edition of the case study, he writes (incorrectly, as he later reports) that she was later married to a putative lover who has appeared as an association to one of her dreams, as if this masculine destination is the obvious one for her. In a long footnote, however, Freud reveals the feminine determinants of Dora's desire – her love for Frau K. Freud is unable to see this clearly, to rid himself of his status as subject and understand how he might be an object in an economy of feminine desire; as Jacobus (1986: 42) puts it, he is 'blinkered when it comes to a triangle in which the man mediates between two women, as he himself mediates between Dora and the (m)other woman'. Dora, however, asserts her own positive status and resists Freud to the end, treating him like a servant, giving him two weeks' notice, then going. In this sense, she uses him and remains free of him, as he fails to fully recognise her desire. On the other hand, there is some evidence (see Bernheimer and Kahane 1985) that she stayed a hysteric all her life. Who 'won' the battle between her and Freud is therefore a debatable point; but something which is clear is that Dora was not simply possessed by him, that she was able in analysis to create some kind of space of her own.

Hysteria, then, cannot be thought of simply as the 'female malady', because it makes questions of sexual difference, identity and power problematic. Its meshing together of body and word removes the clarity of vision so prized by the masculine order. As the man looks in upon it, trying to maintain his distance, so he gets drawn in, becomes one of the characters in a tale he thinks he is writing from outside. Something about the kind of feminine subversiveness that is 'hysteria' is absorbing and tricky, and when it is faced with the equally absorbing and tricky procedures of psychoanalysis, it produces a space in which established and accepted boundaries become unstable and partially dissolved. This may be why hysteria, according to Freud as read by Kristeva, is regarded as a malady of space.

It was suggested earlier that psychoanalysis is in part about recovering an appreciation of time, submitting to its mastery – something which at its most abstract seems to be a masculine association. But if that is so, then the literature reveals there to be quite a tussle going on over the nature and meaning of time. This tussle has the following, complexly interwoven form: one, the feminine is the domain of space, the masculine of time; two, time and space flow into one another, journey without end. Kristeva (1979: 190) writes, 'when evoking the name and destiny of women, one thinks more of the *space* generating and forming the human species than of *time*, becoming or history'. Taken at face value, this is a conventional and familiar gendering of things: the feminine, because of the womb and the maternal function, is associated with space, both in the sense of a place from which something is produced, and one in which something is received, enclosed and held. The masculine dimension, however, is active: the male does things, creates history, writes books and speaks words that have an effect. However, no such simple differentiation can be sustained in Kristeva's work. We have already encountered her saying that the analyst can combine maternal and paternal, holding and differentiation, in a sense combining these stereotypic illuminations of space and time. Now Kristeva complicates any easy identity of masculinity with time and femininity with space – having versus holding – by arguing that it is not that women *are* space rather than time, but that their time is like space; it has space-like qualities.

In 'Women's Time' (1979: 191), Kristeva claims that, from amongst the 'multiple modalities of time known through the history of civilisations', female subjectivity essentially retains two forms: 'repetition' and 'eternity'. The former is seen in those aspects of femininity which have a cyclical and rhythmic quality and hence a relationship with nature which is both regular and exhilarating – both pleasurable in its stereotyped patterning and subversive in its link with 'what is experienced as extra-subjective time, cosmic time'. Women's time as 'eternity' takes a somewhat different form: 'the massive presence of a monumental temporality, without cleavage or escape' – something sombre, unscalable, unmoveable in its solidity, something always present.

Thus, women's time is not men's time, but it is time nevertheless. It is rather like Irigaray's 'natural cycles', but it seems more frightening than that: it is hysterical time in that it is akin to the movements of the body, 'cyclical or monumental'. It opposes masculine time, named by Kristeva as 'time as project, teleology, linear and prospective unfolding; time as departure, progression and arrival – in other words, the time of history' (1979: 192). This masculine time sounds like narrative time, story time; Kristeva writes that 'A psychoanalyst would call this "obsessional time", recognising in the mastery of time the true structure of the slave' (192). Ironically, Lacan was expelled from the psychoanalytic movement on the grounds of his fiddling around with the boundaries of time – varying the length of sessions according to his patients' needs, or his own. Indeed, with all the emphasis psychoanalysts place on creating a therapeutic space, it is the very fixed boundaries of a certain time limit (usually fifty minutes) that offer the strongest definition of what that space is about.

Of course, none of this is straightforward. Kristeva claims that feminine time is both cyclical and monumental. It is, therefore, akin to analytic time: time that carves out a space. There is no beginning or end, therapy is not about doing but about staying – in the Kleinian vision, for instance, it is about the capacity of the analyst to remain a constant and surviving figure in the face of onslaughts from the patient's projected destructive emotions or aspects of self. In this respect, there is a significant, gendered dimension to the Kleinian development: where Freudian analysis is characterised by an orientation towards reconstruction of the narratives of the past, and hence emphasises the cognitive and developmental dimensions of insight, Kleinian analysis is immersed in an ever-unfolding present, where the here-and-now interchange of highly charged emotions is of primary concern. Masculine time, masculine therapy, is the time of doing, of first and second and last; feminine time and therapy is that of being, of waxing, waning and waxing again, of holding.

But therapy is not simply masculine or feminine in this sense, not *either* revealing the history of a complaint *or* exploring the emotional context in which symptoms currently exist. Therapy is a space, but not just a safe haven; it is a *generative* space in which a struggle occurs for the production of new meanings – as in 'Dora', a stream of signifiers producing difference. Therapy therefore encompasses both masculine time and feminine time, but how, and in what form, and to what degree may depend on the dimensions of difference present in the room. Are we talking here of a feminine space punctuated by masculine insertions; or of a struggle for mastery held in bounds by the caress of a containing temporal structure; or perhaps of a hysterical dissolution of 'masculinity' and 'femininity' into an imaginary bisexuality? Or, perhaps, we are back with the imagery of giving birth: a space that, over time, produces something new. Here, as ever, as in Kristeva's reading of Winnicott, we are in the realm of the mother.

SPEAKING WITH THE MOTHER

Irigaray is once again relevant here. 'Woman always speaks *with* the mother; man speaks in her absence' (1989: 134). However, this 'speaking with' is not necessarily to be construed as something positive, as a mode of care and containment – the idealised version of Winnicott and even, in some readings, of Klein. For Irigaray, maternal space, in the absence of a symbolic account of the mother which is not constructed from the masculine position – that is, which is not constructed from within the Lacanian symbolic – is an untheorisable space always threatening to turn into engulfment. Her focus is on the significance of the early, pre-Oedipal mother–daughter relationship, but her argument is that this cannot be symbolised properly under patriarchy, given the way the phallic nature of the symbolic order intervenes. That is, the mediation of all symbolic activity by the lens of phallic discourse wipes out the mother as woman, contributing to misogyny and distorting the account of the pre-Oedipal Imaginary so that it becomes impossible to discover the meaning of the feminine in its own terms. As Wright (1989) points out, this is true of psychoanalysis itself, which tends to examine the mother in terms of *whose property* she is – whether the child or the father, but not in terms of her own positive content. Irigaray herself notes that this lack of a language of the feminine, lack of a true symbolic of difference, leaves the daughter and her mother always absorbed within each other.

> The mother always remains too familiar and too close. The girl has the mother, in some sense, in her skin, in the humidity of the mucous membranes, in the intimacy of her most intimate parts, in the mystery of her relation to gestation, birth, and to her sexual identity.
>
> (1989: 133)

Speaking 'with' the mother means not being able to represent the mother–daughter relationship, in terms both of intensity (which is genuine) and difference (which is potential). So the masculine difficulty in separating from the mother while remaining in contact with her, is matched by the feminine difficulty of becoming a subject at all.

According to Whitford (1989), Irigaray reads women's ontological status in this culture as *déréliction*, 'the state of abandonment, described significantly in the same terms (*un fusionnel*) as the psychoanalytic term for women's failure to individuate and differentiate themselves from their mother' (1989: 112). By contrast, men have a kind of space which is truly their own: 'the fundamental ontological category for men is *habiter* (dwelling), whether in a literal or a figurative sense: men live in "grottoes, huts, women, towns, language, concepts, theories, etc."' (112). So here it is men who have a room of their own; the woman's space turns into absorption in the mother – once again, in a way, she ceases to exist. But there is something else as well, a way in which this non-existence can make itself felt. Grosz (1990b: 174), building on Irigaray's work and emphasising the way phallocentric discourse has made the woman's voice unattainable, draws out the consequences of this non-existence in terms which suggest both its pathology and its potential for subversion.

> As the sexual other to the One sex, woman has only been able to speak or to be heard as an undertone, a murmur, a rupture within discourse; or else she finds her expression in a hysterical fury, where the body 'speaks' a discourse that cannot be verbalised by her.

This idea of an undertone, a murmur, will be returned to later in connection with Kristeva's notion of the semiotic; it suggests something alluring and threatening, something holding the possibility of overturning the dominant order of things.

For the moment, however, let us ask the question of what kind of space it is that men inhabit and offer, what might be the nature of this speaking in the absence of the mother that is also a kind of possession, a dwelling in the woman as well as in the Symbolic. Certainly, men work in culture: rationality rules, but at a substantial price, that of disowning most of what exists. One possible critical route here would be to take up the work of Cixous, to emphasise the subversiveness of the feminine disruption of all forms of continuity, to elaborate the rhythmic and the emotional, to speak the 'jouissance' of the unconscious. All these are meant to be characteristic 'feminine' modes of activity, not necessarily excluding individual men, but built on the premise of a relationship with the body from which most men are very distant. Then this stable place of the man, this *habiter*, ceases to look so promising, and the dereliction of femininity is no longer a state of absence. This also would have the effect of revalorising Irigaray's reading of feminine sexuality as plural – always at least two, not pinned down to the monolithic and imaginary masculine unity. Indeed, Irigaray's general point concerning multiplicity is important here, and presents a vision which is shared by many feminist writers. This is that the monolithic nature of the masculine sexual economy, symbolised by the penis = phallus equation, is one built on a reasoning process in which polarities are constructed (either/or) and then one pole is repudiated (male/female). Gallop (1988: 97) comments, 'Irigaray seems to be advocating a female sexuality that replaces the anxious either-or with a pleasurable both: vagina and clitoris.' Why should there be only one thing at a time, and why should that thing always be male?

As Gallop (1988) points out, the destabilising process set in motion by the prospect of non-phallic modes of sexual identity has a critical impact on the Lacanian assertion of the primacy of the phallus and of its distinction from the penis. The importance of this claimed distinction ('the phallus is not the penis') is that it cuts across genderedness to make both female and male castrated in language – neither are the source of power, the true originators of meaning. Yet, as Gallop emphasises, the phallus is not in fact some totally arbitrary symbol, it is built up on the model of the penis, and it is masculinity which is associated both with the phallus and with power. Lacan himself is unable to counter this identity persuasively, limiting his positive rendering of the nature of the phallus to the slogan that 'the phallus can only play its role as veiled' (1958a: 82). Veils usually connote the feminine, but try as Lacan might to desex the organ, the phallus/penis relationship does seem stubbornly resistant to denial. In her inimitable style, Gallop forces her way through all this:

> The Lacanians' desire clearly to separate *phallus* from *penis*, to control the meaning of the signifier *phallus*, is precisely symptomatic of their desire to have the phallus, that is, their desire to be at the centre of language, at its origin. And their inability to control the meaning of the word *phallus* is evidence of what Lacan calls symbolic castration.
>
> (1988: 126)

The Lacanians are here hoist on their own petard, revealing their desire, but unable to enforce it because of the slipperiness of desire itself. In an important sense, this failure represents a rupture in the Symbolic order; that which appears to be in control is actually at a loss when faced with the positive challenge of language, the unconscious and femininity. In this way, femininity breaks through with the beginnings of its own Imaginary; the possibility is raised of an alternative frame of reference and way of thinking about the relationship of sexual difference to the phallus – that is, to 'power, language and meaning'.

This brings this discussion back to Kristeva. Kristeva is not willing to dispense with the idea of the Symbolic; on the contrary, she argues that sexual difference will only ever become intelligible through the development of theory concerning this order of representation. However, what Kristeva does do – in a move which is as much as anything her most substantial contribution to feminist psychoanalytic thinking – is to re-examine the relationship between pre-Oedipal and Oedipal registers, or rather, between the Imaginary and the Symbolic, to produce an account of their interweaving which is more subtle than that offered by Lacan, and which leaves open many more possibilities for movement and development. Simply put, this re-examination involves the revising of the Imaginary as an order of 'semiotic' functioning, which (as with the Imaginary) is surpassed when the Symbolic comes into being, but which is also a necessary and continuing precondition of the Symbolic – and a source of opposition to, and disruption of, symbolic functioning.

Grosz (1992: 195) offers one of the clearest available descriptions of the notion of the semiotic, and its relationship with the Symbolic, as follows:

> The semiotic (mythically, retroactively) precedes and exceeds the Symbolic, overflowing and problematising its boundaries. In the broadest terms, the semiotic is the input of the undirected body, while the Symbolic is the regulated use and organised operations of that body

in social production. It is only through the Symbolic that we can have access to the semiotic; the former provides the latter with a voice and a mode of representation.

It will be apparent from this that the semiotic is more easily regarded as a register for femininity – bodily, chaotic, made marginal by the operations of the Symbolic order, unable to speak its own name. In addition, in a simple developmental sense its location as pre-Oedipal makes it, within the conventions of psychoanalysis, something concerned primarily with the mother and hence with the sphere of femininity. But there are some important qualifying points to make here. The first is that, while acknowledging the shared marginality of the semiotic and the feminine, Kristeva is categorical in maintaining the Freudian and Lacanian assertion that sexual difference is an Oedipal acquisition, making itself felt retrospectively once the castration complex has been enacted, but not actually operating in the pre-Oedipal period itself. Thus, although many writers, particularly feminists, emphasise the genuinely feminine associations of the semiotic, and although it is also true that in Kristeva's account the semiotic is founded on the primeval space of what she calls the maternal 'chora', emphasising its bodily, enveloping and female quality; nevertheless, her insistence on its pregendered nature makes it a possible site of resistance and subversion in all subjective experience – male as well as female. Moi (1985: 165) states the position simply as follows: 'Any strengthening of the semiotic, which knows no sexual difference, must therefore lead to a weakening of traditional gender divisions, and not at all to a reinforcement of traditional notions of "femininity".' *All* subjects are infiltrated both by the Symbolic and the semiotic, subjected to the law but also 'ruptured by the boundless play of semiotic drives' (Elliott 1992: 222). As long as the semiotic exists – which will be always, for the Symbolic would have no materials out of which to be constructed, were it not for the bodily drives rhythmically expressed in the semiotic register – there is a prospect for heterogeneity and disruption within every subject, male or female, and this prospect can always be glimpsed somewhere, whether in art, in language, in madness or in dreams.

In developing this argument, Kristeva retains a vision of what masculinity might contribute to the developmental and social process. Perhaps showing her indebtedness to Klein (a presence rife in *Freud and Love*, as Jacobus (1990) shows), this contribution appears in a kind of pre-Oedipal triangulation, in which the father has a position even prior to his appearance as representative of the law. But whereas Klein incorporates this potentially outside other *within* the relationship with the mother – for instance making the paternal penis part of the phantasy of the maternal breast – Kristeva strives to renounce the narcissism involved in this kind of fusion. Lacanians might say, if forced to be categorical, that the first desire is to be the desire of the mother; Kristeva's point is that even that desire is already directed outside the self–other circuit, towards a separate space. For Kristeva (1983), thinking about Freud and love, this space is that of the 'father of individual prehistory'.

We plunge, here, into transference. Absorbed in the discourse of the patient, the analyst discovers something else, some other speaking presence, some other point towards which the narrative is moving, or, rather, against which it is making itself heard. In Lacan's rendering of this experience, it is patently Oedipal: it is not the immediate relationship with the other but instead the 'big Other' which is at the heart of meaning. That is, there is some outside element, usually theorised as the cultural Law or the imperatives of language, that 'guarantees' meaning by structuring the possibility of all other relationships, whether they take the form of 'analysis or love' (Forrester 1987: 71). Thus, not just the one-person relationship, of self to itself, but also the two-person relationship, of self to other and of infant to mother, is an intrinsically narcissistic one if left unmediated by a Third Party – the big Other or structures of the Symbolic law. The target of psychoanalysis is to bring the patient into contact with this outside voice, to show how it operates on her or his own history.

In large part, it is transference that promotes this process. Transference experiences, at least when they are interpreted, move the subject away from her or his narcissism towards an insertion into the Symbolic, into the discourse of the Other. In this respect, Kristeva takes up a very radical position. One effect of the Lacanian structure is to make problematic the refusal of mastery which is supposed to be at the heart of Lacanian theory – a contradiction which is perhaps inextricably linked to the masculine orientation of the theory and with the speciousness of much of its apparent flirtation with femininity. In calling upon the big Other of the law a certain amount of fetishising of the Symbolic takes place and a phallic theory is created – the notion that, in therapy, we are positioned by reference to something outside us slides easily into an attempt to uncover and identify that something, to unveil the phallus. Kristeva insists that this is too static a position, that the transference relation must be kept dynamic and meta-phoric, understood as 'the crystallisation of fantasy' (1983: 247) but nevertheless also as something in motion, a 'movement towards the discern-ible, a journey towards the visible' (247). With everything always in motion, phallic turgidity – full knowledge and mastery in the transference, for example – gives way to that fluidity in terms of which femininity always couches its challenge, and in terms of which the semiotic order is conceived.

Psychoanalysts working in a range of different schools of thought have suggested the importance of primary narcissism for the construction of a stable and secure personality (see Frosh 1991). This is challenged by Lacanians, for whom primary narcissism is itself set up on the model of a fantasised relationship with an imaginary object – the reflection in the mirror. For Kristeva, it is more subtle still. Primary narcissism is structured as a 'parry', as a means of escaping emptiness and the horror of dissolution. 'Narcissism protects emptiness, causes it to exist, and thus, as lining of that emptiness, ensures an elementary separation' (1983: 242). The 'abject' is the term applied by Kristeva to this elementary, presubjective separation; abject because it is marked by horror, because the subject, constituted as an experiential emptiness, always tends towards falling into a space of nothing-ness. 'If the object secures the subject in a more or less stable position', comments Grosz (1992: 198), 'the abject signals the fading or disappearance, the absolute mortality and vulnerability of the subject's relation to and dependence on the object.' The fragility of this early subject/object boundary is extreme, making this first motion of the subject-to-be one that can be overwhelmed, producing a state of genuine abjection, of being devoured – of what might be described as 'borderline'. Without mediation, this is precisely what happens in the relationship between the desiring mother and the despairing infant; that is, if the mother's desire is turned towards the child, there is no possibility of a truly maternal 'space', for space disappears and boundaries dissolve.

Here, partly parenthetically, is an element in Kristeva's critique of Klein, and a moment to reintroduce the space created by the father. Taking up the Kleinian assumption that the mother incorporates all the phallic elements of the father, making them her own, Kristeva argues that while the pre-Oedipal mother is certainly phallic in the sense of being the focus of all the infant's desire, there is something outside her from the start, something towards which the mother can look, preventing her from falling into total absorption in, and identification with, her child. For Kristeva, this 'something' is termed the 'father of individual prehistory', or, as a riposte to Klein, the 'archaic inscription of the father'. Kristeva states, 'The archaic inscription of the father seems to me a way of modifying the fantasy of a phallic mother playing at the phallus game all by herself, alone and complete, in the back room of Kleinianism and post-Kleinianism' (1983: 259).

The father of individual prehistory is presented by Kristeva as an entity in the Imaginary sphere – as something therefore clearly operating differently from, but also in conjunction with, the Symbolic father of the Lacanian Law. Kristeva conjures the archaic father in the pre-Oedipal context of the mother specifically as an object towards which the mother can look, turning her desire away from the infant, and so creating a space into which that infant can grow. Kristeva writes that,

> The loving mother, different from the caring and clinging mother, is someone who has an object of desire; beyond that, she has an Other with relation to whom the child will serve as a go-between ... Without the maternal 'diversion' towards a Third Party, the bodily exchange is abjection or devouring.

> (1983: 251)

In Lacanian thought, the mother is always partly structured by the law of the father, as an entity both in and at the boundaries of the Symbolic. But here, Kristeva is working with an idea of the mother as *subject* in a different sense – as having something which is her own (a desire for another) that offers her a space which is also her own (she is not defined solely in terms of her mothering function) and that also makes it possible for the infant to resist being submerged in her closeness and immediacy. This is a theoretical move that opens onto a much more fluid field of gender possibilities than that made available in traditional psychoanalytic thinking. In relation to the father, Kristeva opposes the Lacanian implication of a fixed Symbolic order defined by the phallus and by the paternal 'Non' (fatherhood as a prohibiting function). Instead, she argues for the importance of a more heterogeneous experience of the father – both in the Imaginary and in the Symbolic, something that both creates a supportive space and that makes symbolic regulation and expression possible. This can be seen, for example, in her argument concerning the way the use of symbols (a talking cure, perhaps) promotes the 'triumph over sadness' necessary for recovery from depression. What makes this possible, she writes, is the ability of the individual to identify with something other than the lost object – a traditionally Oedipal scenario. However, this outside figure or 'Third Party' is enabling rather than prohibitive, preventing the subject from being engulfed by the lost object. As such, the Third Party is the 'imaginary father' who functions in exactly this way – creating a space for the infant's subjectivity – in earliest development. Nevertheless, writes Kristeva (1987: 23–4),

> it is imperative that this father in individual prehistory be capable of playing his part as oedipal father in symbolic Law, for it is on the basis of that harmonious blending of the two facets of fatherhood that the abstract and arbitrary signs of communication may be fortunate enough to be tied to the affective meaning of prehistorical identifications, and the dead language of the potentially depressive person can arrive at a live meaning in the bond with others.

As Grosz (1992) points out, this is an image of the imaginary father as something embodying love – contributing to the ambivalence with which many feminists regard Kristeva, as it seems to suggest that the father is superior to the mother in this respect, and more generally that subjectivity can only be secured with the assistance of a patriarchal structure protecting the child against maternal engulfment. But Kristeva's approach here can equally be seen as a plea for a reinstatement of a different notion of fathering from that defined solely by domination – and she specifically notes that it is only in the combination of the imaginary father with the Oedipal one, that symbolic activity can become truly alive.

This is what, ideally anyway, is experienced in the transference during psychoanalysis, and is hinted at in the notion from Kristeva, quoted earlier, that the analyst can hold both 'maternal' and 'paternal' positions. But there is something more at work here, both therapeutically and developmentally, that expresses the openness of the possibilities created by Kristeva's work. If, as in Lacanian theory, the 'Third Party' representing the father and the outside world operates only in the Symbolic as a structure of law and prohibition determining signification, then desire has a kind of closure around it. However much it is constituted in and by lack, it always has its answer in the phallus – something distinct to which it is directed. Lacan's strategy of calling the phallus 'veiled', so making it a slippery and ultimately undefinable entity, does not fully protect it from becoming caught up in the gendered realities of the penis and male power. For Kristeva, however, despite her own argument that the father is always a phallic figure, the imaginary father of individual prehistory is not defined so much by a process of turning *towards* something, but of the mother turning *away*, establishing both her own entitlement and that of the infant. Consequently, the nature of the original Third Party, by being denied Oedipalisation, is left literally questionable.

> The most archaic unity that we thus retrieve ... is that of the phallus desired by the mother. It is the unity of the imaginary father, a coagulation of the other and her desire. The imaginary father would thus be an indication that the mother is not complete but that she wants ... Who? What? The question has no answer other than the one that uncovers narcissistic emptiness: 'At any rate, not I.'
>
> (Kristeva 1983: 256–7)

The direction of the mother's desire away from the infant makes it possible to create a maternal space; in that way, the emptiness of the subject-to-be can become filled, or at least 'blocked up', and turned into 'a producer of signs, representations and meanings' (Kristeva 1983: 258). The presence of, and identification with, this Third Party, this 'Father of Individual Prehistory', makes all the later history of the subject possible. But this is not a matter of uncovering a real Other – a real father or masculine position, as phallocentric theory and practice might suggest. The gesture that saves the infant subject is the mother's desire being turned away from it (from her or him), the realisation that this desire is for something other than the 'I', something 'not I' – but what that is, is an unanswerable question. The restlessness of desire is what matters most; the certainty indeed that, being desire, it has no resting place. So the position of the Third Party, the 'father', is not that of some distinctly and necessarily gendered positivity, despite its association with both ideal love and Symbolic Law. It is, rather, the creation of a space outside into which the subject can look – making it possible both to be with the mother and to develop. In the process, this involves an enormous range of nurturing and symbolic activities, from the most complex manifestations of subversive semiotic irruptions into oppressive discourse, to the most ordinary question one might ask during a therapeutic transaction: 'What is it that you want?'

'THEIR SYMBOLIC EXISTS'

Reading Kristeva as a movement away from Lacan, what is most impressive is the fluidity of the symbolic processes which she uncovers. These build on the genderedness of space and time, and of psychoanalytic practice, but they also produce an idea of how this genderedness can be surpassed – of how the Imaginary dimension of experience can both disrupt and elaborate what is given by the Symbolic. There is no symbolic activity without semiotic processes; moreover, that which appears to be the defining characteristic of the Symbolic – the presence of the paternal Third Party – can be found in a different form in the most archaic moment of development, when the infant subject first breathes in the prospect of separate being. All of this makes the Symbolic appear much more amenable both to subversion and to enlargement.

Applied to the process of therapy, this work suggests that therapeutic scrutiny should produce an awareness not so much of the maternal container, but of the production of a certain kind of other-directedness necessary for survival. None of this is meant to imply that the patriarchal organisation of the contemporary cultural Symbolic is easily overcome: even a cursory look around at the extent of continuing male domination, as well as of the limitations of masculinity, is proof of the difficulty of that process. 'Their "Symbolic" exists', notes Cixous (1976: 255), in a rather different context that asserts the continuing opposition of masculine and feminine, 'it holds power – we, the sowers of disorder, know it only too well'. Recognition of the material reality of this is a necessary precondition for activity. In therapeutic terms, it also confirms that the provision of 'maternal space' – 'holding' space – is an important procedure in its own right, making it possible for external power to be bracketed out while exploration of the patient's internal possibilities is undertaken. But what is being argued here is that this morbid opposition – maternal containment, paternal power – is unnecessarily restrictive, however often it is replayed in everyday life. There is always also some other space, a desire that moves outside the mother–infant, therapist–patient orbit. Without this movement, all space collapses – there is no difference. So the conventional distinctions and oppositions – feminine space versus masculine time, holding versus doing, repetition versus narrative, hysteria versus obsessionality – get taken up into something else, some other intersection of masculinity with femininity, of the Symbolic with the semiotic 'sowers of disorder'. Narcissism is self-protective, it creates a space for growth; but it can only exist when already premised on structures given from outside. So too with this feminine and this masculine, this 'Ladies and Gentlemen' in Lacan's famous image: they protect us, these categories, against the dizzying ambiguities of the fluid unconscious. Neither category, however, is truly 'outside' or other, neither is the 'One' that creates an empathy or a substantive difference. Each is built in relation to the other; perhaps, like space and time, they are 'really' the same thing. As they intersect, and particularly as what is more formal and rational becomes interrupted by what is more disruptive and irrational, some space for imagination and change can be made.

NOTE

1   This chapter is an edited version of Chapter 6 of *Sexual Difference: Masculinity and Psychoanalysis* by Stephen Frosh (1994, London and New York: Routledge).

15

SUBJECT TO CHANGE WITHOUT NOTICE

psychology, postmodernity and the popular[1]

Valerie Walkerdine

Phil Cohen (1992) writes that 'most theories have a strong, if disavowed, autobiographical element in them' and 'most of the general theories have rested on a very slender and sometimes non-existent, empirical base'. But what if the autobiographical element is made to stand in a clearer light and the general seen to be very particular indeed, what then? Sherry Turkle (1992) argued about the mixing of personal and theoretical in the psychoanalyst, that the idea that the analyst who is revealed to have particular problems in a specific area (the most notable being Melanie Klein's relationship with her daughter, highlighted by the play about her) must be said to be biased, her vision clouded by pathology. Instead, Turkle argues that indeed, her very difficulties in this area made her especially sensitive to the issues involved. Of course, aspects of her personal biography drove her obsessions, but this had to be understood as quite opposite from the idea that this perverted and distorted an objective search for scientific truth. It was precisely what she knew, was sensitive to, had problems with, that gave her work strength in a particular direction.

We all have trajectories which implicitly or explicitly fuel our research, but mine, which covers a working-class provincial childhood, primary teacher training (a good job for a woman: 'you can always go back to it'), to teaching, psychology, a Ph.D. in developmental psychology, teaching in education departments, researching cognitive development, gender, mathematics, subjectivity, making art and films, moving to an art department and then Media Studies, must at least rank as one of the more unusual! There are some issues that I want to draw out of this trajectory to make some links between the past and the present. It was the popular imagination that fuelled my growing up and has a special place for me in attempting to explore the issues at stake for me. There has long been, and I want so much to talk about it, a sense of the relation between the masses, the working class, the popular, mass consumption, communication, media, as bad. The masses are seen as bad and the markets and media make them even worse. So we have an endless stream of psychological research aiming to examine the 'effects' and 'uses' of television and other media. There has been, in both psychology and media and cultural theory, a constant seesawing dynamic of good/bad, reactionary/progressive between the mass and the media. But if I was formed as a woman who grew up as one of the post-war mass, the grammar school-educated proletariat, the working-class girl who was shown only the pathologising romance, how come I am a professor today? And how can we examine the place of the popular in the making of the subject? The popular-low, working class, women's – how do we view its place, a place where fact and fiction blend?

This chapter is about me because I am one of its subjects. It is about the possibility of recognition that the traditional boundaries between subject and object have broken down and that this means that our own subjectivity is formed like that of those we research. The implications of that alone are vast. Just like the place that France and the French had in the awakening of my adolescent longing, the longing for the Other, to be Other, someone else, somewhere else, exotic, foreign, so French theory had its place in that other imaginary space, the space of the British Left and emerging 1970s feminism. France was the place where 1968 had happened. There may have been no revolution, but at least it seemed to the eager English imagination that they had been near to one. After all, there had been barricades, riot police, endless attempts to account for the failure of the moment. But that moment was especially important to a group of young radicals who felt trapped, as did many others at the time, by the empiricism and positivism of British psychology, by the failure to take on board the lessons of European Social Theory, the lessons which put theory on an agenda because we wanted to explain the constitution of the subject and its intertwining with the social, the refusal of the idea of a pregiven subject who is made social through a process of socialisation that left the dualism of individual and society intact. It is this work, of which our group formed a part, which refused the split between individual and society, and thus between psychology and sociology, which helped to inaugurate particular forms of media and cultural theory and which rehabilitated psychoanalysis to a place in the British academy, which has been so important to me personally and to many others.

SUBJECT TO CHANGE

James Donald (1991) recalls the politics of the time well when he remembers that in the wake of 1968 the failure of the Left to have a theory of the subject seemed very important: in understanding why the workers had not joined the students in great force, why a revolution had not happened. One of the central issues here that I'm going to return to in the course of this chapter is the place of the (never quite delivering the goods) working class.

It was felt that economistic models failed to engage adequately with the production of subjectivity and the place of this in both the production of the social and social change. In Britain, this was played out in critiques of the old New Left, especially the empiricism of notions of shared experience producing working-class identity, as in the work of E. P. Thompson for example. Thompson had argued that working-class consciousness was produced out of shared experience of oppression. Instead of the idea of shared experience constituting identity, Althusser posited an entirely different relation between ideology and consciousness. While class consciousness had always been central to Marxist thinking, Althusser argued that the traditional models of true or false consciousness linked to accounts of ideology as a process of distortion of perception, an inability to see the true state of oppression and exploitation, were too crude and that much thinking on the Left was too economically determinist. He argued that the realm of ideology was relatively autonomous from the level of the economy and indeed went as far as proposing that it was only determined by the economy in a 'last instance', an instance that, he argued, never comes. Althusser's theory was not supported by a Cartesian account of experience, cognition or perception but by the work of the French psychoanalyst, Jacques Lacan, the man who had been responsible for a structuralist and semiotic reading of Freud. It is his work that became especially important for the argument that I am going to develop. First, in using Lacanian psychoanalysis, Althusser presented the necessity of a theory of a psychologically complex subject as a central aspect of the analysis of the social world, and moreover, an account of the subject taken from psychoanalysis and not from psychology. The particular version of psychoanalysis that Althusser chose also had its own complex version of the social produced in fantasy through the motor of desire. This got over the problem of individual/social dualism and so the split between psychology and sociology (and accounts of a pregiven individual to be made social), but it helped inaugurate the serious study of ideology in its own right. This was enormously important to British film studies, which did much to promote psychoanalytic work in this country by publishing, in the journal *Screen*, a huge body of psychoanalytically inspired film theory. In addition, cultural and media analysis was much influenced by this work, the work that developed from it and the work of Italian Communist Antonio Gramsci, which interpreted ideology through his work on hegemony. In all of these ways studies of the relation between ideology and consciousness in Britain, central to social and cultural theory, took off. Feminists were arguing for the importance of psychoanalysis as well, especially after Juliet Mitchell's influential *Psychoanalysis and Feminism*. Psychoanalysis began to enter the stage of serious academic debate, but it was not any old psychoanalysis, but as Donald put it, 'a feminist rereading of a Lacanian rereading of Freud' (1991: 2). This version began to flourish in sociology, media and cultural studies, literature, anywhere it may be said, except psychology!

It is at this point then that a group of young psychologists and sociologists began to publish a journal aimed at psychologists and social theorists, called *Ideology and Consciousness*. Later, more enamoured of Foucault than Althusser, we changed the name to its initials, *I and C*. In Britain this formed part of the development of new kinds of work in the realm of the psychological, work on subjects and subjectivity, inspired by structuralism, poststructuralism and psychoanalysis. We saw it as a profound critique of the positivism and empiricism of Anglo-American psychology and it has taken until the 1990s for a greater body of work to begin to be established in both countries in traditions that have become variously known as poststructuralist, deconstructive, discursive and postmodern psychologies. One of the first books in this wave was *Changing the Subject* in 1984.[2] This work attempted to go beyond Althusserian structuralism in producing a theory of the subject utilising the work of Michel Foucault, for whom the split between science and ideology, retained by Althusser, was gone beyond to examine the place of the human and social sciences (in my case, especially psychology). For Foucault, psychological stories were not false or pseudo-science but fictions which function in truth, scientific stories whose truth-value had a central place in the government and regulation of the modern and postmodern order. To cut a very long story short, Foucault argued that the individual was not the same thing as the person or the subject of psychology, but a historically specific form of the subject. In this account, the individual was understood as *produced* by means of a set of apparatuses of social regulation, management of populations in which scientific knowledges about what the social and subjective was were fictions which were central in the production of a management which sought to regulate through self-regulation. By producing discourses and practices in education, law, medicine, social work, etc., it was argued that the subject which was so painstakingly described, was actually created.

This work had a number of consequences which, put briefly, were about this subject, not a pregiven entity or essence, but produced in the fictions and fantasies which make up the social world. For Foucault, 'the child', 'the woman', were fictions created in the practices of regulation. And our 'fit' with those stories, how we came to embody them, was what was at stake here. It had a profound impact upon the social sciences and upon literary theory, though its impact on psychology was much slower to take hold. It is an irony that this kind of work on the subject and subjectivity was far more widely known and respected outside psychology. However, it is also the case that, to this day, media cultural and social theorists are apt all too easily to dismiss psychological work in their field as reductionist (viz. Morley 1993), thereby ignoring the psychological altogether, while maintaining an apparent ignorance of the growing body of critical psychological work.

I do not have time here to discuss the particular ways I used this body of work to intervene in debates about developmental psychology (especially cognitive development and language) nor in debates about gender, rationality and education. But suffice it to point out that I argued that subjects are produced within discursive practices and that this is strongly critical of accounts of universalist models of development, for example, or work which understands 'the child' or 'femininity' outside specifically historically and culturally located practices in which subject positions are produced through the interchange of signs.

SUBJECT TO CHANGE: WHO NOTICES?

One of the major issues with this approach was how to understand the relationship of the subject in Foucault's terms to how subjectivity is lived, both in relation to historicity and materiality, and how a non-unitary, non-rationalist subjectivity is held together. This subjectivity cannot be reduced to Thompson's 'lived experience', but the problems of how to understand it were forcibly brought back to me when I came to the recognition that there was something that both I and the theory and politics needed to come back to: the popular and what the French call the popular classes. While I started to work on popular culture, returning to issues that had been important to me as a child (children's literature, girls' comics, for example) something else was happening. Class came back to me with a jolt, not as a theoretical issue, even and perhaps as we shall see especially through the Left, but as a profoundly personal one. Psychoanalysis was bringing back my childhood, or at least my fantasies and memories of it, and with that, a lot of pain. It was a fertile period in my work but also a time in which I was dealing with a deep depression, a terrible anger, which came out in some of my writing, most notably 'Dreams From an Ordinary Childhood' (Walkerdine 1984) and *Democracy in the Kitchen* that I wrote with Helen Lucey (Walkerdine and Lucey 1989). But the depression and anger allowed to come to the surface issues around class and the popular which I now want to explore.

She was always such a good girl, a goody-goody, even. Good at school. But the longing, the desire to get out, to travel, glamour, all the things that girls in her position were set up for. Her mother did her best, but the ambition to be an artist never really took off, even though she loved art more than anything else. Art for her signified the capture on paper of that fantasy (she drew scenes and glamorous women, while Brenda Orton made copies of the blue ladies they sold in Boots department store in Derby, the exoticised blue-tinted blue oriental women). But an artist and art college? All those paint-splattered wild-looking girls in duffle coats: no, a primary school teacher. This was, after all, the respectable working class, that group that came to be so dismissed by the New Left. But she insisted on London. Why did nobody tell her that you could go and study a subject because you liked it? It meant nothing to her or to her friends, like the time when she and Carolyn Hales decided that it was better to do art at training college because it only took three years, whereas going to art school and then teacher training would take five, so you would be able to be adult, work for a living more quickly if you went to training college. And somehow, the little rebellions were never much, and wouldn't have been understood as rebellions by those intellectuals, the highs and lows, the isolation, the ignorance, with the romance of poverty and dirt locked firmly inside the fantasies of the Left itself. When she first wrote about the dreams of her ordinary childhood one reviewer called her life stultified. It hurt and brought once again to the surface that immense well of hate.

But there was that entry into the longed-for space, the glamorous intellectual Left where she felt as though in a masquerade – the splitting, the not belonging, the fear of being found out to be stupid – the parties where people talked of being in the Young Communists at 14, when what she remembered was South Pacific, Radio Luxembourg *and the Methodist Youth Club.*

She felt stupid, frightened, like the time when granny said that Mum had shown her up on a coach trip by eating her fish skin during a fish and chip supper. Or when, having learnt to put the peas on the back of her fork, later as a Ph.D. student, she watched a professor's daughter in her twenties stick her finger in a chocolate mousse or others lick plates (licking tea in saucers, dipping biscuits in tea, were definitely practices to hide, to like but to be ashamed of). This shame didn't start when she joined the intellectual Left, but long before. To be respectable was not to be like the rough children or the families with a dad in prison. It was to wear clean underwear in case you were knocked down and taken to hospital or to polish the silver in case the Queen might call. A vigilant self-regulation was always necessary to avoid being the object of external regulation or pity or charity and you hadn't to want too much either: everything in moderation
much wants more
manage
cope
don't break down
don't get into debt

Sure. And we can find all of those forms of population management which formed my family in that way. But just then, at that moment when I looked for it, for some place in which that history of which I was trying to speak was being spoken about, I found nothing. And perhaps because of psychoanalysis I could no longer split and keep one thing in one place, another in another. The best, the cleverest, beat the poststructuralists at their own game so that they couldn't throw me out, back to the provinces, babies, depression, sinks, coping, moderation and yet the overwhelming need to make a Left and a feminism which refused to look in this direction, take some notice.

WORKING-CLASS SUBJECTS: WHO'S NOTICING?

For feminism, class was often presented in a debate about capitalism versus patriarchy, class versus gender, as though it were possible to be either one or the other and always, as usual, as though class only referred to one class: the pathologised Other, not the normalised middle class. In addition, it was often taken to be the case that working-class girls and women were too feminine or less feminist. For the Left, increasingly in the 1970s and 1980s, the respectable white working class had become the source of the problem, not the hope for the revolution. They were positioned as a problem in all popular Left movements, like the politics of the GLC or Left councils, where the respectable white working class were also viewed as the biggest problem in the implementation of anti-racist and anti-sexist policies. As Franco Bianchini admitted (Bianchini 1987), in the politics of the GLC, for the white working class nothing was done. I couldn't, it seemed, have chosen a worse moment to want to talk about the respectable white working class, precisely the moment when not only was the issue completely out of favour but had come to be associated with the epitome of reaction.

But I want to argue that while in one way the Left appeared to have abandoned the white working class, class having seemed to disappear from the agenda, the proletariat, the mass, has been an obsession, a central if sometimes silent figure during all the debates from modernity through to postmodernity. Indeed, we might say, following Foucault, that stories about the masses circulate endlessly. The issue is not then so much that they have disappeared, but a question of where and how they are talked about, what kind of object they become. And, in all of this, the popular has a particular place.

NOTICING THE MASSES

Let us go back again to that Althusserian moment, when for the British Left the thing to be explained was not the possibility of class consciousness, but failure. Theories of ideology were to explain not a subject whose vision was clouded, but a subject produced in ideologies, in media and other texts. For Althusser, the working class was constructed not in the real relations of production but in a set of imaginary relations in which bourgeois fantasies, especially those of the mass media, had produced the very mirrors in which the workers' identity was formed. By referring to Lacan's psychoanalysis, the way this work was taken up was to clearly imply an account in which working-class identity was an ideological product down to the very unconscious meanings of the original fantasies. Lacan's Imaginary built on Freud's idea of imaginary wish fulfilment. The infant, argued Freud, deals with the terror that it feels when food and warmth and human comfort are inevitably not on tap twenty-four hours a day, by 'hallucinating the absent breast'. Freud later saw this as the origin of fantasies of wish fulfilment and the organiser of psychic life. The phantasy space, unconscious, was the one to be filled with phantasies of plenty and presence. For Lacan then, the Imaginary Order is the order of wish fulfilment fantasies, of an impossible reunion with the lost mother. This eventually Oedipal fantasy could only be solved for Freud by the castration complex and for Lacan the move to the Symbolic Order in which the desire to be the object of the mother's desire is crosscut by a deeply competitive patriarchy, one which is no less a fantasy but a fantasy of control through which the social world is organised. For this, Lacan made reference to the structural anthropology of Lévi-Strauss. In analyses that followed, much work in film theory using this model concentrated fruitfully upon Oedipal analyses of Hollywood movies and, following a very important paper by Laura Mulvey in 1975, on the place of

Hollywood in constructing a patriarchal fantasy of woman, a woman who was not a distorted stereotype, but who did not exist except as symptom and myth of a male fantasy. A fantasy constructed in the Dream Factory itself. This meant that the working class increasingly came to be identified as being totally formed in ideologies, in mass media, trapped in a Hollywood which played upon their most infantile fantasies, constructing a patriarchal fetishisation of women and a sexist and infantilised working class, the very working class constructed in the fantasy of the New Left. I want to argue that this paved the way for not only the dropping of class from cultural analysis but also the idea that by the 1980s the working class no longer existed as a viable entity.

But to explain this I want to go back, to at least the beginnings of social science, to the modern period of grand metanarratives, the grand stories of psychology and sociology, the stories which claimed to tell the truth about the human condition, the stories of, among others, Darwin, Freud and Marx. Darwin's story of evolution charted civilisation as a narrative of survival and adaptation, taken up as social Darwinism in which capitalism and industrial competition, the rise of the bourgeoisie, were explained using an evolutionary discourse, with the white bourgeois male at the highest point, the most civilised with a series of others, those closer to the animals, less evolved: children, women, colonial peoples, the proletariat. The proletariat: the mass, the mob. Marx took it one way, le Bon, another. A civilised proletariat, one better evolved was understood as central to the emergence of the possibility of effective government. In these accounts the state of the proletarian mind was thought of as central to their transformation from a mass or a mob into either docile bodies, law-abiding, well-regulated subject or to that entity 'The Working Class' that would recognise its true mission through the production of the appropriate form of revolutionary consciousness. What Le Bon feared in mob rule, dark anti-democratic forces, threatening the bourgeois order, a threat only lessened by individuation (a theme which was to be central of accounts of the mass from media to football fans) was countered by Marx's modernist proletariat, who had to be able to see the world as it really was and understand the state of its alienation and exploitation in order to make the revolution. The working-class mind seems to have become a heavily contested space. But what if this proletariat, this white, rough and respectable working class is not a fact of modernity, but a fiction, a fantasy, one created in the imagination of the bourgeoisie? A fiction, in Foucault's terms, functioning in truth, very powerful truths that constitute and regulate modern forms of government. In this scenario, the working class always exists as a problem, to be transformed one way or another. It begins to be 'endlessly' described and monitored in every detail. When I say that it is a fiction, I do not mean that poverty, oppression and exploitation do not exist or that class does not become an important designation through which we recognise ourselves, but that the way that the working class is created as an object of knowledge is central to the strategies which are used for its creation as a mode of classification and regulation. These strategies tell us about the fears and fantasies of the regulators, the bourgeoisie, for whom the proletariat forms an Other, to be feared, desired, directed, manipulated. In this sense I am arguing that this truth is constructed inside the fertile bourgeois imagination, an imagination that sees threat and annihilation around every corner because of its shaky position in between the aristocracy and the proletariat. The truth about the working class then is the mirror of the fears and hopes of the bourgeoisie. In these fantasy stories the proletariat become everything which Darwin described as lower, more animal, less civilised, less rational. The mass has to be tamed. It is mapped and classified and found wanting. It is pathological to the bourgeoisie's normal. But it can be made normal: managed, policed to become normal like the bourgeoisie. It can be educated, tested, its intelligence monitored, its mental health, its mothering, fathering, cleanliness, work habits and on and on. This class is endlessly described. And that other class? Only in so far as it is presented as the norm. As Helen Lucey and I described in *Democracy in the Kitchen* (Walkerdine and Lucey 1989), the bourgeoisie is no less regulated, the women no less oppressed, but their oppression inheres in the very normality of which they are presented as guardians.

These stories of course have their heroes and villains, the good and the bad. There are the salt-of-the-earth working class, the hard workers as well as the feckless drunken poor, the bad mothers. Is bourgeois desire enshrined in these fantasies? A desire for a more equitable world in Marx, versus a smooth working capital in liberalism? But are they any less fantasies for that? Just as Edward Said (1988) argued in *Orientalism* that those western stories of the Orient told us more about the fantasies of the West than anything about the East, might not those stories of the working class, endlessly recirculated and enshrined in the everyday regulation of the population as if to make them true, might not they too tell us more about their creators than those so ardently described, so liberally, nay humanistically, regulated? Yes, from the solidarity of the Welsh pit village to C4 and Essex man, might they not all be fictions imbued with fantasy? Shouldn't we be looking at whose stories these are, how they came to be told and what effect they have in the constitution of actual working-class subjects – subjects designated by that very classification?

In fact, if we look back to that moment of the constitution of the mass, I want to argue that it paved the way for modernity's look at the media. In 'The Future of an Illusion' (1927b) Freud wrote that the 'masses are lazy and unintelligent; they have no love for instinctual renunciation, and they are not to be convinced by argument of its inevitability' (186). For Freud then it would be impossible to dispense with control of the mass by the minority. What, in his view, had to happen was the provision of good leadership, which would induce the masses 'to perform work and undergo the renunciations upon which the existence of civilisation depends'. In Freud's view, therefore, civilisation is against the mass. It is the mass which is closer to the body, to pleasure, to animality. Bad leadership, stressing deprivation and leading potentially to fascism with an easily swayed mass, who is closer to their emotions than to rationality, is understood as one side of the coin of which the mass media and consumption are the other. Precisely by catering to the easy pleasures and not the necessary privations, the mass media and markets, in this view, work against civilisation.

CREATING THE POST-WAR WORKING CLASS

I want to move to the 1950s with its idea of the meritocracy, of social mobility, through the tripartite system of education designed to find the bright among the working class, but also a period when the newly consuming working class appear to be deviating from their historic mission: they are taken to be becoming bourgeois. It is the beginning of the mass market and mass media. Indeed, the problem of the mass as proto-fascist reasserts itself in the discourse of the Frankfurt School, which locates the causes of fascism in authoritarian childrearing. In empirical analyses of authoritarianism, using empiricist variants of psychoanalysis which were to have a profound effect on social psychology, authoritarianism is charted by means of projective tests, Likert attitude scales, Rorscharch blots. It is precisely this position that is taken up by the Frankfurt School in the post-war period. It is important to me that at this moment, the moment of my childhood, a number of issues come together in the regulation of the masses: the tripartite system of secondary schooling, leading to the expansion of higher education, the mass market, media and communication. So, Adorno, Horkheimer and others place easily together prejudice, proto-fascism, authoritarianism and the uncivilised pleasures of the mass. The mass then that is at once becoming more educated, is in danger of swamping the world with its easy consumption, its authoritarian parenting, its passive television viewing, its escapism. So this mass is also in danger of swamping the civilised world with the easy pleasures of the uncivilised. It is my view that social and psychological research has a particular place at this point in the surveillance and regulation of the masses, a point which becomes clear in relation to psychological research on media audiences.

Perhaps the love/hate fantasy about the working class always said more about the desire of an intellectual Left for the masses to do the transforming, the dirty work as usual, while they could write, think, lead. In whose fantasies were we constituted and how did we grow up inside those different fantasy scenarios?

The monitoring of the working-class family takes a new turn. The danger of the consuming working class is a turn to reaction, a reaction understood as being central to mass media, with the propagandising appeal. The pathological family joined by the pathologising media. And the way that the one is watched by the other becomes a test of proto-fascism, of abnormality in the social psychology of family viewing. Social theorists begin too to assert that the working class, with its penchant for consumer goods and its wage settlements, has lost its way. It is being caught by the mass and enticed away from its revolutionary goal. Already its decline as a class is mapped: the fantasy is in danger. Simultaneously then, working-class people are being presented with home ownership, consumer goods, holidays, education, the possibility that for the first time their sons and daughters may not have to face the same tiring, poor, soul-destroying jobs as them. The class becomes a place to leave. And why on earth would you not want to leave it for the life that is being offered? Why should anyone see a romanticism in back-breaking work or poverty? Why, having faced so many defeats, would you want to try again? But the injunction to be 'true' and the urge to consume, to better oneself, to move out, constitute the working-class subject as the object of hopelessly contradictory discourses. To want to move out is to sell out and not to sell out is to remain stupid, animal, reactionary, pathological, anti-democratic, take your pick. I don't like the choice very much. YOU can, after all, succeed in education. From that period is *My Fair Lady*, in which a flower girl, living in poverty, can be educated to pass for a princess. To be educated however, is first of all to be maligned as a dirty animal.[3]. The violence of this inauguration into being a lady never struck me when I first saw it as a child. I remember only the songs and the transformation of Audrey Hepburn into someone who could pass for a princess. What she was, what I was, was presented as so very sordid, so very worthless compared with what was on offer: rags to riches, pauper to princess. Glamour, excitement, exotic Otherness. New worlds of wealth and glamour and plenty. So you too can get out, but beware, authoritarian families lurk, families who bear the responsibility for success and failure of grammar school boys, fathers who are too strict for child-centredness, so anti-democratic and not progressive, mothers who deprive, fail, don't talk or stimulate their offspring, who produce delinquents, criminals. Progress is now taken to be in the hands of the liberal middle classes, who allow their children to grow up towards autonomy. The mass is one of the problems: the market, manufacturing and the media. Ah yes, the media.

It is the 1950s which sees the Frankfurt School and other social psychologists such as Henri Tajfel begin to look at the mass media, at groups and intergroup conflict. The early research on media effects begins here. But interesting it is at this time too that these early researchers comment on the power of the new mass markets and media to produce new forms of social and psychic life. C. Wright Mills argued in 1956 that mass communications created a pseudo-world of products and services, but also lifestyles inherent in buying those products and services. Two American anthropologists, Horton and Wohl, talk about the way in which television brings simulated communication into the living room. Interestingly these sentiments are ones that we associate more with the 1980s than the 1950s, and with post-modernity than modernity. However, what is visible here already is a version of the mass subject with an identity defined by that mass consumption. The fear, the danger understood as lurking inside is the production of a proto-fascist mass of consumers, living in a bubble. But who is in the bubble? By the 1950s, is this endlessly to be watched mass of consumers the ones who have lost their way? Of course, we should have known. As usual, the normal middle classes are all right because they see through mass consumption, they talk to their children about television, they buy healthier foods and, of course, but nobody seems to remember this, they have more money and they have access to a culture which they regard as infinitely superior to the one that the poor unfortunates are dragged into. The avant-garde in relation to the popular, but that's another story.

Well, here we are again then back in the 1950s with me and my dreams. Not a grammar school boy or angry young man. Descriptions of me fade. But wait. I had thought that the stories of that time were all about boys and that girls were left silenced as usual. But I was wrong. Quite wrong. What I've discovered while researching this chapter is that there is a whole post-war narrative about girls growing up into upward mobility, the very narratives which so fired my imagination. These narratives, found in *My Fair Lady*, *Gigi*, Walt Disney's *Cinderella*, build upon prewar narratives also featuring girls: Shirley Temple movies, Orphan Annie comic strips, Judy Garland in *The Wizard of Oz*. I am not going to discuss these, except to point to the central place of girls in movies about poverty, wealth and the depression. Here the girls are poor and often orphaned and like Judy Garland they dream of a place where wishes are granted through the intervention of good fairy godmothers, thwarted by bad witches, to reach a place where men can grant ultimate wishes which are about turning poverty to wealth and poor men into fine ones. But by the 1950s, the story of the girl is a story of rags to riches transformation through education. Here, the girl does not just intercede for others, she may actually be shown to move out of the horror that is herself towards a transformation both to adult womanhood and to wealth, glamour and romance. While these movies certainly present wish fulfilment, have strong Oedipal elements, to describe them only in these terms is to miss a central point. The girls in these movies are not constituted only in a sexual wish fulfilment. That narrative only makes sense in relation to a historically specific story about upward mobility, a move to be a lady, through an education leading to the possibility of betterment through a marriage to a person from a higher class. I would say that these films signal a particular trajectory which incorporates education, respectability, glamour, romance and upward mobility through marriage. This story also relates to and builds upon others told in other places for girls, like girls' comics that I have analysed elsewhere. But I think that this is not simply about Althusser's version of Lacan's Imaginary, nor is it Gramscian hegemony. The unspoken and unanalysed elements are poverty, class exploitation and oppression and how women get out of these at a moment at which becoming a 'princess' is shown as the glamorous, perhaps the only way. I would say then that these films constitute a certain truth about class and mobility at a moment at which certain paths and fantasies are open to poor women. Nor do I think that they are, in any simple sense, bad. As I have tried to show, they, far more than the culture of school, helped to get me to the place in which I am today. Without the possibility of those dreams higher education would have meant nothing to me at the age of 14. Contradictory as that message was, it cannot simply be condemned out of hand. It has to be understood in terms of the conditions of my subjectification and as resistance to the life that was accorded to my mother. Why would I want to be a housewife when I thought that I might become a princess? (Or at least something more glamorous, even if that glamour was more circumscribed – actually, rather air hostess or bi-lingual secretary than princess!) But there is something else here too. I think that the glamorous option has to be seen as a defence, a defence against the Other that it hides. Neither the mother nor the father is shown as adequate, rather in the stories I have talked about they are poor, exploited, uncouth, animal, dirty, reactionary, depriving, nasty and sometimes exploiting. What is presented as the feared place, to be defended against at all costs, is a return here. But it is the bourgeois fantasy which constitutes this inadequacy and places it as a grid for the girl to read her own history. That those mothers and fathers struggle to do what they can in the circumstances they find themselves in cannot be contemplated in this scenario. While the working class is endlessly described, very particular stories are being told and some issues do not even get a mention, as I shall demonstrate later. But of course, those working-class women are spoken about everywhere from the 1950s to the present. They stare out of every developmental psychology, education, social work textbook. They are the bad or potentially bad mothers. So while social democracy struggles to reform our mothers, a door opens and a few of us are let in (ashamed, afraid ever to be like that again, defiant). No material for a revolution here. Only a story about how come they (I can't say we now, having escaped the fate worse than death) came to be like this. Failed again.

But, come to think about it, it is not very surprising that the erstwhile middle-class students of 1968 should have missed the class narratives inside Hollywood and opted only for a world of Oedipus. For it was those students, the sons and daughters of the bourgeoisie who, in their own revolt, reacted against the conformist privilege of their parents. Such young people must have found it virtually impossible to identify with a respectable aspirant working class. The young women who so desperately wanted to get out of the despised place and into glamour could hardly claim to be or want to be part of the romanticised and fetishised working class that the rebellious bourgeois youth imagined. That bourgeois resistance simultaneously created a desirable working class that contained everything that they wanted as opposed to the despised parents. This working class was not respectable. They, like the new communes, had dirty kitchens, away from bourgeois housework, a far cry from the incessant cleaning, in case the queen might call, the respectable and tidy houses that I remembered. No, if that working class did not live up to the romantic expectations it would have to be cast aside in the Left's dreams to find a truly revolutionary constituency, one in which it could be imagined that there were no anti-revolutionary deviants: blacks, women, were the next on the fantasy list. And again I am suggesting that what was being created here too was an impossible object, one that like the white working class, could never live up to all the expectations and fantasies placed upon it. And the respectable white working class got dropped while the new theorists of mass consumption went shopping.

POSTMODERNITY AND THE POPULAR

During this time, psychological studies of the mass media tended to be concentrated in two paradigms: so-called 'effects' and 'uses and gratifications' research. The theoretical trajectory of both may be seen in relation to the historical and theoretical trajectory of which I have spoken. This is particularly true of post-war American research and later British work. While the idea of a hypodermic injection of media into the person was abandoned as too simplistic, nevertheless, researchers' main concerns in one way or another depended upon the early psychoanalytic work, even if transformed out of all recognition by empirical social psychology. How the media gratified the mass, the effects on the mass, its uses in their lives, how it related to psychological needs, all played upon this underlying fear of the inherent dangers of mass communication, linked to the already dangerous classes. While the Frankfurt School was understanding the mass as caught inside mass consumption in a pessimism that prefigured Baudrillard, Screen Theory looked to Lacan. This theory had the masses even more tightly caught than the social psychologists, who had left a certain room for voluntarism. Here, the subjectivity of the masses was formed, right down to the unconscious, in the media (an unconscious then not gratified by easy pleasures, but actually formed in the signifiers that make up ideological signs). In Screen Theory the grip appeared so tight that there seemed to be little escape, except to move beyond the imaginary fulfilment of an impossible desire to the Symbolic Order. And it was this which was later resisted in some sociological work, and which led to a rejection of psychological and psychoanalytic paradigms in cultural studies. At the time, certain theorists of subculture and the popular, especially using Gramsci, were apparently defending the working class. I say apparently because work which proceeded from *Resistance through Rituals*, in the 1970s CCCS, understood subculture as arising out of alienation and being a mark of resistance and therefore of proto-revolutionary activity (Hall and Jefferson 1975). It has been well documented that such work was mostly about young men and that it presented only certain groups as working class. There was no place here for the respectable or older or female working class. Later, Fiske (1986) argued that popular culture and especially television, was not a medium which created the identities of the working class, because they as audience were able to raid it for progressive meanings and even to make resistant readings.

There are some important insights in this work, which I do not have the time nor the space to go into here. However, I wonder if there is not a defensive optimism in the way in which these authors see a working class that can make 'progressive' readings, that has not wholly been taken in. In following Fiske and others down this road, there has been a general opposition to the overdeterminist and pessimist psychoanalytic readings in which total identity is produced in the media. Judith Williamson (1986a) argued that her media studies FE class managed to deconstruct an advert at five paces, but insisted that they liked the fantasies presented. They resisted her attempt to take away their pleasure. She later added about Fiske, that it is all very well to defend working-class raiding and progressive moments taken out of more reactionary narratives. In a sense that could be one view of what I took out of *Gigi* and *My Fair Lady*. But I think that would be wrong. Elsewhere, Williamson (1986b) asserts that redemptive readings of what the masses make with the popular are a problem when they have access to only one code. This attempt to suggest that audiences make active meanings in their consumption and are not either passive consumers or have identities totally determined by the text invokes an American discourse of empowerment, of voicing and authentic creation. But I think that it is not only wrong but patronising. It is a defence (again) of the working class as equal but different. Look folks, they are not taken in and they are actually bright enough to make their own meanings! But neither they nor the readings are equal but different. It makes working-class readings seem like the consumption of pick-and-mix sweets in a postmodern shopping mall. But while I think that it is correct to assert that people make what they can of what is available, we seem to move from determinism to voluntarism with no idea how to produce an understanding of subjectivity which is not at either end of these poles. Watching *Gigi* and *My Fair Lady* certainly helped fuel my dreams, but these were dreams that were already being produced in the complex relations set up in the practices of upward mobility, of family class and sexuality, into which I was inscribed. They were also centrally about oppression, exploitation, poverty, something which appears on neither the determinist nor voluntarist agendas.

If I feel patronised by all the equal but different arguments it is because I do not think that the difference is equal at all. A reading of oppression as pathology leads to particular practices through which that pathological subject is to be formed, only to be corrected. Of course in that process we make what we can of what we can find, but that does not mean that there is not a complex psychodynamic at work, nor that the only discussion to be had is about a push-me-pull-you will-they-won't-they reactionary/progressive seesaw. I want to change the agenda. In none of the above analyses can I find any reference to how oppressed peoples are formed and live under oppression. I cannot find it precisely because the agenda has been set elsewhere. If the masses have been the central pivot of analysis the aim has been to account for their place in the making of revolution, democracy, totalitarianism. So busy have some of the intellectuals been in creating the stories of the working class to fit their fears and their desires that they do not seem to have been the least interested in addressing questions about the constitution and survival of oppressed peoples. Indeed, by the 1980s, some cultural theorists in the endless missionary position had come out from socialist puritanism and discovered that if the working class (who no longer existed anyway) liked to go shopping then it was OK to admit you liked it too. So writers in *Marxism Today* variously waxed lyrical about shopping in Camden Sainsburys and the food-court at Euston Station. And opined that if young blacks liked to buy expensive suits then this was a signal that it was OK for the Left middle classes to admit to liking this too, all the time failing to see that the two were not the same thing.

Meanwhile, Baudrillard had caught a sharp dose of Frankfurt School pessimism. Repeating and extending ideas from the 1950s he envisaged a scenario in which media simulation had created an atomised and silent mass, a mass reduced to a physical entity. The mass here is indeed off the streets and in their homes, watching television or playing video games, but the threat of the eruption of violence is still blamed upon them, as in the Bulger case, for example. Now, I think that some of Baudrillard's pronouncements about the masses, while they certainly build upon fantasies of which I have already spoken, are not as wild as they have been presented. And he does try to document the subjectification of a mass which no longer is understood as having an authentic voice, since it is no longer the true revolutionary class. And he does recognise that the class is both endlessly defined and resisting definition, endlessly asked to be autonomous while endlessly asked to conform. And he refuses to read them using those grand metanarratives. He believes that the modernist democratic project of calculation, education, policing, definition of the masses is finished, because the masses refuse it. He has been accused of being pessimistic, but it seems to me that the optimism/pessimism dichotomy is part of the fantasy of the bourgeoisie about the place of the mass that I have been trying to go beyond.

SUBJECTIVITY AND OPPRESSION

The project which I want to signal here is critical of those grand meta-narratives that have endlessly described and defined away the respectable white working class. While it might be said that the atomised and indi-vidualised poor now have no place to turn except the imaginary communities created on the screens in their living rooms, the communities and organisa-tions which were their strength having been crushed, I wonder if that too does not hark back to a romantic reading which was certainly not true in my own childhood. No, I want to try to construct a different story about what media fantasies mean in the lives of oppressed peoples. Here I want to draw on work on oppression and psychodynamics which has helped me and then to go on to look briefly at one example. I do not think that we can explore the constitution of this subjectivity without examining how poverty, pain, oppression, exploitation are made to signify. The popular as escape indeed, the longing, the hope.

I'm trying to construct a way of analysing the production of subjects in practices as a way of getting beyond the dualism of the media effects work and the oversimplification of the text-based work. My aim then is to account for subjectivity and the place of the popular in making oppressed subjects now, not audience research *per se*. I can't go into all the details of how I am approaching the idea of subjectivity in practices, but just let me suggest a few pointers. The practices in which subjects are produced are both material and discursive, but the relation is not one of representation but signification. Indeed, if fictions can function in truth then fictions themselves can have real effects. Subjects are created in multiple positionings in material and discursive practices, in specific historical conditions in which certain apparatuses of social regulation become techniques of self production. These are imbued with fantasy. We cannot therefore separate something called 'working-class experiences' from the fictions and fantasies in which life is produced and read. What is the relation between those fictions and fantasies and the psychic life of the oppressed? What gaps and silences are there in the fictional discourses, like the fact that they may speak of pathology, of difference, of poverty even, but rarely oppression? How then is oppression lived and is it spoken? If so, how? And how is the absent material a relation in this subjective constitution?

I made a documentary called *Didn't She Do Well* (Metro Pictures 1992), about a group of working-class women, all of whom have gone through higher education at some time in their lives. In the Women's Therapy Centre in London, these women talk about their lives. The toll of pain, suffering, survival and courage is almost overwhelming. But they begin to speak about the specific historical constitution that I referred to. I found little help in understanding this discourse from a traditional psychoanalysis that cannot handle materiality, nor from a Lacanian reading. Oppression simply does not enter discussions about the psychopathology of working-class women, who appear on the scene not as upwardly mobile girls, but as pathological mothers.

What I am talking about here are patterns of defences produced in family practices which are about avoiding anxiety and living in a very dangerous world. Work on the holocaust and torture and survival in Latin America (see for example, Puget 1988) has made it perfectly clear that certain defences may be necessary to survive danger and that one cannot assess those defences on a scale of normal to pathological. But it is possible to examine the place of those defences in constituting the very practices in which subjectivity is produced. Just as I suggested that there was something to be defended against in the fantasy of upward mobility, notably the oppressed, poor animal working class, so I want to suggest that such defences are part and parcel of the constitution of the lives of the oppressed and that we can look at the popular in this postmodern order as part of that defensive organisation, as something that makes life possible, bearable, hopeful, but cannot be understood as either good or bad, without locating its place in the conditions and survival of oppression. Gail Pheterson (1993) points out that the defensive structure incorporates all subjects embodied in relations of domination, complex as they are. Class domination then does not just touch the working class, as I have tried to show, but is central to the fantasy structures and defences of the bourgeoisie.

Witnessing humiliation and exploitation acts differently for those who have a cleaner than for a woman who works as one. Middle-class people often only see the working class in relations of service or as frightening others in areas of town that they do not want to enter. Their defences are cross-cut by the way in which the Other is made to signify and the fictions in which they are inscribed. When Ronald Fraser (1984) let a beggar into the manor house in which he lived as a child, he learned painfully that his parents were not pleased, that there are some people who are not to be welcomed into one's home. When the 4-year-old Sarah (Walkerdine and Lucey 1989) looks out and asks her mother why the man cleaning the windows has to be paid for his work, she understands a different relation to work and service and money than the young working-class girl who is told she cannot have new slippers because money is scarce, that her father earns the money at a factory he cannot leave until he is allowed to do so and again from the young child who watches her mother being humiliated in a Social Security Office.

Psychodynamic forces: the wishes, drives, emotions, defences are produced in conflicting relations, in a context in which materiality, domination and oppression are central, not peripheral. But accounts of psychodynamics rarely include these issues as central to the account and as we have seen, they disappeared entirely from the post-Althusserian debates. So the working class, the gradually disappearing class were locked inside ideologies in infantile wish fulfilment because of a refusal to engage with the psychodynamics of oppression. In addition, as Pheterson argues, there has been a reluctance, even a refusal, to call into question 'normal' or 'normative' relations of domination, the 'normal' everyday designations of Otherness, the defences. The consequences of this are enormous and make all accounts very one-sided. Indeed, she goes further and argues that perpetrators of abuse, for example of racism, are understood sociologically and their victims psychologically, as in need of therapy.

This is overwhelmingly the case with the distress witnessed in the upwardly mobile working-class women in my film. I made the film precisely because I wanted to contest the view that this pain is an individual pathology that needs to be corrected, the result of inadequacy or inadequate families. Rather I wanted to make public the psychic effects of living in and under oppression. Oppressed groups, such as the working class, have to survive but survive in a way which means that they must come to recognise themselves as lacking, deficient, deviant, as being where they are because that is who they are, that is how they are made, an insidious self-regulation, while individual effort is allowed to those clever enough to plan an escape, an escape only to be pathologised by those others who romanticise the oppression in the first place. As Pheterson remarks, genocidal persecution is not required to elicit psychic defence; daily mundane humiliation will do.

What then are the consequences of living that daily humiliation and for children to grow up watching their parents face it? How do they live watching parents do without, face hardship, be hurt or killed at work, never stop working, become drudges, old before their time and so forth? Why are not these the questions that are being asked? Bergmann and Jucovy (1982) report that responses to natural disasters have less lasting psychic effect than continuous systematic and organised assault on a people singled out as less than human. It becomes clear then that if we look at the effectivity of the media in the constitution of subjectivity in this way, what is at stake changes dramatically. For indeed, the five women in the film tell us clearly and courageously what that continuous systematic and organised assault is like, what it means to witness the routine humiliation of one's parents and to long to leave, to not be like them, but to feel the terrible guilt of leaving, of survival. A survival which defensively may have to be many hundreds of miles away in another place, which cannot bear to see the pain, the humiliation that has been escaped and to feel the shame both of having been like that and also of getting out when others are still there, and have no obvious means of escape. That is what I want to talk about and it makes equal but different and the trite stories of finding progressive elements in the mass media, trivially offensive.

The five women tell of the shame, the pain of watching parents do without, of fathers who were injured and killed at work or who died prematurely because as one doctor put it to Christine's father, 'I'm sorry Ernest, there's nothing I can do for you, you're worn out.' Or Diane's mother with eight children who she never remembers seeing sitting down. These are the stories, and the identifications and defences become clear in the film, clear that they are means of survival. Fiona McLeod tells about her fears that she might not survive all the pain she has gone through even though she is now a well-paid social worker and lives five hundred miles from her family on an Edinburgh housing estate.[4] Diane Reay tells us of the time she went to a union dance when she first went to university. I have chosen this example because the dance is so redolent of all those balls, the balls in the films in which one had to learn to pass as a lady. Diane is afraid that her masquerade has not worked well enough.[5] Diane took the protective step of marrying the middle-class man who first befriended her, saving her from the men who wanted to constitute her as other, as oversexed and easy. This could be interpreted as a defence against something unbearable, not an ideological failing, an over-femininity of a working-class woman who cannot see beyond patriarchy, as has often been suggested.

Diane wants to be able to be sexual, but to have that sexuality read as animal, dirty and deviant is likely to produce complex conflicts and defences. I am trying to begin to tell a story, one which I can't elaborate here, about the practices of survival in which such defences are not only produced but are necessary. Necessary, but not without contradictions. Seen in this light, wishes for a glamorous upward mobility, a new happy bourgeois family presented in the media portrayals take on a different light. They tell us about what is being guarded against and how practices incorporate stories told to make that survival, escape, hopeful, bearable. These practices must in fact be passed into family practices themselves, and down generations, as complex cultural resources, ways of being and belonging. How does all this relate to working-class families watching television? Families who are the object of all that regulation. Eliana and her sisters watch the musical, *Annie*, on the video. Her father is Maltese, her mother from Yorkshire. Aged 6, the middle of three sisters, Eliana and her sisters decide to put on the video that daddy has bought for them. They play in front of it, their mother in the kitchen, their father out. I want to share with you one tiny little bit of a much longer analysis. *Annie* presents us with a dispossessed and unworking proletariat. Annie is an orphan. She is adopted by the rich and self-made armaments manufacturer, Daddy Warbucks. The only way out for the little orphan is to charm your way into the rich family. Here the working-class family and community is not deficient, it is non-existent. There is only one solution: escape. This version of going to the ball gives not a prince but a family, happiness, servants, plenty; oppression is taken away, defended against. Pain is removed. This magical solution could be especially appealing to Eliana and her sisters. They make reference to only one part of the film and they address their remarks to me, sitting recording in the living room. The remarks are about a sequence in the film in which Miss Hannigan, the drunken woman in charge of the orphanage, is swaying in front of the camera, with a bottle in her hand, apparently drunk.[6] Eliana tells me that Miss Hannigan is only acting: 'She's supposed to be drunk, but she ain't ... 'cos it's water.' Two minutes later the little sister Karen emerges from the kitchen to tell me 'mummy's drunk' which prompts the mother to deny it to me.

So in the film there is a woman who looks drunk but is in fact acting, but at home there is a mother who does not look drunk but is proclaimed to be, all for my benefit. Eliana's mother is systematically being beaten by her husband, a husband who according to her is having a relationship with another woman, a woman he takes the children to visit. Eliana is apparently coping, but is displaying distressing symptoms at school, from docility to the point of apparent stupidity to fits of hidden rage where she breaks the heads off dolls. I want to argue, to cut a long analysis short, that *Annie* provides her with a narrative framework into which she can envisage an escape from the daily misery which she watches her mother endure. The drunken Miss Hannigan is the bad working-class mother, in fact the nearest to a mother that the orphans get. She is contrasted with the beautiful secretary to Daddy Warbucks, Grace, and to Daddy Warbucks himself, who is so charmed by Annie that his hard surface softens to reveal a soft father. So the happy family can be found by dint of the efforts of the resourceful orphan. Daddy is the one with the money, the home, the happiness, to be contrasted with the drunken mummy. In Eliana's life then I am presented with a drunken mummy who is responsible for her own oppression and who comes off badly when compared with the escape offered by daddy and the other woman. Indeed, there is a metonymic relation, since Eliana's daddy bought the video of *Annie*. He is the one who offers the fantasy of escape. But *Annie* offers no narrative of the oppression suffered by the mother, nor of her possible escape. Neither is there any model for the father's cruelty except that which can be tamed by an alluring little girl. But the mother does use the film to provide the girls with her account of what is happening. She addresses her remarks to me in front of them. She talks of the difficulties of her life, her suffering and why she gets angry with the elder sister's siding with her father. The conversation and the film therefore act as a vehicle through which she can refute the *Annie* version of events. The video offers dreams of escape, but its presence cannot be judged outside some understanding of the conditions under which the family lives and the practices that they have produced to cope with these. Fantasy and escape then have to be understood as part of a whole ensemble of defences against the pain of that routine oppression and humiliation of which I spoke earlier.

SUBJECT TO CHANGE ...

So how is it possible to produce a new kind of psychological work on the masses and the popular? The issues which fuelled the debates of modernity have not disappeared or dissipated, now that the fragmented poor watch the television behind the net curtains on the fifteenth floor of a crumbling tower block, while it is being turned into a non-working class. Indeed, in some ways, the fears grow more desperate, the concerns as potent. Is it possible to work in another way on the relation of the subject to the popular? I have turned to work on myself, precisely because I have been remade as a professor out of the feared mass. Is the process of my civilisation then, the move to bourgeois culture, also one which allows me to work as both insider and outsider? Is there a new kind of knowledge that can be constituted in this way? Surely we must be able to tell some new stories. Sometimes the cultural theorists miss the wood for the trees, because they are so busy charting resistance or raiding that they seem to miss the way that routine humiliation, the present forms of management, constitute the subjectivity, defences and coping practices of most of the population. So busy looking at a progressive/ reactionary dichotomy and working with, not taking apart, this fiction which functions in truth, they seem not to see the ways in which subjects cope, produce defences against extreme conditions that frankly sometimes are not very nice. How they long for things that never seem to enter the intellectual imagination because the latter seem not only not to know how to look but because they are so busy talking about working-class fantasies, without ever analysing their own. Yes, intellectual work and personal histories do come together. And I do not think it is simply about objectivity or bias, but about how certain stories get told and how they too can fulfil fantasies, be defences. I want to address the questions which seem to be left out of the constant descriptions of the majority of the population. Such descriptions help to build a fortress simulacrum to keep them out because the romantic fantasies have failed, to construct only a defence against its opposite – the nightmare. Puget's (1988) work on psychoanalysis in Argentina during the dictatorship documents that defence well. The feeling that it will always happen to somebody else defends against the terror that it will indeed happen to oneself. So I want a happy ending, you know. Like those happy ever after stories of my childhood or my mother's soothing 'it'll be alright chicken'. But, neither that nor Gramsci's optimistic will seem appropriate. We need to look in a new way at our daily lives and recognise that the end of grand metanarratives of 'The Working Class' is not to discard oppression. Indeed, quite the reverse. But the professor has an injunction to speak, to profess, tell certain kinds of stories with an authority vested in her position. The little girl was so quiet, trying so hard to say the right thing, to be loved. To the schoolgirl, the goody-goody, the teacher, the silent student. But there was always a rage underneath all that goodness, the nurturant loving of the children in the progressive classroom, always the deep fear of being thrown out for her ideas, for opening her mouth to spit out all that anger, the anger that she found lurking even beneath the engaging feminine smile of her childhood.[7] The dilemma of the angry powerful woman is enacted at both conscious and unconscious levels, but it is also lived historically and socially.[8] It has long been women who have had an injunction to speak about the personal, to tell their secrets, just as it has always been the working class who have been asked to tell of their lives, to explain their pathology, while the fact that it takes two classes to tango appears to have escaped the notice of those who constantly ask us to tell it like it is. It does not seem surprising then that the injunction to speak about it has become one of the modes of regulation of the modern age, to bare all, to allow the natural to emerge, only to be better regulated. In our understanding of the regulation of the postmodern order, we need to examine the place of that voicing and where it appears, on television, films, the radio, the popular; who is being made to speak, what to and for whom? No, we have to create some other stories, which face the present and confront it, write new songs and begin to sing them.[9]

DEDICATION

The chapter is dedicated to the memory of Jo Spence, 1926–92.

NOTES

1   This paper was given as an inaugural lecture and included a mixed-media presentation. It has not been possible to preserve the entire flavour of the visuals and sound in the text. I note where visuals were shown and what they were. A videotape of the lecture including the visual material is available from the author.

2   By Julian Henriques, Wendy Hollway, Cathy Urwin, Couze Venn and myself (1984).

3   At this point in the original paper I showed a clip from *My Fair Lady*, with Audrey Hepburn and Rex Harrison. The clip shows Hepburn as Eliza being inaugurated into the regime by which Professor Higgins is to change her speech patterns to make her pass for a lady. Higgins tells Eliza that she is very dirty and the whole scene is quite violent.

4   Fiona McLeod talks in the clip about her feelings about her father, who died in a fire in their flat. She felt that he wanted to be a giant but feared that in reality he was a dwarf. She sees herself as like him now and finds it hard to believe in herself. Like him, she fears that she might not survive.

5   Diane tells the group that she went to a union dance and was pursued by a man who said that he thought he'd seen her in Woolworths. She interprets this as meaning that he thought

that she was a girl from town, who had just happened to get into the dance. She did not want to be seen as this because the male students treated such women badly, seeing them as having 'easier virtue'. She concluded that her masquerade was not working well enough and that she felt that she had to try harder.

6　I showed the clip of this on video at this point.

7　This has been documented in my installation 'Behind the painted smile' and discussed in the piece of the same name in *Schoolgirl Fictions* (1991), where my period as a primary school teacher is also written about.

8　The tightrope walked by many women in the academy, who have been patronised and envied, rejected and passed over, made to work harder and for less reward than their male colleagues, continues to be a testimony both to the courage of women and to their continued oppression.

9　The original paper ended with the song 'Coming' from Sally Potter's film *Orlando*. On the screen were end credits, thanking the Department of Media and Communications for support and technical assistance with the lecture, especially Colin Aggett, June Melody, Joanne Donovan, Helen Pendlebury and Joanne.

16

MAKING SPACE FOR THE FEMALE SUBJECT OF FEMINISM

the spatial subversions of Holzer, Kruger and Sherman

Gillian Rose

The troubled relationship between the subject and the feminine has long concerned feminist writers. Freud's puzzled question, 'What does woman want?', has often been taken as paradigmatic of the refusal of subjectivity to femininity in phallocentric discourse; many feminists have argued that the qualities of rationality, consciousness and agency attributed to the subject in western humanist traditions are qualities also attributed by those same traditions only to masculinity. Femininity is thus at once entirely unimportant to the project of the (hu)man subject and yet also central to its fear of and desire for its Other, the non-subject, the abject. In the face of such erasures and fantasies, feminists have insisted that women are indeed subjects. They insist on the difference between Woman – the feminine as it is imagined in phallocentric discourse – and women as subjects only partly and problematically positioned through the interpellations of Woman. As Teresa de Lauretis (1986a) argues, 'subjectivity' is thus central to feminist politics in at least two senses. First, there is a concern for the ways women are subject(ed) to masculinist definitions of femininity; and second, there is the search for women's resistance to those disciplining processes, a search for women as subjects on their own terms.

The focus of this chapter is on one particular feminist notion of subjecthood which acknowledges both relations of power as they constitute identity, and feminist efforts to elude those relations. It explores an aspect of a particular feminist argument which describes subjectivity as a political project, and one of the most eloquent exponents of that argument is de Lauretis. Her account of 'the female subject of feminism' (de Lauretis 1986a: 14) has become an influential statement of a subjectivity constructed for political ends. 'The female subject of feminism' is:

> the concept of a multiple, shifting, and often self-contradictory identity, a subject that is not divided in, but rather at odds with, language; an identity made up of heterogeneous and heteronomous representations of gender, race, and class, and often indeed across languages and cultures; an identity that one decides to reclaim from a

history of multiple assimilations, and that one insists upon as a strategy.

(de Lauretis 1986a: 9)

The 'female subject of feminism' is a way of thinking about subjectivity. It offers a subject conceptualised as complex and contradictory. This subject is multiply structured through a diverse range of shifting, mutually mediating and conflicting discursive interpellations: gender but also class, 'race', sexuality and able-bodiedness, to name just a few. It is an identity fractured by the gap between the conscious and the unconscious, between language and desire, between discourse and its excess. It is a subject which is both produced by discourses of identity and destabilised by the contradictions and failures of those same discourses and their affects. It imagines 'the female subject [as] the site of differences' (de Lauretis 1986a: 14). Its moments of solidity and certainty are rare and fleeting, contingent and strategic; moments of alliance and coalition with other subjects are not assumed in advance to be possible but are worked and struggled for. And this female subject of feminism is theorised in these terms for reasons which are above all political.

This notion of subjectivity as a political project is political in various senses. It is the result of struggle and debate within the feminist movement, for example, since its emphasis on complexity and difference is a consequence of the critiques made by black and lesbian feminists, among others, of the overgeneralised accounts of 'women' which appeared in many early, white, straight feminist arguments. Those generalisations occurred at least in part because their feminist authors were implicitly relying on a humanist notion of the subject as the innocent foundation of knowledge, and so generalised from their specific experiences to all women. In consequence, differences among women were neglected. For Judith Butler, writing more recently, however, 'the task is to interrogate what the theoretical move that produces foundations *authorizes*, and what precisely it excludes or forecloses' (Butler 1992: 7). The more recent stress on the complexity of subjectivity is due then in part to a concern to avoid the exclusionary universalism implicit in such generalisations about 'women'. This fear of exclusion has also encouraged a hesitation in relation to what are perceived as essentialist interpretations of subjectivity. For, if in many recent feminisms the subject is no longer innocent but problematic, then so too are the knowledges in which the subject is embedded and through which it is constructed. The effort to think through a subject position in terms of difference, contradiction and instability is connected to the effort to situate the production of knowledge, including knowledge about the subject, in a highly complex, shifting and power-ridden world, and to render any action on the basis of such knowledge both accountable to a specific position and vulnerable to other interpretations (Haraway 1991; Rich 1986a). In that sense, as Probyn (1993) argues, the question of subjectivity in feminism is not only the question, 'Who am I?', but also, 'Who is she?' This pairing of questions works towards refusing the structures of erasure of and fantasy about the Other from which so many women have suffered yet which have been replicated within the feminist movement itself. This feminist notion of subjectivity as a political project is thus also political in that it seeks to intervene in the relations of power/knowledge by redefining subjectivity; as Butler argues, 'to deconstruct is not to negate or dismiss, but to call into question and, perhaps most importantly, to open up a term, like the subject, to a reusage or redeployment that previously has not been authorized' (Butler 1992: 15). As well as being the result of political struggles, then, this feminist notion of subjectivity is also political because it intervenes in the power/knowledge nexus in order to reconstitute the subject in explicitly relational terms.

The discussions surrounding this notion of the 'female subject of feminism' have been complex and diverse, but perhaps two main emphases can be detected. Both exist within the work of de Lauretis. First, there is an insistence on the inadequacy of language as we now have it to articulate the desires of the feminist female subject. Arguments are made about the need to escape the signifying limits of phallocentrism:

> To deconstruct the subject of feminism is not, then, to censure its usage, but, on the contrary, to release the term into a future of multiple significations, to emancipate it from the maternal or racialist ontologies to which it has been restricted, and to give it play as a site where unanticipated meanings might come to bear.
>
> (Butler 1992: 16)

In this emphasis, there is a turn to the unconscious 'as a *resistance* to identification' (de Lauretis 1990: 126), or to discursively unrecognisable transgressions (Golding 1993b), or, more unusually, to a careful reflection on experience (Probyn 1993), as guarantees of an escape from the limits of phallocentrism. Alongside this insistence on what is unnameable in phallocentric language, however, there is also a search to construct that feminist female subject through an effort to name a new identity, or to articulate aspects of identity which have previously been ignored or taken for granted (this is what I assume de Lauretis means when she talks about reclaiming a self from assimilation). One of the most cited of these efforts has been that of Minnie Bruce Pratt (1984). In her essay 'Identity: Skin Blood Heart', Bruce Pratt recounts her gradual realization of just what her gender, sexuality and 'race' mean in terms of who she is, who she wants to be, and how that means she relates to others. Here there is a sense that the 'political consciousness' of the female subject of feminism is precisely conscious (de Lauretis 1987: 137): a notion of subjectivity as self-conscious agency.

Neither of these two emphases – on what is beyond representation and what can be represented, or on the unconscious and the conscious – are neatly distinguishable from the other. Bruce Pratt, for example, sees her efforts to articulate her position as a continual effort motivated by a desire for that which is not yet realisable. De Lauretis herself advocates both positions, and has argued that in fact an oscillation between these two strategies is typical of feminist discourse. And she articulates this oscillation in spatial terms:

> Now, the movement in and out of gender as ideological representation, which I propose characterizes the subject of feminism, is a movement back and forth between the representation of gender (in its male-centered frame of reference) and what that representation leaves out, or, more pointedly, makes unrepresentable. It is a movement between the (represented) discursive space of the positions made available by hegemonic discourses and the space-off, the elsewhere, of those discourses: those other spaces both discursive and social that exist, since feminist practices have (re)constructed them, in the margins ... of hegemonic discourses and in the interstices of institutions.
>
> (de Lauretis 1987: 26)

De Lauretis is far from being the only feminist to speak of feminism in these spatialised terms (Rose 1993a), and I assume that such spatial images resonate because space itself is bound into the power/knowledge relations addressed by the notion of a female subject of feminism. This means that space is also central to subjectivity.

The connection between subjectivity and spatiality elaborated most fully by feminists is probably the mutually constitutive link between 'the master subject' – that is, white, heterosexual, middle-class masculinity – and the view of everywhere from nowhere which hopes to construct a transparent space in which the whole world is visible and knowable (Haraway 1991). Trinh Minh-ha, for example, describes the claims to knowledge which constitute the master subject as 'territorialized knowledge':

> It secures for the speaker a position of mastery: I am in the midst of a knowing, acquiring, deploying world – I appropriate, own and demarcate my sovereign territory as I advance – while the 'other' remains in the sphere of acquisition. Truth is the instrument of a mastery which I exert over areas of the unknown as I gather them within the fold of the known.
>
> (Trinh 1990: 327)

This transparent space depends on, and reproduces, a subject position which imagines it is 'looking down like a god' with a 'lust to be a viewpoint and nothing more' (de Certeau 1984: 92); it captures its Others as passive figures in a landscape, and while constantly gazing at their Otherness never specifies itself. In this scopic regime, subject positions and spectating positions merge through a specific spatiality: a space which is produced by, and reproduces, the fantasy of the (potentially) all-seeing, all-knowing humanist subject (Deutsche 1991). To push the terms of this argument further, I want to suggest more generally that particular imagined spatialities are constitutive of specific subjectivities. Identities are constituted in part by the kind of space through which they imagine themselves. In contrast to the effort to stabilise the master subject through transparent space, for example, just as Probyn

(1993: 1) imagines the multiple and shifting dimensions of feminist subjectivity as like layers and layers of acetate transparencies, their lines mobile, fusing with and repelling each other, so I imagine several spatialities entangled, contradictory and shifting, mapped by and mapping each female feminist subject.

This chapter explores some possible dimensions of the spatialities of a female subject of feminism (in the urban West) by looking at the work of Jenny Holzer, Barbara Kruger and Cindy Sherman.[1] As Jones (1991) remarks, the work of these three artists has often been examined by feminist critics, and they have often been praised for challenging masculinist interpretations of Woman. It is thus possible to make a direct parallel between their work and the project of the female feminist subject: both aim to resist the figure of Woman as the material of masculinist representation. However, as Jones (1991) also comments, this praise is usually offered only in the terms set by Holzer, Kruger and Sherman themselves. Here, I want to consider their work particularly in relation to the argument just sketched about subjectivity and spatiality. I want to argue that the possible effects of the critique of subjectivity and sexual difference in the work of Holzer, Kruger and Sherman vary, and that spatialities are central to these variations. As the first section of this chapter will argue, their earlier work invites interpretation in terms of the 'space-off' of masculinist discourse because it tries to destabilise both phallocentric systems of meaning and their everyday space and subject by implying a space and a subject beyond them. I try to imagine what this 'space-off' does to everyday space and everyday representations of the subject. But in thinking about the spatialities of their work I also want to explore the risks of invoking that elusive 'space-off' and its enigmatic subject, and it is with these risks that the first section of this chapter ends. In the second section I suggest that more recent work by Holzer, Kruger and Sherman can be interpreted as an effort to counter that risk. I suggest that in their later projects, they offer more specified notions of both subjectivity and space, as alternatives to the master subject and his everyday space. Their work thus shifts, in that oscillation identified by de Lauretis as typical of feminism, between implying the non-representable and offering a more positive alternative. In so doing it creates a complex series of statements about power, subjectivity and spatiality, which offer some strategies for thinking about the space in which it might be possible to map the female subject of feminism.

DISRUPTING THE EVERYDAY, FRACTURING THE SPACE

De Lauretis locates feminism as a political struggle in a specific space: the everyday. She comments on 'the epistemological priority that feminism has located in the personal, the subjective, the body, the symptomatic, the quotidian, as the very site of the material inscription of the ideological' (de Lauretis 1986a: 11). The everyday is the 'ground where socio-political determinations take hold and are real-ized' (de Lauretis 1986a: 11–12), because the everyday is where identities are constituted. It is the arena of 'the sociocultural practices, discourses and institutions devoted to the production of men and women' (de Lauretis 1987: 19). Images which both constitute and represent sexual (and other) differences proliferate in this everyday space. Owens describes these images as stereotypes, and comments that 'while the stereotype enjoys an unlimited social mobility – it must circulate freely if it is to perform its work – it must nevertheless remain fixed, in order to procure the generalized social immobility which is its dream' (Owens 1984: 101). Both masculinity and femininity are performances made meaningful in relation to these stereotypes, more-or-less disciplined performances for an everyday audience which expects certain displays and rejects others. For many women, as Russo (1986: 213) remarks, the threat of everyday 'mistakes' – too much rouge, a dingy bra strap showing, a voice too shrill in laughter – is threat of misperforming, of becoming a grotesque spectacle. This audience of and for gender, complex as it is, is made by and reconstitutes 'the sociocultural practices, discourses and institutions devoted to the production of men and women'; the everyday is the space of the construction of masculinity, femininity and the heterosexual contract (among other subject positions). And because so many of its disciplinary strategies are visual, the everyday constitutes a particular kind of space. More specifically, it is a space constructed through a voyeuristic relationship between looking and being seen. Voyeurism is an act in which 'gratification is obtained without intimacy' (Deutsche 1991: 11); it is a controlling and distanced way of looking, and it is the voyeuristic distance between the performer and audience of everyday gazes which constructs everyday space.

Clearly this everyday space is extraordinarily complex. The proliferation of stereotypes on which Owens (1984) comments produces contradictions in the everyday: posters for slimming aids directed at Woman are displayed next to appeals for money for the passive-ised Others of the 'third world'; brutal documentary images are used to sell upmarket clothing (Back and Quaade 1993). Although it is the space of the culturally recognisable, it also includes the apparently culturally unacceptable, since it is the territory of both the Same and its Others. The space of the everyday thus has both its 'euphoria of the panorama' (Kruger and Mariani 1989: ix), and also its spaces around which a certain knowledgable ignorance centres – the slum, the inner city, the closet (Stallybrass and White 1986; Keith and Rogers 1991; Sedgwick 1991) – and this too produces instabilities. Nor is its heterosexuality guaranteed (Fuss 1992). Yet images of being caught and confined in just this disciplinary space proliferate in feminist writing of all kinds (Rose 1993a). Feminists emphasise what Owens (1984) describes as 'social immobility' by arguing that women are positioned in a particular way within the space of everyday glances and performances. 'In a world ordered by sexual imbalance, pleasure in looking has been split between active/male and passive/female. The determining male gaze projects its fantasy onto the female figure, which is styled accordingly' (Mulvey 1989: 19). Woman is defined as the sexualised object of the masculine gaze, and de Lauretis (1990: 119) hazards from this that 'this constitutive, material presence of sexuality as objectification and self-objectification ... is where the specificity of female subjectivity and consciousness may be located'. Hence the importance of various notions of 'masquerade' in discussions of femininity (Iversen 1991); femininity is seen as a performance for a masculinised audience. Despite its complexity, then, feminists have argued that everyday space 'is always already an illusion produced by specific technologies of representation that are not recognized as such in order to naturalize specific structures for ideological reasons' (Wigley 1992: 386).

The early work of Holzer, Sherman and Kruger was located in just this everyday territory. All three depend for much of their work on the images and language of one of the main sources of gendered stereotypes: the mass media. All engage with the mass media both to challenge those stereotypes and to make their work accessible to a wide audience. All share Sherman's feeling that 'I wanted to make something that people could relate to without having to read a book about it first. So that anybody off the street could appreciate it' (quoted in Nairne 1990: 132). Holzer and Kruger display much of their work not in galleries but in public spaces: as posters, on electronic billboards, as flyers, on trucks, as T-shirts. As Holzer says, 'from the beginning my work has been designed to be stumbled across in the course of a person's daily life' (1985: 64). They share a strong sense both of the power relations embedded in the everyday and of the instabilities in those relations, and one of their strategies is thus to utilise the contradictions of the everyday against itself. All three depend for their critique on strategies of mimicry and parody. Sherman says, 'I wanted to imitate something out of the culture, and also make fun of the culture as I was doing it' (quoted in Nairne 1990: 132). And Kruger has described this parodic strategy in terms of the 'space-off' described by de Lauretis, suggesting she is at once 'outside' discourse and utilising its resources: 'We loiter outside of trade and speech and are obliged to steal language. We are very good mimics. We replicate certain words and pictures and watch them stray from or coincide with your notions of fact and fiction' (Kruger, in Foster 1982: 89). Their strategies of mimicry, then, are designed both to echo and to displace hegemonic ways of seeing and knowing. In this first section I want to suggest some ways in which these strategies also engage with the spaces and subject positions of the everyday.

Both Holzer and Kruger seek to find and amplify the instabilities which already exist in the everyday: its contradictions, its excesses, its outrages. Kruger's efforts at displacement are perhaps the more obvious, at least at first glance. In the early 1980s she produced a series of images in which blocks of text were superimposed onto black and white photographic images salvaged from magazines or newspapers (see Figure 16.1). Kruger's work depends on citing the visual stereotypes found in these sources; men appear in business suits, women as passive poses and objects (Owens 1984). The terse text and the stark design are often explained by referring to the years Kruger spent as a magazine designer. In content, though, the slogans are far from the

Figure 16.1 Barbara Kruger, untitled (you are not yourself), 1982. Courtesy Mary
Boone Gallery, New York

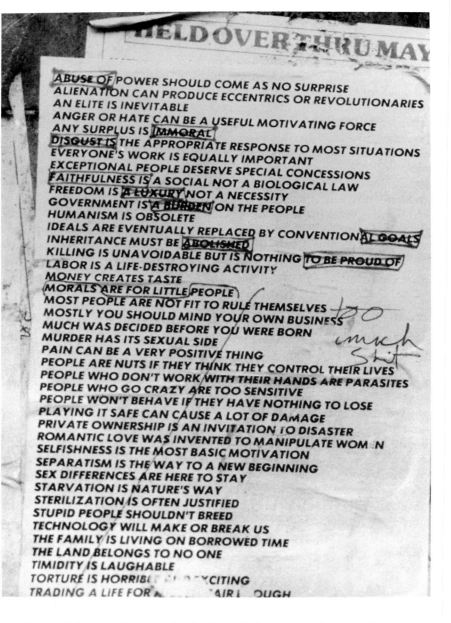

Figure 16.2 Jenny Holzer, selection from *Truisms*, 1980. Courtesy Barbara Gladstone Gallery, New York

reassuring language of advertising. Over a photo of the eyes of a man peering through some kind of optical instrument, Kruger states *surveillance is your busywork*. A photo of the head of a woman seen through frosted glass says *you thrive on mistaken identity*. Placed in city streets, on billboards and in subway carriages, these images and pronouns seem directly to address the webs of everyday gazes and to render their subject positions material: as Owens says, 'Kruger appears to address *me*, this body, at this particular point in space' (1984: 98). The spectators of the everyday are materialised, and they are materialised in much of Kruger's work as masculine. Over the silhouette of a naked woman surrounded by pins, for example, she writes *we have received orders not to move. Your gaze hits the side of my face* appears over the profile of a stone female head. The words *your body is a battleground* are superimposed on a photograph of a woman's face, half an ordinary photo and half a negative. In another image, Kruger writes *I am your reservoir of poses* across a photograph of a huge sun hat. Kruger insists on the power of masculinity to gaze, and the position of femininity as that which is looked at. But the texts of her work also disrupt that gaze's ability to know through seeing and understanding the stereotype; 'she suspends their masculine pleasures with the impertinences of superposed texts' (Linker 1990: 62). Kruger both draws on and defies hegemonic visual and textual codes.

Holzer's strategies are also impertinent, and funny. In her series of *Truisms* and *Inflammatory Essays*, and her *Living* and *Survival* series, which she developed between 1977 and 1987, Holzer used the language of advertising to displace its own authority. She polished phrases resonant with the certainty of dominant discourses, pervaded with a faded sense of cliché, and listed them one after another, on posters which appeared in New York streets (see Figure 16.2). In March 1982, the Spectacolor sign in Times Square beamed some of her *Truisms*: *protect me from what I want, money creates taste, you are a victim of the rules you live by, often you should act sexless* (see Figure 16.3). The result was 'verbal anarchy in the street' (Foster 1982: 88), as Holzer's own truisms competed with each other – *children are the cruelest of all, children are the hope of the future* – and with the surrounding advertisements (Graham 1981). Although this work only rarely addressed the question of sexual difference explicitly, like Kruger, Holzer constructs language as a site of conflict. And since the languages she pulls apart form part of the codes constructing everyday space and sexual difference, Holzer's banal confusions also heighten the fragmentation and contradiction of everyday space and gender identity. As Holzer said, referring to a more recent installation, 'I wanted to have as many things working as possible, up to the point of it being ridiculously confused, purposeless and grotesque' (Holzer 1990: 36).

Sherman's mimicry focuses on the visual language of the mass media. Her most famous parodies are her *Film Stills* of the late 1970s. In this series of about eighty fairly small black and white photographs, Sherman assumes a repertoire of poses and disguises, and pictures herself as a diverse range of figures from advertisements, Hitchcock and Godard films, Hollywood

Figure 16.3 Jenny Holzer, selection from *Truisms*, 1982. Courtesy Barbara
Gladstone Gallery, New York

B-movies and photo-magazines. At first, Sherman's repetitive manipulations
of the familiar iconography of femininity lure the spectator into a certain
complicity with her images; this is how we expect women to be seen, as
figures, often passively waiting or dreaming, or interrupted by something
happening outside their own scene (Williamson 1986). Women are represented through particular visual codes, sometimes looking like objects
'composed for the outside world and its intrusive gaze' (Mulvey 1991: 141),
but more often caught as if unaware of the camera's distant gaze (see Figure
16.4). We voyeuristically scan the images in the same way that we are voyeurs
of the everyday. And then a certain unease may begin. The cleverness of these
images consists in their repetitious provision of accurate signs of feminine
dress, demeanour and location. (In Sherman's work, femininity is constituted
not only through the figure but also through her location: women caught in
kitchens (#3, #10), women snapped in gardens (#47), vulnerable women
dwarfed by skyscrapers (#21), women glimpsed through the forecourt of a
concrete building (#63, #83).) The spectator can easily decode these signs, too
easily, and the signs become evident as such. The spectator's growing
awareness of this encoding of femininity reveals the complicity of us all with
everyday gazing. Sherman's manipulation of the visual construction of
femininity then disrupts claims to know what women really are by their
dress and location by revealing the constructedness of femininity. In

Figure 16.4 Cindy Sherman, untitled #54, 1980. Courtesy Metro Pictures, New York

Sherman's work, the everyday itself is examined. She destroys the transparency of its gazes and materialises its (masculine) voyeurism at (feminine) figures.

Although the work of Kruger and Sherman seems more able to carry a feminist interpretation than that of Holzer – it is easier to spot their engagements with the problematic of sexual difference – none the less their work is also much more complicated than a simple opposition between masculine and feminine. Although Sherman and Kruger clearly draw on Mulvey's analysis of 'woman as image, man as bearer of the look' (Mulvey 1989: 19), their work also confuses that neat division of scopic labour. In much of their work, for example, it is not quite clear on reflection what position the spectator is supposed to occupy. None of their work offers a simple or innocent alternative viewing position, and critics of all three artists have commented on the confusion their work engenders for the spectator. The usual positions are refused; the gaze which usually makes sense of everyday things is refracted or foiled. The masculine gaze is deflected, the feminine object lost. Holzer's enigmatic truisms put the onus of interpretation onto the reader, who is asked to negotiate their mutually exclusive claims to truth: a task only possible if coherence is abandoned. Sherman's work, in its manipulation of the positions of voyeur and and looked-at, artist and model, active and passive, offers no stable subject position; as Mulvey (1991: 142) says, there is 'no resting point that does not quickly dissolve into something else'. Even Kruger's preremptory slogans are ambiguous when their forms of address are considered (Owens 1984). If *surveillance is your busywork*, who is speaking? And who to? Is it addressed to men? Or does it comment on women's complicity with the dominant gaze? What then am I to make of the same image with the text changed to *surveillance is their busywork* (see Figures 16.5 and 16.6)? The uncertain subject position constituted by their work means that 'meanings shift and change their reference like shifting perceptions of perspective from an optical illusion' (Mulvey 1991: 147). A disorientation, a displacement, occurs for the spectator of Sherman's photographs or of Holzer's or Kruger's posters, even as they are materialised as spectators. Everyday space is confounded and its efforts to stabilise subject positions through stereotypes is disrupted.

I want to suggest that this disruption offers a 'stimulating aether of the un-named' (Sedgwick 1991: 63). It is a refusal to be assimilated into the figure of Woman and an implicit reminder of the heterogeneity of the female subject of feminism. I also want to suggest that it implies a spatiality other than the territory of everyday phallocentrism. In her discussion of mimesis, by which she means a mimicry of femininity which does not reduce a woman to that which she mimics, Irigaray (1977: 84) suggests that 'if women are such good mimics, it is because they are not simply resorbed in this function. *They also remain elsewhere.*' If everyday space and masculinist subjectivity is constructed through the visual regime of the stereotype, then Irigaray implies that the effort to occupy a different subjectivity also requires a different spatiality, the space-off or 'the elsewhere of discourse here and now' (de

Figure 16.5 Barbara Kruger, untitled (surveillance is your busywork), 1985.
Courtesy Mary Boone Gallery, New York

Lauretis 1987: 25). Stumbling across a poster by Kruger or Holzer, or absorbed by Sherman's manipulations, I imagine that occupying the uncertainties of subjectivity they offer is also to glimpse the possibility of another spatiality, to feel a crack in the grids of the everyday, to sense an aporia in the disciplining webs of gazes and performances. It is to dream of a space which is not the territory of phallocentrism. In this early work, neither Holzer nor Sherman nor Kruger specify the qualities of this other spatiality; indeed, as it is outside meaning they cannot. But it is there as a possibility: an emptiness fracturing the proliferation of stereotyping in everyday space.

These disruptions to spectating subject positions and the everyday space of spectatorship are what I want to emphasise in my commentary on the work of Holzer, Sherman and Kruger; they subvert the power of Woman and

Figure 16.6 Barbara Kruger, untitled (surveillance is their busywork), 1988.
Courtesy Mary Boone Gallery, New York

thus of Man. However, there are other ways to understand these images.
Several critics have used some very traditional terms in order to elaborate the
rejection of Woman by these women artists; for in refusing to represent
women in culturally intelligible ways as Woman, they can be accused of

presenting femininity as Man's enigmatic Other once more. Their masquerades may lead to enigmatic displacements, but that enigma is not necessarily
unassimilable to masculinist discourses; indeed, as many feminists have
argued, it is central to them (Jardine 1985). In western cultures, for example,
the Sphinx has long since stood as a symbol of feminine mystery, and reading
Holzer's work has been described as 'an encounter with the Sphinx' (Foster
1982: 91). Moreover, that mystery is also often understood as dangerous and
horrifying, and Kruger's work has been described as enacting 'the medusa
effect' (Owens 1984): a castrating femininity. In her discussion of Sherman's
project, Mulvey (1991) argues that Sherman progressively reveals this
misogynistic dislike of the mystery of femininity on its own terms. She
argues that in the *Film Stills*, 'an overinsistence on surface starts to suggest
that it might be masking something or other that should be hidden from
sight, and a hint of another space starts to lurk inside a too plausible façade'
(Mulvey 1991: 141). Here Mulvey is commenting on the space through which
femininity is imagined in phallocentric discourse: 'the phantasmagoric space
conjured up by the female body, from its exteriority to its interiority'
(Mulvey 1991: 139). By the mid-1980s, Sherman was still photographing
herself, but had changed her mask from that of acceptable femininity to
grotesque animal masks, and then to blasted body parts dismembered by the
photo frame, and Mulvey sees these images as a visualisation of that
'phantasmagoric space'. Sherman's photographs are understood as the
articulation of masculinist fears about the feminine enigma lurking behind
everyday masquerades; once Sherman refuses the mask, all that is left is a
horror of the unspeakable.

Mulvey is suggesting that Sherman's work follows the logic of femininity
as a masquerade for a masculine audience to a masculinist conclusion. Hers
is an important point to consider, since Sherman's co-masqueraders, Holzer
and Kruger, have also turned increasingly to the bodily in their more recent
work, and critics have repeatedly described their feelings of physical
revulsion and disgust at that work. Holzer's installation at the Guggenheim
Museum in New York has been described as like being in the head of
someone insane (Holzer 1990: 36); Sherman's work as being really frightening (Danto n.d.: 11); Kruger's as provoking fear, disgust and denial (Squiers
1987: 85). Is this disgust a masculine disgust at femininity, as Mulvey
suggests? Is the 'elsewhere' I have described as an exhilarating absence not a
stimulating aether at all, but just another glimpse of mysterious femininity as
a gulf into which Man stares horrified? Is it perhaps not a feminist 'elsewhere'
at all, but in fact an all too familiar topography of the feminine?

It is impossible to adjudicate between these two possibilities. Work that is
addressing sexual difference is quite likely to produce (at least) two rather
different responses from an audience itself constituted in part through that
same difference. The space implied in the work of Holzer, Sherman and
Kruger may be interpreted both ways. But here I want to pursue my feminist
reading to suggest that, in more recent work by these three artists, there is
a rather different relationship posited between subjectivity and space. Instead

of an un-namable disruption, a hostility to Woman, Man and everyday space, there is an effort to articulate some of the conditions of female feminist subjectivity. There is a move towards the bodily not in horror but with tenderness, and an exploration of a different spatiality through which to constitute that body in those tender terms. Their representation of the bodily provides a space through which the constructed and relational qualities of the female subject of feminism can be articulated.

RE-PRESENTING THE BODILY, DEMANDING A TENDERNESS

I have argued that the process of representation is central to everyday space and to the en-gendering of subjects in that space. Bodies and identities are disciplined largely through the kinds of images Owens (1984) describes as stereotypes. But bodies do not become stereotypes; as Kruger (1991: 446) says, 'the stereotype exists where the body is absent'. In their more recent work, Holzer, Kruger and Sherman have all engaged with the bodily in an effort to assert a material dislocation from the interpellations of the stereotype. If the previous section examined their earlier work in terms of their enigmatic disruptions which implied a 'space-off' beyond phallocentrism, then this section examines their more positive reworkings of representational practice and the positioning of the viewer in relation to the bodily; after all, now 'we have discovered the coercions of the media; we must develop the means by which we would confront them' (Linker 1990: 87). This is not to suggest that Holzer, Kruger and Sherman see the brute body as something on which to ground an alternative representational system, however. The body is central to the feminism being explored in this chapter, not because it offers a safe haven from the representational ravages of phallocentrism, but precisely because it is central to the constitution of Woman; as de Lauretis (1986a: 12) argues, 'the body is continually and inevitably caught up in representation'. Rather, Holzer, Kruger and Sherman are suggesting that seeing the physicality of the self in a particular way may provide moments of resistance for thinking new selves, new relations, new subjects of feminism:

> As my self does all the complex and mundane manoeuvres required of it, the sound of other sexed selves beckons and empassions. Wrought in the experiential and in the theoretical, these selves carry with them the movement of bone, of body, of breath, of imagination, of muscle, and the conviction of sheer stubbornness that there are other possibilities. These selves are made to speak of transformation, refracted they provide glimpses of other positions, lodged in the terrain of the social they rearticulate a geography of the possible.
>
> (Probyn 1993: 172)

In this section, I want to argue that Holzer, Kruger and Sherman are subversively engaging with the bodily in order both to confront the viewer with the costs of stereotyping and also to assert the need to care for the body in the face of the kind of disgust discussed by Mulvey and other forms of harm. In this work, the body is represented as a site for imagining the construction of the complex and relational female subject of feminism. And the representation of the bodily in this way also involves a spatiality which is not that of the everyday.

Holzer's recent work has shifted from billboards and Spectacolor signs to installations in art galleries. Although this work is less of a direct intervention into everyday space than many of her earlier projects, it still addresses one of the central issues of the everyday: the ways bodies are categorised, disciplined, manipulated and hurt. One of her most sustained considerations of this theme was her installation at the Dia Art Foundation in New York in 1989, called *Laments*. *Laments* consisted of eleven marble sarcophagi and eleven light-emitting-diode columns in a dark room; the words engraved on the marble were also transmitted by the LEDs and by tapes of voices. The texts are elaborations of the conviction that *death is the/modern issue*. The texts and voices explore disease, war, wounds, murder, ecological death, aggression.

> The new disease came.
> I learn that time
> does not heal.

The language is terse and the voices lonely and flat.

> It became too hot.
> The black dirt's
> heat made the
> air wriggle . . .
> I cannot go
> where it is
> cooler because
> people there
> are awake
> and armed.

There are no images of bodies in this installation, despite the intense viscerality of the texts; the bodily is invoked by words written and spoken. But the embodied subjectivity of the spectator is certainly rendered problematic. Indeed, 'spectator' no longer seems the appropriate term, with its connotations of distance and control, since the spectating gaze is foiled in *Laments*. This is mainly because of the space Holzer has created in the installation. The space is disorientating, the darkness punctured by the LEDs, their electronic messages travelling in different directions, by the soft spotlights on the sarcophagi and by the different voices. In this space there can be no panorama. Here it is difficult to achieve the distance of the everyday from the implied bodies; speaking about *Laments*, Holzer said she hoped that 'you will completely lose the space' (Holzer, in Waldman 1989:

19). That everyday space is here displaced; instead of a cool and distanced gaze at bodies I feel pulled into an intimate relation with the bodily, made to care for it, to feel horror at the violence perpetrated on it. In *Laments* I feel less of a spectator and more of a participant, placed in close proximity to the dying by a particular kind of organisation of space, and made to ache for the bodily. In insisting, then, on a relation to the bodily which is not one of violence and/or disgust, Holzer also argues for a different kind of space, constructed not through voyeurism but through intimacy and care.

While Holzer addresses physical kinds of violence, Sherman continues to focus on what Gayatry Chakravorty Spivak describes as 'epistemic violence'. Her most recent photographs turn to the western tradition of representing the naked female body (Avgikos 1993), and show nudes – of a kind – in the traditional poses of that genre, poses which constitute the masculine gaze which voyeuristically scans the female figure and also tries to dominate the everyday. The spatial organisation of these photographs is the space of the everyday gaze at a landscape with figures: a panorama of the female body. Through this space, Sherman addresses the possessive and eroticised scenography of the naked Woman painted for the pleasure of the masculine viewer (Berger 1972), and also comments on the exotic Orientalism which often pervades such images in western art – a brief comment on the racialised as well as the gendered dynamics of such images. But the bodies in these photographs have lost the veneer of humanity which even Berger insisted would exist in an image of a naked female when the heterosexual contract between painter/husband and model/wife was one struck between equals (Berger 1972: 57–8). For the bodies in Sherman's photographs are made up from bits of broken dummies, medical equipment, remnants of fabric, cheap wigs, junk, masks. Instead of the proliferation and plenitude of everyday stereotypes, these images offer brutalised fragments of bodies, bodies made from objects. Sherman here is articulating the effects of everyday voyeurism on the body. She is visualising the objectification of bodies; as the voyeur fetishistically glances at hands, legs, hair, breasts, she materialises each part of the feminised body as an object disjointed from the rest. Here the costs of everyday voyeurism to those so subjected are being counted.

Given the spatiality of these more recent images, I want to return to Sherman's photographs from the mid-1980s and suggest that some of them visualise something other than masculinist disgust at the enigma of femininity. For, as Mulvey (1991) herself notes, those photos from the mid-1980s have a spatiality rather different from that of the everyday. They are overwhelming and fragmented. Sherman used a much larger size of photograph for this series than she had previously, and the resulting image cannot be encompassed in a glance. This reduces the spectator's feeling of control over her work. The viewpoint of these photos does not offer the euphoria of the panorama, but looks closely and down onto bodies and earth, gravel, lichen. These images of bodily decay offer no commanding panorama but rather insist on a careful study of all the image, all its details. Sherman did not invite a disciplining surveillance of these images – the kind of gaze her anti-

Figure 16.7 Cindy Sherman, untitled #153, 1985. Courtesy Metro Pictures,
New York

nudes of the 1990s expect and reject – but rather an attention to the bodily which is a sort of fascination at its extraordinariness and vulnerability (#177). There is a wonder at the bodily in this work, which, together with the violence to which these bodies have been subjected, invokes a kind of tenderness for the body (see Figure 16.7). And as Holzer laments, such tenderness is all too rare in this era of commonplace torture, mass starvation, war and the abandonment of those with AIDS. I'm suggesting that, like Holzer, some of Sherman's photographs from the mid-1980s posit a different relation to other bodies from that of everyday space. Both Sherman and Holzer are insisting on the care of the bodily. And they are doing this by displacing the voyeurism of the everyday and the stereotype with a space which invites attention to and care of the bodily detail.

Viscerality is a path Kruger has also followed. In her more recent work, she has begun to use coloured images of bodies or body parts, with text, and one image asks *do I have to give up me to be loved by you?* across a depthless and enveloping image of a pumping red heart. Here again, the voyeurism of the everyday has been eschewed in favour of bodily detail, and the body itself becomes the site of a demand for subjectivity. The nature of this demand is elaborated in other recent work by Kruger, which uses lenticular screens. These screens invite the viewer to move in order to see the two images they contain. In one sense, these screens continue Kruger's earlier efforts to challenge the assumed wholeness and innocence of the humanist subject and everyday space, since the same image can produce two contradictory messages for any one spectator. The words *My hero!* across a photo of a barechested man looming out of the frame, for example, shifts into *You can dress him up but you can't take him out*. This renders the spectator's position fragmented, even unstable. But because the viewer has to move to make this happen, bodies themselves become the site of contradiction. For Kruger, the body is a site of multiplicity; 'the body's multiplicity has a strategic aim, for in multiplicity is the key to mutability, to social transformation, and to change' (Linker 1990: 87). The bodily in her work thus becomes the location of a demand for a particular kind of subjectivity which is mutable and thus open to (re)construction. Kruger's lenticular screens also embody the relation between viewer and seen, and this reminds me that to think of the feminist subject of feminism is not only to (re)think myself but also to think about the other woman, other women. As Probyn (1993: 112) says, feminists must 'remember that selves always work with other selves within discursive events'. An installation by Kruger in her New York gallery in 1994 emphasised this point. Its critique of nationalistic, (neo-)fascistic, fundamentalist oratory surrounded the visitors (or rather, again, participants) with images, text and sound, and problematised their relation to the other visitors by problematising their relation to the audience, to the crowd and to the social. For Kruger, the bodily constituted through an intimate space is not a way of specifying an essential identity but rather a sign of the constructed, relational and above all political (feminist female) subject.

In this section I have argued that in their more recent work, Holzer,

Kruger and Sherman have become less elusive and more prescriptive. Instead of disrupting phallocentrism from an enigmatic 'elsewhere', I have suggested that their newer projects articulate the material costs of phallocentrism (and other oppressions) more directly, and that they also offer alternative ways of seeing and space. By focusing on contemporary forms of violence, their work insists on the centrality of embodied subjectivity to relations between people, and on the need to rethink the dominant form of subjectivity. Their representation of the corporeal as a non-determining site for redefining subjectivity for political ends also emphasises the complexity of subjectivity and the need for tender forms of intersubjectivity. These concerns parallel the ways in which I suggested that the female subject of feminism was political at the beginning of this chapter. And in the spatiality of their more recent work, Holzer, Kruger and Sherman also suggest a more intimate space in which to map that feminist female subject.

CONCLUSIONS: SUBJECTIVITY, SPATIALITY, BODILY

In this chapter, I have argued that everyday space is not simply a container within which the various technologies of gender work, but that it is itself produced by those technologies. In this territory, phallocentrism tries constantly to establish its hegemony, and in particular to regulate identity through the circulation of stereotypes. For many feminisms, the space of the everyday is 'the space of recognition' (Adams 1993: 131), a space in which women are recognised as Woman by a voyeuristic gaze searching for images of its Other, hoping to recognise itself in and against its Other, attempting to make its meaning by seeing the stereotype, at a distance, the distance of everyday space. In this argument, everyday space is a constitutive element of the contradictory relation between the phallocentric subject and his Other; but, imagined differently, space could also constitute a different relation, a different subject, a subject which could imagine a relation between subject and subject, the female subject of feminism.

If, then, as Kruger comments, *your body is a battleground* because in part a battle over the definition of the subject, it is clear that the kind of space in which that battleground, that body and that subject might be mapped is not separate from the battle itself. This battle is also being waged over the meaning of space. For feminists dreaming of new kinds of subjectivity, new kinds of space must also follow. And indeed, the chapter has argued that spatialities are central to the feminism implicit in the work of Holzer, Kruger and Sherman. I have tried to argue that their work explores different spatialities which are connected in complex ways to different notions of subjectivity. For example, they have addressed the dominant subjectivity – that of white, middle-class heterosexual masculinity – and shown its connections to an everyday spatiality of surveillant voyeurism. They have also tentatively begun to suggest another spatiality which is constituted by, and itself constitutes, a different relationship to the body and the subject from that of everyday space and the master subject. Against the distance and the objectification of everyday voyeurism, Holzer, Kruger and Sherman have offered a different kind of space. That space allows a relation between subjects rather than between the subject and his Others. It refuses the stereotype in favour of a certain bodily detail. It replaces the violence to which so many bodies are subjected (not just female bodies) with a tenderness and care. Careful attention to the bodily becomes a space of resistance.

Yet this is not a simple reversal of terms. The work of Holzer, Kruger and Sherman oscillates between the imperative to describe an alternative to the disciplines of the everyday and a desire for the unrepresentable. None of these artists suggest simply replacing the distance of the voyeuristic panorama with the closeness of the intimate detail; neither do they suggest replacing the visual with the tactile. They are neither surrendering the gaze nor abandoning space, because their larger *oeuvre* also insists on an unrepresentable spatiality and imagery. Their work implies that 'elsewhere', the space-off beyond discourse, and that 'elsewhere' is central to their project of 'trying to construct another kind of spectator who has not yet been seen or heard' (Kruger 1991: 435). This 'elsewhere' renders their turn to bodily intimacy provisional; it undermines any essentialism which their turn to the bodily might otherwise imply.

Parveen Adams (1993) argues that it is politically and subjectively vital to keep a sense of a space empty of recognition as a sign of the vigilance and hope with which a politics of care must conduct itself. I will conclude by suggesting that to think about the geography of the female subject of feminism is not to be able to name a specific kind of spatiality which she would produce; rather, it is to be vigilant about the consequences of different kinds of spatiality, and to keep on dreaming of a space and a subject which we cannot yet imagine.

NOTE

1   For the work of Holzer, see Holzer (1990) and Waldman (1989); for the work of Kruger, see Linker (1990); and for Sherman, see Avigkos (1993), Danto (n.d.) and Kellein (1991).

17

ETHNIC ENTREPRENEURS AND STREET REBELS

looking inside the inner city

Michael Keith

A friend of mine wants to be a parliamentary candidate in a constituency in the East End of London where about 60 per cent of the Labour vote in the last General Election came from the Bangladeshi community. The community originates from the region of Sylhet and has been present in the area of East London near to the docks for many decades but is largely the product of the last wave of post-war boom British Fordist migrant labour in the late 1960s. Or perhaps not. Given that this particular fraction of New Commonwealth settlement was deployed principally in the rag trade and in the restaurant and catering business – case studies in the flexible labour process – Sylheti settlement in the East End is perhaps better understood as the first British case of post-Fordist labour migration. Either way, the vagaries of both trades have contributed to an instability in the racialised labour market which paradoxically parallels the East End's history of casualised labour, and almost normalises the 'exceptional' case of Bengali settlement.

My friend comes from a particular subdistrict within Sylhet but grew up principally in the London Borough of Tower Hamlets and has made a reputation fighting the racism for which the area is so notorious, rendering the term East Ender itself at times a metonym of white English working-class identity. Another friend, articulate, middle class, in Tower Hamlets for less time but likewise with a long and honourable record of support for progressive left politics also wants the same parliamentary seat. Commonly, representations of him, instrumentally choosing to exceptionalise his identity, point out that whilst he comes from Bangladesh he is from Dacca and a more affluent background than most Sylheti people locally. I ask my first friend why he should be preferred in the parliamentary selection process, expecting a similar line. But instead of restating this categoric form of the global/local, the subdivision of Sylheti/non-Sylheti in London, he says to me: 'Michael, this is a poor working-class inner city area. We need an Eastender to be the MP here. That is why I should be chosen.' For a moment the whiteness of the East End itself is not only challenged but also rhetorically colonised. The moment, though on one level insignificant, stays with me. It is an exemplary case of the manner in which for all the sometime fluidity of a new politics of cultural difference that takes race, class and gender as mutable, the much celebrated hybridised identities that result, defining themselves as speaking subjects, invariably come to rest in a moment of closure that creates, however temporarily, an inside that lends meaning to 'community', that defines a structure of sensibility without which trans-cultural communication cannot begin. It is an inside invoking complex notions of spatiality that is subject to endless reinvention, (mis)appropriation and (mis)representation. It is also quite clearly an inside that cannot be measured within a straightforward metric of correspondent truth. More specifically, the incident, though anecdotal, highlights the manner in which a particular set of places – in this case the East End, Bangladesh, 'Asia', and most of all, 'the inner city' – do not so much bracket identities as become constitutive features of them. In this sense the mapping of subjects is constituted by the invocation of place as much as the genealogy of placing.

On one level there is something quite unproblematic here. We make sense of the world by the stories we tell ourselves, by the urban narratives and tales of the inner city through which characters come to life, sometimes exemplary, other times exceptional, sometimes didactic, other times mundane, sometimes reassuring, other times horrifying, though as always this horror itself rests on the sublime conflation of disgust and desire.

Such characters owe their life to the narrative forms through which they are allowed to emerge. The racialised city has a full cast list, the process of 'mapping the subject' is about how such subject positions or 'subjectifica-tions' are made politically, epistemologically and aesthetically visible. In this chapter I want to look briefly at the manner in which two subject positions of racialised otherness draw on the historical genealogies of representations of 'blackness' to define the parameters of policy thinking and urban policy practice that may on the surface appear to be free of such culturally specific traces.

The two characters are those of the *ethnic entrepreneur* and the *street rebel*. These two figures share an historical provenance that is far too complex to outline in great detail here, though both become organising themes through which race, gender, class and sexuality are invoked to make sense of the inner city. Thus read they become sublime personalities with iconic status. The ethnic entrepreneur is the assimilationist hero, the street rebel the bourgeois nightmare. They are characters who have made sporadic appearances throughout industrial history but whose realisation is always historically and geographically specific, from the Huguenot weavers of the seventeenth century to the Asian shopkeepers of the twentieth; from the hooligans of the nineteenth century to the black rioters of the 1980s, almost a hundred years later.

For the ethnic entrepreneur the inner city market is a mysterious place that generates its own protocols of institutional behaviour and generates their 'placed' forms of expertise, for the rebel the city provides a stage for transgression, a place where the 'shouts in the street' can be heard. 'Race' echoes through both characters, though creating quite distinctively racialised subject positions. Both are quite clearly absent presences in the imagined cities that inform contemporary urban policy.

We know that the inner city is not the straightforward product of a uniform set of processes working through the production of spaces in a sovereign political economy to manufacture identifiable tracts of urban decline. It is, among many other things, the product of the political imagination which generated popular understandings of an inner city problem, a mythical space that is at once both inside and outside our own society (Keith and Rogers 1991), that owes its own genealogy to the very cultural roots of a fascination and fear of the city itself.

In such a context I take it as unexceptionable that, *inter alia*, policy initiatives in the inner city rest on narratives of urban decline and salvation, replete with aestheticised and deeply politicised representations of contemporary urbanism. These stories of the inner city require and define their moral cast list, an array of exemplary characters who can serve as both the suitable objects and subjects for techniques of state intervention that are sometimes characterised as urban policy. This chapter is about how such characters are imagined, written and enacted. For the purposes of this volume I am interested in teasing out the forms and techniques of government in the inner city which render racialised subjects visible at some times but not at others.

In small area urban policy the very act of mapping at once both erases and reinvents, sometimes creates and defines 'a place' from a disparate set of census tracts. Political spaces are transformed into natural objects. This involves a process of erasure and a process of reinvention. The vagaries of cartographic boundary lines confer upon an area of the inner city a unity that is entirely governmental. It may be rationalised in terms of an array of pathological or material indices of deprivation that maps 'a population' or it may be marked by a particular potential for economic growth but it defines a 'suitable case for treatment', a proper place for inner city government. It is not the case that by making visible the fabricated governmental root of this inner city geography there is an implicit alternative natural or organic geography of authentic urban areas. It is instead the case that implementing agencies create a space of governmental practice which, thus legitimised, becomes the natural object of regulatory practice.

In this chapter I want to draw on two such mapping exercises in London, one in the East End, the other in Deptford. Both are cases of governmental mapping of the official inner city through the City Challenge initiative, launched in 1991 by Michael Heseltine when Secretary of State at the Department of the Environment. It was an initiative characterised by an auction of victimhood; local authorities had to compete one against another to prove that both their 'inner city' was more deserving of additional resources and that their techniques of urban renewal were the most impressive. Two London Boroughs, Lewisham and Tower Hamlets, were among the eleven local authorities out of seventeen invited to bid, that won City Challenge initiatives for their locality in the first generation of the exercise.[1]

Some basic features of this creation need to be restated. People in the northern wards of the London Borough of Lewisham are generally poorer than most other people in the capital city, have worse housing conditions, have poorer health, have less chance of being in employment, greater chance of being poorly, or seriously ill or dead by tomorrow morning than people living in other parts of London (Deptford City Challenge Evaluation Project 1993). The clichéd litany of socio-economic deprivation indices that identifies the area itself as an area of lack may be the product of particular techniques of quantification and may be the potential source of tales of urban pathology but they are no less grim for being so. It was in this area, which receives most attention in this chapter, that Deptford City Challenge (DCC) was established in 1991 and began operation in April 1992.

In the Spitalfields area of the East End a similar list of socio-economic characteristics, coupled with a number of derelict land sites with clear development potential, near to the City of London, prompted the government to invite the London Borough of Tower Hamlets to locate its City Challenge bid in the Spitalfields area rather than in any other part of the borough.

Attenuating in one place, emerging in the other, our two racialised characters appear quite differently in the two sites, generating quite different forms of political action. In one case the character is progressively erased, the ethnic entrepreneur so favoured in the free market world of the Thatcherite 1980s looks progressively out of place in the corporatist 1990s. In the other case the spectre of inner city uprisings offers the Faustian bargain of political salience and state criminalisation to the previously unheard and unseen mobilisations of young Bengali men. The purpose of this chapter is neither to valorise nor to denigrate the agency involved in both forms of action, only to make visible the manner in which such political manoeuvring depends on, however momentarily, inhabiting such subject positions; occupying the inside of characters generated by an institutional landscape that defines the inner city.

ETHNIC ENTREPRENEURS IN THE INNER CITY MARKET

In the wake of the uprisings of the early and mid-1980s, and Margaret Thatcher's election-night pledge to 'do something about those inner cities', British urban policy went through one of its periodic changes of direction, this time closely steered by the libertarian right's invocation of the natural social relations that emerge from the free market-place, that utopian site of social order. Old and new policies were bundled together and presented in the form of the 'Action for Cities' programme (Atkinson and Moon 1994). One element of this new portfolio was the prominence given to the potential for enterprise to resolve the crisis of the inner city. There had long been a fascination with the sometime remarkable success of migrant minorities in the economies of metropolitan receiving countries on both sides of the Atlantic, an interest occasionally endorsed and reified by academic study of a phenomenon of *ethnic minority enterprise* (Ward and Jenkins 1984). For the designers of urban policy, the technocratic engineers of the imagined city, the ideological power of such a subject was considerable.

As a 1980s character the ethnic entrepreneur was given prominence through an unholy alliance between political expediency and academic fashion. Institutionally recognised and defined agencies were created to seek out these racialised saviours of the inner city. Specifically, the government created enterprise agencies with a brief to enhance the potential of black and ethnic minority businesses in the inner city, and charities such as the Prince's Trust were quick to follow suit with similar initiatives, whilst the newly created TECs were also allocated old Section 11 monies in the form of the Ethnic Minority Grant, part of which was to be directed at 'enterprise services'. In the recognised 'inner city' area of the London Borough of Lewisham the Deptford Enterprise Agency (DEA) was established in 1988, funded by the DTI and by Section 11 monies, whilst the local authority also created its own STEP Business Club with a brief to support 'ethnic enterprise'. In part strategically linked to such developments, many different ethnic minority groups have formed their own associations across the country, though rarely working across areas that match exactly the imagined geographies of 'inner city government'. In the case of South London the Afro-Caribbean Business Association worked across a broad swathe of the area, not just within the Deptford City Challenge boundary, and were a powerful political lobby locally.

Deptford City Challenge were obliged to present to government a 'face' of the private sector to represent capital in the structure of 'partnership'. Through a logic of 'forums' created to represent fractions of the partnership the local authority, in establishing the implementing agency that was to become City Challenge, created a 'business forum' to 'represent' the private sector interests within the partnership. In its early form the body attracted little interest once it became known that monies of the new agency – some £7.5 million annually – were not to be hypothecated to each forum. The intention instead was that these representative bodies would elect members of the board who would, in a non-executive fashion, steer the strategic direction of urban regeneration. In short the business forum did not control resource allocation, although it did create a rubric through which 'business' in all its forms, from major corporate players to stall-holders in the local street market, was to be made institutionally visible. Significantly, it homogenised fractions of capital through a democratic structure that defined the very different elements that constitute the 'private sector' in terms of what they had in common, mediated through a mandate that awarded each properly acknowledged 'member' of the business forum a single vote. Thus the 'local' private sector was institutionally defined within the overarching corporate framework.

The case for developing a support framework for local business, an element of the programme that the DoE had already pointed to as weak in the first-year review of DCC, was strengthened by an external consultant's report which made the case for the development of an independent local business association. From late 1993 onwards, once it became known that the 'business forum' would be transformed into a long-term Deptford Business Development Association (DBDA), the representation of the private sector developed into a controversy that split the board and threatened the working of the new regeneration agency. At one point a senior officer of the company, when asked how serious they thought the crisis could become responded:

> it could become extremely serious. I think it could fracture the board at one level. The local business sector could be even more split than it is. And we might not be able to function in this area. It would be very difficult to know how we were going to go anywhere.[2]

At the heart of the issue was the fact that though time-limited themselves to five years, the new City Challenge companies were intended to create a permanent impact on their area, leaving behind sustainable institutions. In the case of the Business Forum/DBDA transformation, sustainability of the new organisation was linked to a £0.5 million 'soft loans' (reduced rate loans) package for inner city enterprise floated by Midland Bank and a possible future as a 'one stop shop' or 'business link' supplying 'enterprise services' for the South Thames TEC. An uncontentious consultative arena had been transformed into a site of political contest and perceived potential financial power.

Although Midland Bank was to control the soft loans scheme it was mooted to be linked to the DBDA, who would administer business 'health checks' to those receiving funds, for which they would be paid. This tied the initiative to funding and resources. Although the link with Midland was subsequently broken – the former, it transpired, could proceed without the latter – the DBDA remained closely linked with the possibility of attracting a 'business link' into the area, another possible funnel for regeneration monies.

The process of urban regeneration is shaped by the political agenda at ministerial level and the manner in which this translates into an institutional environment established in the inner cities by the various departments of government. This is true historically as much as it is the case in contemporary times and it is frequently the case that political fashion changes more quickly than the institutional forms that it creates. The city landscape is characterised by an institutional archaeology. Various initiatives, all the product of their own time, litter an area, all bearing testament to yesterday's vogue notions of inner city regeneration. Hence within a few miles of Deptford it is possible in 1994 to find urban programme initiatives, reflecting the urban policy agenda of the 1970s, an Urban Development Corporation, reflecting the government agenda of the early 1980s, a Task Force, reflecting the responses to the inner city riots of the mid-1980s, various enterprise initiatives reflecting the 1987 election-night promise of the Prime Minister and the development of Greenwich Waterfront, reflecting the preferred left local authority perspective of the same time. And, of course, a City Challenge company.

These initiatives, in their different ways, were created as different solutions to the single, politically defined problem, the decline of the 'inner cities'. Moreover, whilst it would be overstating the case to suggest that there is no lesson-learning process, they cannot be considered to evolve one from another in enlightenment fashion, passing on 'good practice' and building on the accumulated wisdom of their predecessors, at least not in any straightforward sense. They are instead at least as much the product of changing political thinking and definition of the 'inner city problem', mixed with the perennial need of government at ministerial level to launch new initiatives invested with maximum symbolic impact. Consequently, they inevitably involve some level of duplication, one of the other – in funding revenue posts for the voluntary sector, in financing capital improvements to the environment, in training initiatives and in grants and loans packages for small businesses.

All places have their appropriate codes of behaviour of subjects – the notion of 'good practice' in urban regeneration creates a realm of technocratic, value-free government of the value-loaded social world of the inner city. In this sense the encouragement of ethnic enterprise had by 1994 been routinised within this world and the institutional players had been endorsed by several years of central and local state funding. These practices set funding precedents, invented a whole new vocabulary and set of administrative protocols that defined the black business community as a political entity as well as one aspect of the local economy.

However, with the apparent absence of uprisings within the black community in the 1990s, the central government will to address explicitly issues of racial deprivation in the inner cities, never particularly strong at the best of times, is not what it was ten or even five years ago. Partly in consequence, in the market-place of institutional fashions the notion of 'ethnic enterprise' is in government circles now about as avant-garde as yesterday's breakfast.

This leaves the institutional legacies of the 1980s competing to survive in an era when Section 11 funding, one of the major resources for ethnic minority support in the inner city, has been largely abolished, where the ethnic minority grant in the TECs has also been abandoned and where the calculation of the Standard Spending Assessments for local authorities has been changed to exclude ethnic minority presence as indicative of social need and thus cut central government support for local authorities with a large ethnic minority presence.

At the same time, across the country the newly empowered TECs are looking to establish 'one stop shops' for business advice and support, a concept that in its most recent guise is known as the 'business link'. TECs are relatively well supported, and South Thames TEC even ran into some controversy when in the 1992–3 financial year a seven-figure underspend was represented in their annual report as a 'profit'. The apparent affluence of the

TEC, and their increasingly significant role in disbursing enterprise monies, contrasts vividly with the ostentatiously empty coffers of the local authority. It was in this context that the new mediation of City Challenge regeneration ran into difficulties, caught up amidst unsubstantiated but publicly aired accusations of disingenuous manoeuvring, personal and corporate corruption and institutional racism involved in its development. All developments were framed by alternative invocations of 'race' in the inner city and all were underscored by competing and contradictory representations of who were to be the proper subjects of urban regeneration.

At one City Challenge board meeting in late 1993 the DBDA issue erupted into controversy. Objections were raised about the implications of the DBDA in terms of the possible duplication of existing services and it was alleged that the ethnic minority business sector were being excluded from the new package. Two members commented: 'we should treat this report with all the contempt it deserves ... This is a load of baloney. I have never been so insulted in my life', 'This is nothing but a feather in somebody's cap'. Other members disagreed and expressed concern that any uncertainty around the association might jeopardise Midland Bank's commitment to a scheme welcomed by all.

Board Member: '[without DBDA] would we loose the Midland £500,000?'
DCC Officer: 'I think we might. They have said that they are very keen on the Business Development Association to act as a referral agency.'

It was also suggested by another board member that because of the weakness of the company's commitment to the business sector, 'the reputation of City Challenge within the local business community is already at a low ebb'.

The issue went to the vote with the majority in favour of delaying approval of the DBDA pending further investigation by the chief executive. Controversy continued over subsequent months. The business forum, which had approved their own transformation into DBDA at an earlier meeting that was not well-attended, reconvened in January 1994, in a well-attended and rancorous event that eventually supported DBDA but in the terms of one DCC officer 'split down the middle on race grounds', voting fifteen to six in favour of supporting the transformation, with several present deemed controversially ineligible to vote. By March 1994 disagreements were still not settled and for the chief executive the issue had become 'a corporate embarrassment'. A resolution appeared to be imminent in June when a reconciliation launch of DBDA took place, facilitated by an external consultant. However, controversy engulfed the appointment of a manager of the organisation, again focused around the race issue. This prompted the suspension of City Challenge funding of the new business association, pending the resolution of alleged equal opportunities irregularities and the future of the DBDA at time of writing remains uncertain.[3]

At board level the struggle took form rhetorically through the allegation that the new association excluded 'black business', a collective whose very existence was disputed by other members of the nascent organisation.

In this context it is useful to distinguish unsubstantiated charges of corruption from the rhetorical structuring of these charges. Both those vigorously opposed to the new DBDA and the officers of the DCC concerned with promoting it spoke a new language through an old tongue. Agencies that had subsisted on the support for ethnic minority enterprise, owing their funding to Section 11 monies that were rapidly disappearing, were forced to rationalise their behaviour in terms which echoed the protocols, vocabulary and political realities that reflected the normalisation of something called ethnic minority enterprise.

For one board member the DBDA development was a case of racial 'divide and rule' whilst for one enterprise agency locally:

> Well, I mean, we're not talking about political democracy or a plebiscite or whatever it is. I mean, what is democracy – 2 per cent, 3 per cent, 4 per cent? I don't know. But I mean, it certainly wasn't . . . and it didn't have, or it does not have, as far as I'm aware – even as it stands – the confidence or the support of the black and ethnic businesses in any significant degree . . . it will fail because it will not be the only representation in the area. I'm confident that it's not going to have the confidence of the ethnic minority and community businesses in any significant fashion.

Not dissimilarly, one senior DCC officer suggested that the original feasibility study for the association was not as 'inclusive as it should have been' and that 'the community of Deptford is a diverse one and you ignore diversity at your peril'. The suggestion was that there was not a premeditated problem, but still a genuine one: 'I don't know that they were cut out [black businesses] – that is to suggest that it was deliberate – I think there was a flaw in the process.'

Crudely, for some the contest was all about the new agency as source of potential finance:

> Then we decided that we know very little about the DBDA and it was coming into the Deptford area and regenerating the whole area, going to get money from the Midland Bank and from the Prudential Insurance, and they were going to bring investment into the area and all the rest of it and to me it was just a wild idea and it was pie in the sky as far as I was concerned. So we said how can people sit back and say they are going to do that because the local community don't know nothing at all about the DBDA. If you walk round Deptford and say to anyone: 'Have you heard about the DBDA?' they would probably think it was an illegal organisation and that nobody had heard at all about it! So we try as far our best to try and find out what the DBDA was and what they are going to do.

It was in similar terms that at one time a board member suggested that there was no legitimacy in going ahead with the transformation from Business Forum to DBDA, and in reference to the controversial vote in favour of this move suggested that 'There has been a lot of mistrust for whatever reasons in the community. I don't call it a mandate, I call it a ******* stitch-up.' Yet the rhetoric of those involved directly with the new DBDA contested this challenge by disputing the legitimacy of the terms in which it was set. At the heart of the race issue within this debate is whether or not 'the black business community' is a meaningful term and a collective that defines a real 'player' in this equation that can take part in the new 'partnerships' through which the inner city of the 1990s was to be governed. In the terms of one influential figure in the history of the Business Forum:

> I don't know why people like xxx are insisting on – well I don't know what they really want. I mean the kind of logical corollary of what they are saying is they want some kind of black section of the business forum. They want a constitutional guarantee to a certain number of seats ... I will stand against it because I believe in democracy in a way. I mean I do agree with Edmund Burke in that it is sort of rule by the swinish multitude but nonetheless we have to work with it ... I don't believe there is this kind of racial tension within the business community that maybe xxx or xxx or xxx worries about, certainly like xxx. It is a convenience or whipping post I think to whip us with, and when I say us I mean sort of white business people, in order to get a disproportionate share of resources, time, energy, you know that kind of thing. When you look at the business forum, when you go and talk to some of these people, there is not this sort of racial worry that I am trying to be persuaded exists.

Once the DBDA became a contested issue various interests clearly began to perceive, rightly or wrongly, that rival interest groups were trying to seize control of an initiative that some construed as a potential source of future revenue and legitimacy was to be contested in every way, from who was to define the geographical extent of the local business community – should governmental boundaries, corporate entities or racialised business associations define this territory – to a game controlling the membership of the new entity. Though in some respects parochial, this struggle was significant, though not because there was clearly signalled right or wrong on either side. Conventional understandings of racist exclusion are of little help in making sense of the processes at work here. It instead reveals the changing visibility of racial difference within the institutional frameworks that define the inner city.

Politically, some truths become unspeakable. In slightly simplistic terms the visibility of racial deprivation in the 1980s was only rendered acceptable through the nostrums of the free market. Left to choose between working within this frame of reference and receiving no money at all, it was understandable that whilst some would take such notions at face value others would subscribe more pragmatically to the maxims of 'ethnic enterprise'.

Alternative paradigms of labour marginalisation that might have diagnosed racialised poverty in different terms and prescribed alternative remedies to such inequality were denied mainstream political currency in the formulation of 1980s urban policy. The 'black business community' could thus never be divorced from the political symbolism with which it was endowed. This is not to denigrate the historical significance of migrant minorities in innovation and niche marketing within the metropolitan economies, only to contextualise the legitimation of inner city subjects.

At the heart of the new private–public partnerships through which so much of 1990s government policy is mediated is a notion that there is such a 'thing' as an identifiable 'business community'. Private sector interests are institutionalised through novel structures of government which determine the 'players' at the table in this contemporary form of urban corporatism. It is easy in hindsight to suggest that the history of the DBDA illustrates that there is a complexity to the nature of private enterprise that belies the imagery of 'the business community', which at times almost implies a sense of common purpose that does not fit well with the multiplicity of interest groups and the institutional maze that defines the contemporary business environment. Moreover, the affair illustrates how innovations in urban regeneration must negotiate this institutional maze; exemplifying the sort of pragmatic power-brokering that has for long been characteristic of city government.

More significantly, at another level, the example reveals the whiteness of the urban policy implementing agencies through which 1990s techniques of city government are mediated. The new partnerships are, at least in part, about key 'players' in the inner city coming together, a form of government that is unmarked by racial difference. Ironically, normalised institutional roles for public and private sectors endow the ethnic entrepreneur with far less symbolic power than was the case in the free market rhetorics of the 1980s. Yet it is equally the case that such changes are not readily framed within the mainstream left paradigms of the same era. Debates that owe their provenance to 1980s equal opportunities notions of hypothecated resources for minorities, and a focus on the position of ethnic minorities among the personnel of local government, cannot readily come to terms with this new vision of the city as a set of institutional relationships. With the 'community' reified as a singular player at the bargaining table of partnership it too is constructed in singularly deracialised terms.

In this new landscape the shibboleth of the ethnic entrepreneur is discarded even as the reality of ethnic minority businesses continue to struggle against the odds in impoverished urban areas. The racialised subject position of the mythical ethnic entrepreneur is not consonant with the consensually colour-blind and implicitly white form of the new urban corporatism of the 1990s.

In the summer of 1992 in the London Borough of Tower Hamlets, a series of disturbances loosely connected to fights between 'gangs' of young people culminated in clashes between young people in Brick Lane, the symbolic heart of the local Bengali community. Luridly reported in the local press, the events were one of several key passages whose representation has created a constitutively racialised link between Bengali masculinity and the streets of the East End in the last few years.

The demographic structure of the Tower Hamlets Bengali population is largely a product of the timing of Bengali migration. Male-dominated first-wave migration, with family reunification often taking place in the mid-1970s, produced a baby boom generation that came to adolescence and maturity in the late 1980s and early 1990s. This generation of frequently British-born Bengalis has been ill-served by all the agencies of collective consumption. The borough has long been demonstrated to be one of the poorest in Britain; housing conditions reflected extraordinarily high levels of overcrowding and health figures were poor for any and every index of deprivation that could be chosen (Runnymede Trust 1993; Tower Hamlets Homeless Families Campaign 1993). Moreover, the same generation came to maturity at a time when education locally was also badly hit by the teaching disputes of the mid-1980s and Bengali 'underachievement' in schools has long been a fascination of education academics. These trends were compounded by a political culture set by the particular racialised coding of the Tower Hamlets Liberal Democratic Party (Lester 1993) which controlled the borough from 1986–94 and for which the problems of 'young people' readily translated into a bigoted imagery reflecting the asymmetrical age distributions of white and Bengali communities. Youth services were frequently cut back to allow for alternative uses for local government revenue. One study revealed that in Stepney neighbourhood, which covered a large area of Bengali settlement, focusing in particular on the Ocean Estate, grants to the voluntary sector were reduced from £426,065 in 1990–1 to £44,215 in 1991–2, a cut of almost 90 per cent (Mohan 1992). In the same 'neighbourhood' the Community Education Service, which was the principal funder of youth support services, regularly underspent its budget in the early 1990s.

It is only in this context that it is possible to understand the progressive emergence of a series of representations of the social problem of putative Bengali criminality (CAPA *et al.* 1993). The evidence of growing antagonism between police and young British Bengalis is incontrovertible, the creation of new inner city subjects the result of both collective action and the manner in which such actions were framed in the mass media. In 1992 in London there were marches protesting at police tactics, campaigns about specific incidents and arrests and appeals for calm from senior police officers. One national newspaper (*The Independent*, 20 April 1992: 4) went so far as to suggest that the East End Bengali community was a likely focus of an imminent incidence of serious public disorder. Significantly, the story was told through the image of young Bengali men occupying the streets of the East End, challenging police order – the organising trope of narratives of street crime – with the Ocean Estate linked directly to the sites of black uprisings in the mid-1980s. The politics of such representational practices have been addressed elsewhere (Keith 1994) but of central importance for this chapter is the manner in which the street becomes a site through which slippages occur in racist discourse.

It has also been argued elsewhere that through racist constructions of criminality the Criminal Justice System has become a locus of racialisation, manufacturing a criminalised classification of 'race' that coexists with alternative, often contradictory, invocations of 'race' that derive from other racialising discourses. A theoretical analysis of criminalisation helps to understand the manner in which 'blackness' is produced as a sign of criminal otherness (Keith 1993). Geographical references perform organising roles within these common forms of criminalising rhetoric. The racialised other is constructed as dangerous, defined through presence in the public sphere. Hence in the seminal work on the social construction of black criminality in Britain, Stuart Hall and his associates, when deconstructing *the ghetto*, regularly resort to the metaphor of the 'black colony' as both victim of these racist practices of criminalisation and (apparently) social reality (Hall *et al.* 1978) . The lawless black ghetto is a place that is both a racist myth and a site of criminalisation.

This racialised subject position of criminality can envelop British 'Asian' communities as well, most commonly through discourses of gang violence, as experience in Southall and Birmingham demonstrates. There is a barely hidden genealogy here of place and identity, normatively construed. The black body is interpellated through the street. It was, after all, Lord Scarman who so egregiously talked about 'West Indian' people as 'a people of the street' (Scarman 1981), and the representations of Bengali criminality have almost invariably focused on youth, masculinity and the public streets through which 'the youth problem' becomes visible. This is precisely how the slippages of racist discourse work; the tropes of criminality elide 'alien others'. Here we have the corollary of Fanon's 'look a Negro',[4] a successful racist placing of the body of the other in the field of vision. We know these stories. They appeal to a knowledge that predates the moment of representation. They place race in the field of vision in a way in which sense becomes self-evident. Racist representations of dangerous Bengali young men work because the meaning is already self-evident, the product of years of conditioning.

But the street is also constructed as the site of insurrection, and when in Autumn 1993 Quddus Ali was savagely beaten outside the Dean Swift pub in Stepney the demonstration of respect that gathered outside the London Hospital in Whitechapel and developed into serious clashes with the police was rapidly classified in the national media as the riot that *The Independent* had been predicting over a year earlier. This was a moment of confrontation rapidly endowed with extraordinary political symbolism when Derek

Beackon became the first successful BNP (fascist) candidate in a local by-election a few days later. The position of racialised street criminal was found to be, once again, one step away from the arrival of the racialised street rebel.

There is clearly a sense here in which the liberal white left themselves, in a desperate search for the transformative political subject, will cast young Bengali men as the teleological delivery boys; and as Stuart Hall pointed out in a not dissimilar context in 1981, the streets will not only stage glorious insurrection, they will also witness the fact that it is upon young male Bengali heads that the fully armed apparatus of the state will fall; it is they who will be attacked on the streets at the vigil for Quddus Ali outside the London Hospital, and they who will be confronted by and confront the BNP gangs who increasingly conspire to roam the streets of Tower Hamlets to go Paki-bashing. Street politics is easy for the absent.

Now this is undoubtedly a complex story and what is important here is to understand that whilst racist discourse may map criminality through the street and romantic left rhetoric map insurrection similarly, there is a space in which agency takes on board and transforms such interpellation. The subject positions of oppression are regularly taken on board as the vessels that secure movements of resistance – from pejorative black to politically black, from pejorative queer to queer politics – the categories are taken on board to be mocked and subverted in the mimesis of the mirror dance.[5]

The events of 1993 resonated, the problem of 'Bengali youth' became a problem of inner city government, as the local City Challenge demonstrated in microcosm. In early 1991, when the degree of socio-economic deprivation was just about as bad, when the demographics were known, when the realities of everyday life for young Bengali men were not so different from three years later, the original Bethnal Green City Challenge Action Plan barely mentioned the 'problem' of 'youth' at all (Bethnal Green City Challenge 1991). By late 1993, one riot and one fascist later, 'promoting youth empowerment and access to opportunities' had become a key strategic aim of Bethnal Green City Challenge (Bethnal Green City Challenge 1993) and a year after that there was already a focus on a youth forum and even a 'youth representative' on the board of the company, something that cannot be reduced to the vocabulary of co-optation. But such a change is a product of a particular diagnosis of the problem of government in the contemporary East End, a diagnosis that must bring the street to the state.

CONCLUSION: RACE-POWER AND CARTOGRAPHIC SUBJECTIFICATION

Colin Gordon has suggested that in defining the nature of governmental rationality Michel Foucault talks about the problems of contemporary government in terms of

the daemonic coupling of 'city-game' and shepherd game: the invention of a form of secular pastorate which couples 'individualization' and 'totalization'.

(Gordon 1991: 8)

The conduct of conduct at the heart of governmental practice is the move towards simultaneous invention of the social and complete knowledge of the individual. Just one aspect of such trends within the contemporary inner city is the definition of racialised subject positions divorced from ethical self-evidence, something that at one level might appear to deny the possibility of transformative politics. Yet seen differently the characters that emerge in this way raise new questions that disrupt some of the old certainties of political action and that reframe old problems of authenticity, of aesthetics and of the nature of institutional racism.

They demand a contextualisation of the processes of rendering visible particular subjects and the manner in which a variety of discursive practices, including techniques of city government, shape speaking positions, and map the grounds from within which political subjects might speak and trap themselves in their own parochialism.

It is only in so doing that the vagaries of cultural theory can be brought to bear on the concrete realities of institutionalised political practice. It is at such points that there is some sense to the claim that 'the non-synchronous temporality of global and national cultures opens up a cultural space – a third space – where the negotiation of incommensurable differences creates a tension peculiar to borderline experiences' (Bhabha 1993: 218). It is through this imperative to move consciously between positionalities that such a language opens up the vocabulary of inside and outside to a progressive politics, returns us again to the selective appropriation of 'the East End' with which this chapter started, disrupts the race-power cartographies through which subjects are mapped.

The institutional rationalities of the agencies through which urban policy is mediated are structured by the definition of the inner city as a problem of government. But the subjects that both inform and are defined by their techniques and practices both reflect and challenge the characters such agencies script; the ethnic entrepreneur and the street rebel are contingently both the products of hegemonic power and the cast list of political change.

In short the institutional landscape of the inner city creates a cultural reality that in part defines the frames through which mapped subjects are rendered legitimately visible; 'the relation between government and the governed passes, to a perhaps ever-increasing extent, through the manner in which governed individuals are willing to exist as subjects' (Gordon 1991: 47). It is through the interstices of this structural complex that communities of resistance can emerge as political subjects. Living the true lie of racialised subject positions is at least in part about both colonising the normalising whiteness of British government and reappropriating the racialised subject positions such techniques and practices generate.

NOTES

1   Much of the work for this chapter is based on work of the author as part of the Deptford City Challenge Evaluation Project (DCCEP) at the Centre for Urban and Community Research, Goldsmiths College. DCCEP are Elsa Guzman-Flores, Michael Keith, Aileen O'Gorman, Nikolas Rose and Phillippa Superville. Thanks are due to all of the team and the many individuals working within the City Challenge process who have been so co-operative in the conduct of this work.

2   All quoted material in this chapter is drawn from the reports in the work of the Deptford City Challenge Evaluation Project.

3   After the completion of this article a compromise was reached between different parties and the new Business Association launched but the institutional realisation of the subject position of 'ethnic enterprise' remained uncertain.

4   Fanon's opening of *Black Skin, White Masks* captures the subjectification of the black body through the white gaze. Taken together with his axiomatic notion that 'the black man's soul is a white man's artefact' and the memorably epigrammatic comment that 'The Black man is not. Any more than the White man' Fanon's work has become increasingly important in cultural studies projects that trace back the construction of processes of racialisation to the social context in which the formative experiences of identification and identity formation take place (Fanon 1967). It is this *taking place* element of this process that makes an understanding of the spatialities of a sophisticated urbanism indispensable to anti-racist theory and practice.

5   The most impressive exposition of this is surely to be found in Michael Taussig's *Wild Man* (1987), developed theoretically in the more recent *Mimesis and Alterity* (1993).

18

CONCLUSIONS
spacing and the subject

Steve Pile and Nigel Thrift

> Outside and inside are both intimate – they are always ready to be reversed, to exchange their hostility. If there exists a border-line between such an inside and outside, this surface is painful on both sides.
>
> (Bachelard 1964: 217–18)

A concluding chapter usually follows a conventional pattern, drawing up the accounts for a volume by pointing to abuses and exclusions and by setting up a research agenda that makes redress. We are not going to draw up a list of the wounded and the missing. It is a melancholy task which too often confuses accumulation with understanding. Instead, we will briefly point to five different ways in which extant maps have so far failed to produce a feeling for the terrain of the subject, developing the latter way in some detail through different takes on the City of London.

The first of these is that too much of the current literature on the subject is too academic, in that it reads the writings of intellectuals on body, self, person, identity and subject as somehow constituting the histories of body, self, person, identity and subject (for example, was the Cartesian self ever so general in the West as so many of these intellectuals' histories of intellectuals assume?). The result of this academicism is that too much of the literature on the subject is relentlessly self-referential, clinically austere and makes too easy a link between symbolic inversion and resistance. The consequence is that there is still a gulf between theory and much of the work on everyday usages of body, self, person, identity and subject found in anthropology, sociology and historical studies which has only been bridged by a very few studies like Ginzburg's (1980) *The Cheese and the Worms* or Walkerdine and Lucey's (1989) *Democracy in the Kitchen*.

The second way in which extant maps have sometimes failed to produce a feeling for the terrain of the subject is by failing to articulate a clear sense of exploitation and oppression. Talk of relations of power can sometimes obscure the grinding, relentless nature of oppression and the way it forces accounts and choices which may not always be attractive to bourgeois academics. Instead of facing up to this task of description, researchers have often reached for fantasies of otherness which, in classic postcolonial terms, trap the colonised in the fantasies of the coloniser and which therefore play right into the hands of prevailing relations of power by silencing other actual or potential speaking positions. This effect is probably most clearly seen nowadays in the dumping overboard by many ambiguous academics of the 'white working class', a strategy which both closes down the task of description and also avoids 'more difficult emotions. What is it to "be" like that, "how can they?"' (Walkerdine and Lucey 1989: 43).

The third way in which extant maps have sometimes failed to produce a feeling for the subject is precisely in their neglect of emotions. This is perhaps surprising. After all, desire figures large in modern accounts of the subject. Anger is an emotion that is often professed to. Yet emotions are still relatively little studied even though their importance has been acknowledged. Three examples will suffice. First, as we pointed out in the second chapter, affect is now seen as the set of prediscursive states that is elaborated on in practice. In other words affect is what makes practice into what Marx and Engels called 'sensuous human activity'. Second, once the importance of emotions is acknowledged, language can be refigured in a Vygotskian manner as an active rhetorical-expressive form of understanding whose aim is not only to specify emotions but also to produce them, as 'the making of a sense, which works not so much to communicate ideas ... but to prompt in us an affective reaction through which others can feel the movement of our minds, to which they feel they must respond in some way' (Shotter 1993a: 51). In turn, this means that 'social ideologies are not simply implanted into a lacking subject; rather, the reproduction of states of representation depends upon engaging the affective investment of human beings' (Elliott 1992: 264). Third, it is becoming increasingly apparent that emotions vary historically, cross-culturally and geographically. For example, emotions like pride and loneliness clearly vary from culture to culture. Some emotions, like the Japanese *amae*, seem to have no exact correlate in western cultures. Other emotions, like the medieval *accidie*, no longer exist in the western emotional repertoire (Harré 1986). In other words, emotions must be studied because they are a crucial element of how we go on. In turn, such a study also holds out the fascinating prospect of an accord between theories of practice and the psychoanalytic traditions.

This latter point relates to the fourth way in which extant maps have sometimes failed to produce a feeling for the terrain of the subject, and that is in failing to stress the importance of the link between social practices and forms of the unconscious, a link which becomes much clearer once 'the *affirmative* character of psychical production' (Elliott 1992: 262) is posited. This is the kind of programme that Elliott wants to see put in place:

> the problem that must be confronted ... concerns the complex dialectical interplay between the imaginary dimensions of the unconscious and the structuring of states of representation ... An adequate account of the unconscious in relation to the social field ... should focus on three levels, each of which is only methodologically

distinguishable. The first concerns the profoundly imaginary character of unconscious representation. It must be recognised that the unconscious imaginary is the creative work of representation as such. To grasp the primary unconscious is to grasp the imaginary way subjectivity 'opens out' to self-identity, others, reason, society, and political engagement. The second concerns the interlacing of these imaginary forms (drives, image-production, originary narcissistic investments) with broader social influences of pleasures (symbolic representation forms). It must be recognised that subjects are never passively determined by such symbolic forms, but rather actively interpret ... social significations through their representation activity. The third level is that of concrete ideological relations of power and domination. It must be recognised that the imaginary and symbolic forms through which human subjects derive pleasure are structured within culturally specific social and political relations.

(1992: 264)

Such a programme has, until now, probably been most vigorously pursued by Castoriadis (1984, 1987) who has 'socialised' psychoanalysis as a 'practical-poietic' activity of self-transformation. For Castoriadis, the imaginary is an ongoing faculty of 'living signification' which stems from the human ability of

> positing or presenting oneself with things and relations which do not exist, in the form of representation (things and relations that are not and have never been given in perception). We shall speak of a final or radical imaginary as the common root of the actual imaginary and of the symbolic. This is finally the elementary and irreducible capacity of evoking images.

(Castoriadis 1987: 127)

The fifth and final way in which extant maps have sometimes failed to produce a feeling for the terrain of the subject is in the matter of space. As we have seen, figures of space abound in the literature on the subject. These figures interrogate the interrelation of numerous registers of space in the constitution of subjects, including the geo-political, the semiotic, the somatic and the psychic. As Pollock (1988: 68) puts it, 'space can be grasped in many dimensions'. They are used to signal a sense of fluidity associated with current reformative cultural impulses. And they allow the subject to be thought of as both inside and outside. For example, writers like de Certeau and Deleuze clearly view real, external space as a precipitate of the division between the inside and the outside of the subject. Deforming inner and outer space and the relation between them through notions like 'the fold' allows these writers to ponder on the mismatch between 'inside' and 'outside', 'desire' and 'reality'.

But what is interesting about these writers, and many others who use spatial figures to motivate poststructuralist ideas, is how often they remain trapped in a textual world, an inside of their own devising. This is a world which, it is often assumed, has a relatively unproblematic connection with external reality. But, like Kirby, we

> remain unconvinced that the interchange of (academic) discourse and (political) reality can take place quite so easily, simply by wishing or writing it into existence. If the real is nothing more than the accretion of discourses, it has become real due to long practice and popular 'consent'. As academics, we refuse to sacrifice the possibility that changing discourse can change the way we live, but we must delineate their means of interchange more complexly.
>
> (1995: 210)

That such a problem can arise stems in part from assuming that space is in-different, that it acts as a fluid medium in which mobile subjects dwell. But, of course, space is not like this. For example, there is the matter of boundaries. Boundaries are important, both as ways of fixing and displaying the subject by making it impossible to move (a state of affairs of which Foucault has made us keenly aware) and as ways of positively constructing the subject, since they signal when and where one has moved. In other words,

> space offers qualities that seem contrary to the otherwise total fluidity that these critics offer. It connotes difference and distance, location and separation and limitation at the same time that their theories radically foreshorten such solidities, temporally and corporeally.
>
> (Kirby 1995: 208)

This is where much of the writing employing spatial figures still seems curious to many geographers. It neglects the crucial importance of different *places* – performed spaces in which psychical and social boundaries are only too clear, in which resources are clearly available to some and not others, in which physical force makes contact – in fostering difference by generalising different places into in/different spatial figures. In other words, in the process of metaphorisation ground is lost. We index space but we also become lost in it. And we lose a lot thereby. We can no longer get at 'the systematic psychological study of the sites of our intimate lives' (Bachelard 1964: 8). We can no longer 'carve in a being that remains in its place' (Merleau-Ponty 1962a: 292). We can no longer understand what Casey (1993) calls 'implacement', an ongoing cultural process of participating in places and reshaping them. We can no longer understand that 'just as there is no place without body – without the physical or psychical traces of body – so there is no body without place' (Casey 1993: 104). And so on. In other words, as Pratt (1992: 244) puts it, 'we should ... recognise the limits of any metaphor and resist being seduced by geographical and spatial metaphors that are ultimately aspatial and insensitive to place'.

Therefore in the final section of this concluding chapter we turn to an actual place – London – and to two takes on this place which itself consists of a constellation of different geo-political, semiotic, somatic and psychic spaces, each of them intertwined with the others in myriad encounters:

> the city's inhabitants create an exquisitely complex geometry, a geography passing beyond the natural to become metaphysical, only describable in terms of music or abstract physics, nothing else makes much sense of relationships between roads, rails, waterways, subways, sewers, tunnels, bridges, viaducts, aqueducts, cables, between every possible kind of intersection.
>
> (Moorcock 1988: 7)

PLACING THE SUBJECT. TAKE 1: THE CITY OF LONDON AS ELITE SPACE

This section concentrates on the City of London (Thrift and Leyshon 1992, 1994). This small, tightly-bounded area of London has been produced as a coherent space of practices of trade, money and finance over many hundreds of years. The City is a space, like many other spaces of business, in which and from which people exert *power*. Yet the space, the business and the people are not the same as those found in many of these other spaces.

In this section we want to briefly sketch out some of the differences between the City and many other apparently similar spaces of economic power – between the form of power that the City and those apparently similar other spaces exert – as being the result of the kind of powerful subjectivity that is able to be produced in the City because of its network of collective, embodied spatial *practices* which underline Burkitt's (1994: 16) statement that 'social power is conducted through physical presences and absences as well as through linguistic presences and absences'. We must start by pointing to certain general features of the practices of trade, money and finance which are common to many other centres of trade, money and finance but which were and are present in peculiarly concentrated form in the City. These revolve around the production of 'information', time, space and sociality. First of all, places like the City are the sites of power struggles over knowledge; there is a hunger for 'information' (understood in its broadest sense) because it is through gathering information (and quite often keeping that information sacrosanct) that business advantage is gained. Second, time is important because it has been through the manipulation of increasingly long periods of time that business in the City has been able to gain an advantage, from the early bills of exchange to modern derivatives, and from the use of courier mails through to virtually instantaneous electronic communications. Third, space is important since the practices of trade, money and finance nearly all involve practices of ordering of goods, money and knowledge over long distances which in turn have led to the active pursuit of socio-technical innovations, 'immutable mobiles' (Law 1994), like systems of long-distance navigation, the telephone and telegraph, and, latterly, computerised telecommunications. Fourth, sociality is important.

Nearly all the business of trade, money and finance requires some degree of co-operation and trust; the idea of an isolated profit-maximising firm is, like the idea of the isolated sovereign subject, a fantasy. From the earliest days of the Merchant Adventurers, the history of the City has been a history of successive combines of merchants who have been able to successfully splice together information, time, space and sociality.

It is no surprise that the concentration of practices associated with these features over a long period in a particular space has led to the development of specific ways of life. From an early period, those working in the City interweaved with each other, through work-a-day relations and also through specific institutions of governance like the Corporation and the Guilds. In so doing they developed a particular *style of joint action* which may at first have been something of a hybrid but which gradually over time became a sedimented way of life, known for its conservative tinge. This style of joint action is often called 'gentlemanly' (Cain and Hopkins 1993a, 1993b). As a style it has four main elements. First, it was founded in occupations that placed a premium on the organisation of people and finance rather than on processing raw materials. Second, it cleaved to English upper- and upper middle-class values, and most specifically a code of honour which coveted gentility, tended to understatement and was strongly masculine in character (McDowell and Court 1994). Third, it was sustained by personal social networks, dependent upon a particular class background which allowed confidence and trust to be built up. Fourth and finally, it was constantly replenished by histories of the City constructed in such a way that they both declared the worth of the City's practices and gave the City's participants a further means of identification. The gradual calcification of British society over the course of the nineteenth century, and especially the growth in importance of public schools and Oxbridge, only strengthened this style (Cassis 1987).

Such a style of joint action was produced on a day-to-day basis by the embodied practices of the City. These were of a number of interrelated kinds. There were, first of all, particular class-specific modes of embodiment-handshakes, accounts, clothing (especially the suit) and the like. There was a specific form of practical 'memoro-politics' (Hacking 1994) embedded in specialist languages, japes and so on. There were the specific time-space rhythms of the City which meant that the powerful tended to be in the same places at the same times each day (even down to the railway journey to work). Finally, there was the spatial form of the City itself, an enclosed world full of enclosures, a bounded world full of boundaries, which it required specific backgrounds and knowledges to enter. Thus, from an early point in time, and in contrast to similar centres like New York, the built form of the City was as understated as many of its denizens. It often required specific knowledge to read which buildings were the haunts of the powerful. The power of the City was often hidden – a very English structure of feeling (see Thrift 1983b) – behind closed or uninviting doors:

> the inhabitants of [the City] are, for the most part, worthy merchants who are shrewd where their business interests are concerned and who care for nothing but these same interests. The shops, where many of them have made fortunes, are so dark, so cold, so damp that the West End aristocracy would disdain to have them as stables.
>
> (Tristan 1840, cited in Kynaston 1994: 142)

In other words, we can see the City of London as a powerful 'consciousness machine', which was able, through a particular style of joint action, to become a subject-producing space of the first order. Further, it was producing a very particular kind of subject (Douglas 1992).

Nowadays, the City is more open to outside influences. The City's subjectivities are more obviously hybrid (Allen and Pryke 1994). Yet the City still retains a hold on its subjects. What is striking is how much of the gentlemanly style of joint action still remains, a signature etched in the embodied practices of the City and in the space of the City itself. For example, whilst there are new, grandiose buildings, the vast majority of the City's built form remains small in scale and feel (Jacobs 1994), and even in the new grandiose buildings, there are usually specifically built inner sanctums which recall a gentlemanly past through their use of oak lining, special tea services and the like. Thus power in the City is still exercised in a City way, through a politics of practice which is still effective, in part because it is so difficult to see in a society which has become more and more used to seeing the exercise of power. In the City, those who are 'in the know' know. They do not need a map because they know *where* they are (socially, culturally, geographically and so on).

PLACING THE SUBJECT. TAKE 2: THE CITY OF LONDON AS SUBJECT-SPACE

> the city's inhabitants create an exquisitely complex geometry, a geography passing beyond the natural to become metaphysical.
>
> (Moorcock 1988: 7)

Placing the subject has far-reaching consequences for the ways in which subjectivity, society and space are understood. This book has shown that the subject cannot be thought of as being placeless, even – or especially – in those moments where the individual feels 'out of place'. The subject of theory requires a place to be or not to be. However, being or not being in place is not quite so easy to delimit: as we have tried to show, tracing the multiple webs of space, time and power requires a complex and subtle map – aspects of the question at hand may be demonstrated in a case study such as *The House of Doctor Dee* by Peter Ackroyd.

The point of this case study is not to provide a literary criticism of Peter Ackroyd's book, but to make some points about the ways in which mapping the subject invokes thinking about the relationships between subjects and objects, the real and the imaginary, places and subjectivity, powers and knowledges, images and words. The book itself raises specific questions about the past and the present, the body and site, sight and vision, science and magic, memory and time. Though presented as an alchemy of opposing elements, these dualisms should be thought of as artificial and limiting: that is, they are mobilised as, through and by specific moments of the power/ knowledge nexus.[1] Taken together, this is a (select) collection of ingredients which may well make a spell for creating our very own homunculus: i.e., *The House of Doctor Dee* will help reveal a magic which helps understand what happens when the subject-of-theory is mapped into places.

> 'The past is difficult, you see. You think you understand a person or an event, but then you turn a corner and everything is different once again ... It's like this house, too. Nothing ever seems to stay in the same place' says Matthew Palmer to his friend Daniel Moore.
>
> (Ackroyd 1993: 136–7)

Ackroyd explores the intersection of locatedness in space and time with identity and embodiedness through the device of the house of Doctor Dee. Matthew Palmer inherits his father's place in Cloak Lane, Clerkenwell, London. The house has a specific and general history: from basement to upper floors, the building goes through a series of physical and spiritual transformations as each generation alters it to suit their own structure of living. The house thereby comes to represent the sense that 'there is no such thing as history' because 'history only exists in the present' (Ackroyd 1993: 264). Yet the past remains difficult because it haunts the present: 'the past is restored around us all the time, in the bodies we inhabit or the words we speak' (39). So, when Matthew first enters the house, it is made to convey an embodied sense of time: 'When I walked towards the steps, it was as if I were about to enter a human body' (3). As the story continues, Ackroyd places in tension the lives of Matthew Palmer and Doctor John Dee, connecting them via the notion of the homunculus.

Cosgrove describes Dee as a renaissance environmentalist because he believed that 'humans actively transform nature as well as contemplating it (Cosgrove 1990: 347). This can be seen in Dee's attitude towards one form of contemplation: mathematics. Cosgrove argues that Dee outlined 'the ways that mathematics gives access to knowledge of both worlds, temporal and spiritual, and also directs us to the transforming power of the machine' (Cosgrove 1990: 347). Thus, Dee was interested both in architecture as a means of creating the world and in the mechanics of London. One key to Dee's organisation of the power and knowledge was through a specific regime of the visual in which images were reified in a way which gave them 'an aesthetic unity in which language, ritual, spectacle, image and metaphor become active agents in human transformations of nature and the invention of machines' (Cosgrove 1990: 350). Ackroyd uses this notion of vision to incorporate both science and magic, in a way which makes Dee a signifier of what is beyond both science and magic, past and present. Ackroyd has Matthew Palmer feel that

John Dee himself had, in one way or another, belonged to every time. He was in part a medievalist, expounding ancient formulae, but he was also an active agent in contemporary natural philosophy; he was an antiquarian, who speculated about the origins of Britain and the presence of ancient cities beneath the earth, but he was also one of those who anticipated a future scientific revolution with his experiments in mechanics; he was an alchemist and astrologer who scrutinized the spiritual world, but he was also a geographer who plotted navigation charts for Elizabethan explorers. He was everywhere at once and, as I walked about his old house, I had the sense that somehow he had conquered time.

(Ackroyd 1993: 132–3)

Leaving aside a rather narrow definition of geography, amongst Dee's experiments was the attempt to create a 'little man' or homunculus out of baser elements; the ancient formula involving the use of a sealed glass tube from Antwerp, spagyricus, horse dung, four true magnets in the shape of a cross and regular watering. Gradually, Matthew Palmer's exploration of the history of the house leads him to three intertwining histories: those of Doctor Dee, of his father and of his own identity. In the same way that these pasts become perpetually shifting sands, which can never quite be fully grasped, so too his own identity and the identity of the house shift and change – sliding between never completely absent insanity and never fully present others. His presence in that place and at that time leads Matthew to doubt both his mind and body – is he sane? Is he human? When, at the end, Matthew eventually resolves his place in the world, it is no accident that Ackroyd should have him echo his earlier words: 'The past is difficult, you see. You think you understand a person or event, but then you turn a corner and everything is different once again. It's like this house. Nothing ever seems to stay in the same place' (Ackroyd 1993: 276–7).

So far the placing the subject in The House conveys a sense of inter-connectedness between the subjectivity and place, but places cannot be isolated so easily from their surrounds, and – as importantly – neither can subjectivity. In other words, the subject (Matthew) cannot be isolated from the object world (The House, London) in which he finds himself: the outside world is constitutive of his subjectivity and he is an active agent in the constitution of his world. Ackroyd demonstrates this relationship in two places. First, when Daniel Moore describes his search for his identity in these terms: 'sometimes I feel as if I'm excavating some lost city within myself' (83). Moreover, in the story, Matthew, Matthew's father and John Dee go in search of London as a means of recovering a lost identity:

And as I walked through the city, I saw so many houses and streets fading before me that I seemed to be forever treading upon shadows ... I still walked and thought of Doctor Dee. It was believed by him that light descended into matter, and that in the very constitution of the material world would be discovered the great mysteries of the spirit hitherto covered by clouds and darkness.

(Ackroyd 1993: 275–6)

Thus, the house of Dr Dee and the streets of London become means of establishing the place of the subject, but where 'place' cannot be known in advance of subject constitution and where 'place' must necessarily be simultaneously real, imaginary and symbolic. In other words, subjectivity and place cannot be separated without foreclosing an understanding of the located subject and the agency and identity of place. Then again, mapping the subject requires more (or less) than simply multiplying the terms subject and place such that the self is totally fluid and fragmented and a place becomes another location in multiple and shifting social and personal processes. *The House of Doctor Dee* demonstrates that people become enmeshed into their own histories and geographies in specific ways and that these co-ordinates are also fixed through intersecting power/knowledge relationships, including science, spirit, language, house, London, the machine, light and vision.

There is more to be discovered about mapping the subject and some of this may involve casting light into the great mysteries of *places* hitherto covered by clouds and darkness.

NOTE

1   In her book, Rosalind Williams also finds dualisms structuring, and structured by, specific intersections of power and knowledge (1990); while her focus is on the metaphorical significance of the city and 'the underground', ours is on the allegory of the city and 'the House'.

BIBLIOGRAPHY

Abercrombie, P. (1933) *Town and Country Planning*, London: Thornton Butterworth.

Aberdare, Lord (1938) 'Fitness and Fresh Air', *The Nottinghamshire Countryside*, 2(1): 16.

Ackroyd, P. (1993) *The House of Doctor Dee*, Harmondsworth: Penguin.

Adams, P. (1993) 'The Three (Dis)Graces', *New Formations*, 19: 131–8.

Allen, J. and Pryke, M. (1994) 'The Production of Service Space', *Environment and Planning D: Society and Space*, 11: 453–76.

Allport, G. (1954) *The Nature of Prejudice*, Reading: Addison/Wesley.

Alpers, S. (1983) 'Interpretation without Representation, or, the Viewing of Las Meninas', *Representations*, 1: 30–42.

Andersen, H. C. (1900) *Fairy Tales*, translated by H. L. Braekstad, with an introduction by E. Gosse, two volumes, London: Heinemann.

Anon. (1938) 'Strength Through Joy: Suggestion for a Rural Fitness Policy', *The Nottinghamshire Countryside*, 2(1): 15.

Appleton, J. (1986) 'A Sort of National Property: The Growth of the National Parks Movement in Britain', in *The Lake District: A Sort of National Property*, London: Countryside Commission/Victoria and Albert Museum, 113–22.

Arkin, R. and Baumgardner, A. (1986) 'Self-presentation and Self-evaluation: Processes of Self-control and Social Control', in R. Baumeister (ed.) *Public Self and Private Self*, New York: Springer-Verlag, 75–98.

Armstrong, N. (1986) *Desire and Domestic Fiction: A Political History of the Novel*, New York: Oxford University Press.

Ashmore, R. and del Boca, F. (1981) 'Conceptual Approaches to Stereotypes and Stereotyping', in D. Hamilton (ed.) *Cognitive Processes in Stereotyping and Intergroup Behaviour*, Hillsdale: Erlbaum, 1–35.

Atkinson, P. (1985) *Language, Structure and Reproduction: An Introduction to the Sociology of Basil Bernstein*, London: Methuen.

Atkinson, R. and Moon, G. (1994) *Urban Policy in Britain*, Basingstoke: Macmillan.

Avgikos, J. (1993) 'Cindy Sherman: Burning Down the House', *Artforum*, January: 74–9.

Bachelard, G. (1964) *The Poetics of Space*, Boston: Orion.

Bachelard, G. (1981) *La Poétique de l'Espace*, Paris: Presses Universitaires.

Bachman, C. (1990) 'Le Fantasme Urbain des Mickeys', *Urbanismes et Architecture*, 234: 66.

Back, L. and Quaade, V. (1993) 'Dreams, Utopias, Nightmare Realities: Imaging Race and Culture Within the World of Benetton Advertising', *Third Text*, 22: 65–80.

Baden-Powell, R. (1922) *Rovering To Success: A Book of Life-Sport for Young Men*, London: Herbert Jenkins.

Bakhtin, M. (1986) *The Dialogical Imagination*, Austin, Texas: University of Texas Press.

Baldwin, J. A. (1971) *The Mental Hospital in the Psychiatric Service: A Case-Register Study*, London: Nuffield Provincial Hospitals Trust, Oxford University Press.

Baldwin, J. M. (1895) *Mental Development in the Child and the Race*, New York: Macmillan.

Bangay, F. (1992) 'Glimmers of Light', in F. Bangay, J. Bidder and H. Porter (eds) *Survivors' Poetry: From Dark to Light*, London: Survivor's Press, 22–3.

Barnes, P. (1934) *'Trespassers Will Be Prosecuted': Views of the Forbidden Moorlands of the Peak District*, Sheffield: P. A. Barnes.

Basow, S. (1980) *Sex-Role Stereotypes*, Monterey: Brooks/Cole.

Bateson, G. (1972) *Steps to an Ecology of Mind*, London: Intertext.

Batsford, H. (1945–6) *How To See The Country*, London: Batsford (first published 1940).

Baudrillard, J. (1968) *Le Système des Objets*, Paris: Denoel.

Baudrillard, J. (1981) *For a Critique of the Political Economy of the Sign*, St Louis: Telos.

Baudrillard, J. (1983a) *Simulations*, translated by P. Foss, P. Patton and P. Beitchman, New York: Semiotext(e).

Baudrillard, J. (1983b) *In the Shadow of Silent Majorities and Other Essays*, translated by P. Foss, P. Patton and P. Beitchman, New York: Semiotext(e).

Baudrillard, J. (1987a) *The Ecstasy of Communication*, New York: Semiotext(e).

Baudrillard, J. (1987b) *Forget Foucault*, New York: Semiotext(e) (first published 1977).

Baudrillard, J. (1988a) *America*, translated by C. Turner, Verso: London.

Baudrillard, J. (1988b) *The Evil Demon of Images*, Sydney: Power Press.

Baudrillard, J. (1990) *Fatal Strategies*, London: Pluto.

Baudrillard, J. (1990a) *Revenge of the Crystal*, translated by P. Foss and J. Pefanis, London: Pluto Press.

Baudrillard, J. (1990b) *Fatal Strategies*, translated by P. Beitchman and P. Niesluchowski, New York: Semiotext(e)/Pluto.

Baudrillard, J. (1990c) *Cool Memories*, London: Verso.

Baudrillard, J. (1990d) *Cool Memories II*, Paris: Galilee.

Baudrillard, J. (1991) 'The Reality Gulf', *The Guardian*, 11 January.

Baudrillard, J. (1992) *Jean Baudrillard: Selected Writings*, edited by M. Poster, Oxford: Polity Press/Blackwell.

Baumeister, R. (1986) 'Epilogue: The Next Decade of Self-presentation Research', in R. Baumeister (ed.) *Public Self and Private Self*, New York: Springer-Verlag, 241–5.

Bean, P. and Mounser, P. (1992) *Discharged From Mental Hospitals*, London: MIND.

Beck, U. (1992) *Risk Society*, London: Sage.

Bell, D. (1995) 'Pleasure and Danger: The Paradoxical Spaces of Sexual Citizenship', *Political Geography*, 14: 139–54.

Bell, D., Binnie, J., Cream, J. and Valentine, G. (1994) 'All Hyped Up and No Place To Go', *Gender, Place and Culture*, 1: 31–48.

Benedikt, M. (ed.) (1991) *Cyberspace: First Steps*, Cambridge, Mass.: MIT Press.

Benjamin, J. (1990) *The Bonds of Love*, London: Virago.

Berger, J. (1972) *Ways of Seeing*, London and Harmondsworth: British Broadcasting Corporation and Penguin.

Berger, J. (1984) *And Our Faces, My Heart, Brief As Photos*, London: Writers and Readers.

Bergmann, M. S. and Jucovy, M. E. (eds) (1982) *Generations of the Holocaust*, New York: Basic Books.

Bergson, H. (1950) *Matter and Memory*, New York: Basic Books.

Bernheimer, C. and Kahane, C. (eds) (1985) *In Dora's Case: Freud-Hysteria-Feminism*, London: Virago (second edition 1990, New York: Columbia University Press).

Bethnal Green City Challenge (1991) *Bethnal Green City Challenge Action Plan: Working Together to Unlock Opportunities*, mimeo submitted to DoE.

Bethnal Green City Challenge (1993) *Action Plan 1994–95*, mimeo submitted to DoE.

Bettelheim, B. (1978) *The Uses of Enchantment. The Meaning and Importance of Fairy Tales*, Harmondsworth: Penguin.

Bewick, T. (1794) *History of British Birds. History and Description of Land Birds, Volume 1*, Newcastle and London: Robinson.

Bewick, T. (1804) *History of British Birds. History and Description of Water Birds, Volume 2*, Newcastle and London: Longman.

Bhabha, H. (1986) 'The Other Question: Difference, Discrimination and the Discourse of Colonialism', in R. Ferguson, M. Gever, M.-h. T. Trinh and C. West (eds) (1990) *Out There: Marginalization and Contemporary Cultures*, Cambridge, Mass.: MIT Press, 71–87.

Bhabha, H. (1993) *The Location of Culture*, London: Routledge.

Bianchini, F. (1987) 'GLC R.I.P. Cultural Policies in London, 1981–1986', *New Formations*, 1: 103–17.

Bibbings, L. and Alldridge, P. (1993) 'Sexual Expression, Body Alteration, and the Defence of Consent', *Journal of Law and Society*, 20: 356–70.

Birnbaum, S. (1989) *Disneyland: The Official Guide*, Los Angeles: Houghton Mifflin Company.

Bishop, P. (1991) 'Constable Country: Diet, Landscape and National Identity', *Landscape Research*, 16(2): 31–6.

Blumberg, H. (1972) 'Communication of Interpersonal Evaluations', *Journal of Personality and Social Psychology*, 23: 157–62.

Blunt, A. (1994) *Travel, Gender and Imperialism*, New York: Guilford Press.

Boddy, W. (1994) 'Archaeologies of Electronic Vision and the Electronic Spectator', *Screen*, 35: 105–22.

Bogue, R. (1989) *Deleuze and Guattari*, London: Routledge.

Boon, J. (1981) *Other Tribes, Other Scribes*, Cambridge: Cambridge University Press.

Bordo, S. (1990) 'Feminism, Postmodernism, and Gender-scepticism', in L. Nicholson (ed.) *Feminism / Postmodernism*, London: Routledge, 133–56.

Bordo, S. (1992) 'Review Essay: Postmodern Subjects, Postmodern Bodies', *Feminist Studies*, 18: 159–75.

Bordo, S. (1993) *Unbearable Weight: Feminism, Western Culture and the Body*, Berkeley: University of California Press.

Bordo, S. and Moussa, M. (1993) 'Rehabilitating the "I"', in H. J. Silverman (ed.) *Questioning Foundations. Truth, Subjectivity, Culture*, London: Routledge, 110–33.

Boundas, C. V. (1994) 'Deleuze: Serialisation and Subject-formation', in L. V. Boundas and D. Olkowski (eds) *Gilles Deleuze and the Theatre of Philosophy*, New York: Routledge, 99–116.

Bourdieu, P. (1977) *Outline of a Theory of Practice*, Cambridge: Cambridge University Press.

Bourdieu, P. (1990a) *In Other Words: Essays Towards a Reflexive Sociology*, Cambridge: Polity Press.

Bourdieu, P. (1990b) *The Logic of Practice*, Cambridge: Polity Press.

Bourdieu, P. and Wacquant, L. J. P. (1992) *An Invitation to Reflexive Sociology*, Cambridge: Polity Press.

Braidotti, R. (1989) 'The Politics of Ontological Difference', in T. Brennan (ed.) *Between Feminism and Psychoanalysis*, London: Routledge, 89–105.

Braidotti, R. (1994) 'Towards a New Nomadism: Feminist Deleuzian Tracks; or, Metaphysics and Metabolism', in C. V. Boundas and D. Olkowski (eds) *Gilles Deleuze and the Theatre of Philosophy*, New York: Routledge, 157–86.

Bramwell, A. (1989) *Ecology in the Twentieth Century*, London/Yale: Yale University Press.

Breinnes, W. (1992) *Young, White, and Miserable: Growing Up Female in the Fifties*, Boston: Beacon Press.

Brennan, T. (1993) *History after Lacan*, London: Routledge.

Brennan, T. (ed.) (1989) *Between Feminism and Psychoanalysis*, London: Routledge.

Breuer, J. and Freud, S. (1895) *Studies in Hysteria*, Harmondsworth: Penguin (first published 1974).

Bristow, J. (1989) 'Being Gay: Politics, Identity, Pleasure', *New Formations*, 9: 61–81.

Brockway, S. (1989) 'The Mask of Mickey Mouse: The Symbol of a Generation', *Journal of Popular Culture*, 22(4): 25–34.

Brody, H. (1981) *Maps and Dreams*, Harmondsworth: Penguin.

Brontë, C. (1966) *Jane Eyre. An Autobiography*, Harmondsworth: Penguin (first published 1847).

Brown, J. (1986) *Velázquez: Painter and Courtier*, New Haven: Yale University Press.

Browne, J. P. (1992) *Map Cover Art: A Pictorial History of Ordnance Survey Cover Illustrations*, London: Ordnance Survey.

Bruce Pratt, M. (1984) 'Identity: Skin Blood Heart', in E. Bulkin, M. Bruce Pratt and B. Smith (eds) *Yours in Struggle: Three Feminist Perspectives on Anti-Semitism and Racism*, New York: Long Haul Press, 9–63.

Bruno, G. (1993) *Streetwalking on a Ruined Map: Cultural Theory and the City Films of Elvira Notari*, Princeton: Princeton University Press.

Bukatman, S. (1993a) *Terminal Identity: The Virtual Subject in Postmodern Science Fiction*, Durham: Duke University Press.

Bukatman, S. (1993b) 'Gibson's Typewriter', *South Atlantic Quarterly*, 92: 627–46.

Burkitt, I. (1991) *Social Selves: Theories of the Social Formation of Personality*, London: Sage.

Burkitt, I. (1993) 'Overcoming Metaphysics: Elias and Foucault on Power and Freedom', *Philosophy of the Human Sciences*, 23: 50–72.

Burkitt, I. (1994) 'The Shifting Concept of the Self', *History of the Human Sciences*, 7: 7–28.

Burnett, A. and Moon, G. (1983) 'Community Opposition to Hostels for Single Homeless Men', *Area*, 15: 161–6.

Burrell, G. and Hearn, J. (1989) 'The Sexuality of Organizations', in J. Hearn, D. Sheppard, P. Tancred-Sheriff and G. Burrell (eds) *The Sexuality of Organization*, London: Sage, 1–27.

Butler, J. (1989) 'Gendering the Body: Beauvoir's Philosophical Contribution', in A. Garry and M. Pearsell (eds) *Women, Knowledge and Reality*, London: Unwin Hyman, 253–62.

Butler, J. (1990) *Gender Trouble: Feminism and the Subversion of Identity*, New York: Routledge.

Butler, J. (1991) 'Imitation and Gender Insubordination', in D. Fuss (ed.) *Inside/Out: Lesbian Theories, Gay Theories*, New York: Routledge, 13–31.

Butler, J. (1992) 'Contingent Foundations: Feminism and the Question of "Postmodernism"', in J. Butler and J. Scott (eds) *Feminists Theorize The Political*, New York: Routledge, 3–21.

Butler, J. (1993) *Bodies That Matter: On the Discursive Limits of 'Sex'*, New York: Routledge.

Cain, P. J. and Hopkins, A. G. (1993a) *British Imperialism: Innovation and Expansion 1688–1914*, London: Longman.

Cain, P. J. and Hopkins, A. G. (1993b) *British Imperialism: Crisis and Deconstruction 1914–1990*, London: Longman.

Caine, C. (1938) 'Notts. Ramblers' Federation', *The Nottinghamshire Countryside*, 1(4): 19.

Califia, P. (1993) '*Sex* and Madonna, or, What Do You Expect From a Girl Who Doesn't Put Out on the First Five Dates?', in L. Frank and P. Smith (eds) *Madonnarama: Essays on Sex and Popular Culture*, Pittsburgh: Cleis Press, 169–84.

Callon, M. (1986) 'Some Elements of a Sociology of Translation', in J. Law (ed.) *Power, Action, Belief: A New Sociology of Knowledge?*, London: Routledge and Kegan Paul, 196–233.

Callon, M. (1991) 'Techno-economic Networks and Irreversibility', in J. Law (ed.) *A Sociology of Monsters. Essays on Power, Technology and Domination*, London: Routledge, 132–61.

Campbell, A. (1968) 'Population Dynamics and Family Planning', *Journal of Marriage and the Family*, 30: 202–6.

CAPA Legal Advice and Support Group, Centre for Bangladeshi Studies, Queen Mary and Westfield College, Tower Hamlets Race Equality Council (THREC) (1993) *Young Bengalis and the Criminal Justice System*, Proceedings of a conference held at QMW, University of London.

Carlstein, T. (1982) *Time Resources, Society and Ecology*, London: George Allen and Unwin.

Carter, P. (1987) *The Road to Botany Bay: An Essay in Spatial History*, London: Faber and Faber.

Carter, P. (1992) *Living in a New Country: History, Travelling and Language*, London: Faber and Faber.

Cartledge, S. and Ryan, J. (eds) (1983) *Sex and Love*, London: The Women's Press.

Cascardi, A. J. (1992) *The Subject of Modernity*, Cambridge: Cambridge University Press.

Casey, E. S. (1993) *Getting Back into Place*, Bloomington: Indiana University Press.

Cassis, Y. (1987) *La City de Londres 1870–1914*, Paris: Belin.

Castoriadis, C. (1984) *Crossroads in the Labyrinth*, Brighton: Harvester.

Castoriadis, C. (1987) *The Imaginary Institution of Society*, Cambridge: Polity Press.

Cavanagh, J. C. (1981) 'Early Developmental Theories: A Brief Review of Attempts to Organise Developmental Data Prior to 1925', *Journal of the History of the Behavioral Sciences*, 17: 38–47.

Centore, F. F. (1989) *Being and Becoming: A Critique of Postmodernism*, New York: Greenwood.

Chamberlain, J. (1988) *On Our Own: Patient-Controlled Alternatives to the Mental Health Care System*, London: MIND (first published 1977).

Chambers, I. (1994) *Migrancy, Culture, Identity*, London: Routledge.

Chapman, D. (1968) *Sociology and the Stereotype of the Criminal*, London: Tavistock.

Chase, M. (1989) 'This is No Claptrap: This is Our Heritage', in G. Shaw and M. Chase (eds) *The Imagined Past: History and Nostalgia*, Manchester: Manchester University Press, 128–46.

Chase, M. (1992) 'Rolf Gardiner: An Inter-war, Cross-cultural Study', in B. Hake and S. Marriott (eds) *Adult Education between Cultures*, Leeds: Leeds Studies in Continuing Education, 225–41.

Cixous, H. (1976) 'The Laugh of the Medusa', in E. Marks and I. de Courtivon (eds) (1981) *New French Feminisms*, Sussex: Harvester.

Clifford, J. C. (1988) *The Predicament of Culture: Twentieth Century Ethnography, Art and Literature*, Cambridge, Mass.: Harvard University Press.

Clifford, J. C. (1992) 'Travelling Cultures', in L. Grossberg, C. Nelson and P. Treichler (eds) *Cultural Studies*, New York: Routledge, 96–111.

Coburn, O. (1950) *Youth Hostel Story*, London: National Council of Social Service.

Cohen, A. P. (1985) *The Symbolic Construction of Community*, London: Tavistock.

Cohen, P. (1992) *Playgrounds of Prejudice*, London: University of East London.

Collins English Dictionary (1977) *Collins English Dictionary*, London: Collins.

Connell, R. W. (1993) 'The Big Picture: Masculinities in Recent World History', *Theory and Society*, 22: 597–623.

Contemporary Medical Archives Centre (CMAC): A5/162/5: Family Planning Association (UK) in the CMAC at the Wellcome Institute for the History of Medicine, London.

Contemporary Medical Archives Centre (CMAC): PP/RJH/A1/7: Dr R. J. Heathrington papers in the CMAC at the Wellcome Institute for the History of Medicine, London.

Cook, R. (1951) *Human Fertility: The Modern Dilemma*, London: Victor Gollancz.

Cooper, D. (1978) *The Language of Madness*, London: Allen Lane.

Cooper, M. (1990) 'Making Changes', in T. Putnam and C. Newton (eds) *Household Choices*, London: Futures Publications.

Corbin, A. (1986) *The Fragrant and the Foul: Odor and the French Social Imagination*, Cambridge, Mass.: Harvard University Press.

Cornish, V. (1930) *National Parks, and the Heritage of Scenery*, London: Sifton Praed.

Cornish, V. (1932) *The Scenery of England*, London: Council for the Preservation of Rural England.

Cornish, V. (1933) 'Aesthetic Principles of Town and Country Planning', *Scottish Geographical Magazine*, 49: 320–3.

Cornish, V. (1935) *Scenery and the Sense of Sight*, Cambridge: Cambridge University Press.

Cornish, V. (1946) *Geographical Essays*, London: Sifton Praed.

Cornwall, N. and Lindisfarne, N. (eds) (1993) *Dislocating Masculinity: Comparative Ethnographies*, London: Routledge.

Cosgrove, D. (1985) 'Prospect, Perspective and the Evolution of the Landscape Idea', *Transactions of the Institute of British Geographers*, 10: 45–62.

Cosgrove, D. (1990) 'Environmental Thought and Action: Pre-modern and Post-modern', *Transactions of the Institute of British Geographers*, 15(3): 344–58.

Courtine, J.-F. (1988) 'Voice of Conscience and Call of Being', *Topoi*, 7(2): 101–9.

Coveney, P. (1967) *The Image of Childhood. The Individual and Society: A Study of the Theme in English Literature*, Harmondsworth: Penguin (first published 1957 as *Poor Monkey*).

Coward, R. (1989) *The Whole Truth: The Myth of Alternative Health*, London: Faber and Faber.

Crary, J. (1990) *Changing the Observer*, Cambridge, Mass.: MIT Press.

Crimp, D. (1987) 'How to Have Promiscuity in an Epidemic', *October*, 43: 237–71.

Critchley, S. (1992) *The Ethics of Deconstruction: Derrida and Levinas*, Oxford: Blackwell.

Csikszentmihalyi, M. and Rochberg-Halton, E. (1981) *The Meaning of Things: Domestic Symbols and the Self*, Cambridge: Cambridge University Press.

Cunningham, H. (1991) *The Children of the Poor: Representations of Childhood Since the Seventeenth Century*, Oxford: Clarendon Press.

Cunningham, V. (1988) *British Writers of the Thirties*, Oxford: Oxford University Press.

Curry, D. (1993) 'Decorating the Body Politic', *New Formations*, 19: 69–82.

D'Emilio, J. (1983) *Sexual Politics, Sexual Communities: The Making of a Homosexual Minority in the US, 1940–1970*, Chicago and London: University of Chicago Press.

D'Emilio, J. (1993) *Making Trouble: Essays on Gay History, Politics, and the University*, New York: Routledge.

D'Emilio, J. and Freedman, E. (1988) *Intimate Matters: A History of Sexuality in America*, New York: Harper and Row.

Danto, A. C. (n. d.) *Cindy Sherman: History Portraits*, New York: Rizzoli.

Darwin, C. (1873) *The Expression of the Emotions in Man and Animals*, London: Murray.

Darwin, C. (1877) 'The Biographical Sketch of an Infant', *Mind*, 2: 285–94.

Davenport, K. (1992) 'Games without Frontiers', *Leisure Management*, January: 45–6.

Davidoff, L. (1974) 'Mastered for Life: Servant and Wife in Victorian and Edwardian England', *Journal of Social History*, 7(4): 406–20.

Davies, P. (1992) 'The Role of Disclosure in Coming Out Among Gay Men', in K. Plummer (ed.) *Modern Homosexualities: Fragments of Lesbian and Gay Experience*, London: Routledge, 75–86.

Davis, M. and Wallbridge, D. (1987) *Boundary and Space: An Introduction to the Work of D. W. Winnicott*, London: Karnac Books.

de Castro, J. (1952) *Geography of Hunger*, London: Victor Gollancz.

de Certeau, M. (1984) *The Practice of Everyday Life*, London: University of California Press.

de Certeau, M. (1986) *Heterologies*, Minneapolis: University of Minnesota Press.

de Lauretis, T. (1986a) 'Introduction: Feminist Studies/Critical Studies', in T. de Lauretis (ed.) *Feminist Studies/Critical Studies*, London: Macmillan, 1–19.

de Lauretis, T. (1986b) *Feminist Studies/Cultural Studies*, Bloomington: Indiana University Press.

de Lauretis, T. (1987) *Technologies of Gender: Essays on Theory, Film and Fiction*, London: Macmillan.

de Lauretis, T. (1990) 'Eccentric Subjects: Feminist Theory and Historical Consciousness', *Feminist Studies*, 16: 115–50.

de Lauretis, T. (ed.) (1991) 'Queer Theory: Lesbian and Gay Sexualities', *Differences*, 3: whole volume.

Dean, K. (1979) 'The Geographical Study of Psychiatric Illness: The Case of Depressive Illness in Plymouth', *Area*, 11: 167–71.

Dean, K. (1982) 'The Psychiatric Admission Process: A Geographical Perspective', unpublished Ph.D. dissertation, Department of Geography, University of Hull, Hull.

Dean, K. (1984) 'Social Theory and Prospects in Social Geography', *GeoJournal*, 9: 287–99.

Dear, M. (1977) 'Impact of Mental Health Facilities on Property Values', *Community Mental Health Journal*, 13: 150–7.

Dear, M. (1980) 'The Public City', in W. A. V. Clark and E. G. Moore (eds) *Residential Mobility and Public Policy*, Beverley Hills: Sage, 219–41.

Dear, M. and Taylor, S. (1982) *Not on Our Street: Community Attitudes to Mental Health Care*, London: Pion.

Dear, M. and Wolch, J. (1987) *Landscapes of Despair: From Deinstitutionalisation to Homelessness*, Oxford: Polity Press.

Debord, G. (1983) *Society of the Spectacle*, London: Red and Black.

Defoe, D. (1989) *Moll Flanders*, Harmondsworth: Penguin (first published 1722).

Deleuze, G. (1983) 'Politics', in G. Deleuze and F. Guattari *On the Line*, New York: Semiotext(e).

Deleuze, G. (1986) *Cinema 1: The Movement-Image*, London: Athlone.

Deleuze, G. (1988) *Foucault*, London: Athlone.

Deleuze, G. (1989) *Cinema 2: The Time-Image*, London: Athlone.

Deleuze, G. (1992) 'What is a *Dispositif*?', in T. J. Armstrong (ed.) *Michel Foucault: Philosopher*, Hemel Hempstead: Harvester Wheatsheaf.

Deleuze, G. (1993a) *The Deleuze Reader*, edited by C. V. Boundas, New York: Columbia University.

Deleuze, G. (1993b) *The Fold: Leibniz and the Baroque*, Minneapolis: University of Minnesota Press.

Deleuze, G. and Guattari, F. (1983) *Anti-Oedipus: Capitalism and Schizophrenia*, London: Athlone (first published 1976).

Deleuze, G. and Guattari, F. (1988) *A Thousand Plateaus: Capitalism and Schizophrenia*, London: Athlone (first published 1987, Minneapolis: University of Minnesota Press).

Deleuze, G. and Guattari, F. (1994) *What is Philosophy?* London: Verso.

Deleuze, G. and Parnet, C. (1988) *Dialogues*, London: Athlone (first published 1987, New York: Columbia University Press).

Denzin, N. K. (1991) *Images of Postmodern Society: Social Theory and Contemporary Cinema*, London: Sage.

Deptford City Challenge Evaluation Project (DCCEP) (1993) *Deptford City Challenge Baseline Study*, London: DCCEP, Goldsmiths' College.

Derrida, J. (1982) *Margins of Philosophy*, Brighton: Harvester Press.

Derrida, J. (1988a) 'Interview with Jean-Luc Nancy', *Topoi*, 7(2): 113–21.

Derrida, J. (1988b) *Limited Inc*, Evanston: Northwestern University Press.

Derrida, J. (1992) 'Jacques Derrida', in D. Jones and R. Stoneman (eds) *Talking Liberties*, London: Channel 4 Television.

Deutsche, R. (1991) 'Boy's Town', *Environment and Planning D: Society and Space*, 9: 5–30.

Dews, P. (1987) *Logics of Disintegration: Post-Structuralist Thought and the Claims of Critical Theory*, London: Verso.

Diprose, R. and Ferrell, R. (1990) *Cartographies: Poststructuralism and the Mapping of Bodies and Spaces*, Sydney: Allen and Unwin.

Doane, M. A. (1991) 'Dark Continents: Epistemologies of Racial and Sexual Difference in Psychoanalysis and the Cinema', in *Femmes Fatales. Feminism, Film Theory, Psychoanalysis*, London: Routledge, 209–48.

Doane, M. A. (1993) 'Technology's Body: Cinematic Vision in Modernity', *Differences*, 5: 1–23.

Doel, M. A. (1992) 'In stalling Deconstruction: Striking out the Postmodern', *Environment and Planning D: Society and Space*, 10(2): 163–79.

Doel, M. A. (1993) 'Proverbs for Paranoids: Writing Geography on Hollowed Ground', *Transactions of the Institute of British Geographers*, 18(3): 377–94.

Doel, M. A. (1994a) 'Something Resists: Reading – Deconstruction as Ontological Infestation (Departures from the Texts of Jacques Derrida)', in P. Cloke, M. A. Doel, D. Matless, M. Phillips and N. J. Thrift *Writing the Rural: Five Cultural Geographies*, London: Paul Chapman.

Doel, M. A. (1994b) 'Deconstruction on the Move: From Libidinal Economy to Liminal Materialism', *Environment and Planning A*, 26: 1041–59.

Dollimore, J. (1991) *Sexual Dissidence: Augustine to Wilde, Freud to Foucault*, Oxford: Clarendon Press.

Donald, J. (1991) *Psychoanalysis and Cultural Theory: Thresholds*, London: Macmillan.

Douglas, M. (1966) *Purity and Danger: An Analysis of the Concepts of Pollution and Taboo*, London: Routledge.

Douglas, M. (1992) 'The Person in an Enterprise Culture', in S. Heap and A. Rose (eds) *Understanding the Enterprise Culture: Themes in the Work of Mary Douglas*, Edinburgh: Edinburgh University Press.

Dovey, K. (1985) 'Home and Homelessness', in I. Altman and C. Werner (eds) *Home Environments*, New York: Plenum Press, 33–61.

Dreyfus, H. L. (1991) *Being-in-the-World: A Commentary on Heidegger's Being and Time, Division I*, Cambridge, Mass.: MIT Press.

Dreyfus, H. L. and Rabinow, P. (1982) *Michel Foucault: Beyond Structuralism and Hermeneutics*, Brighton: Harvester.

Dreyfus, H. L. and Rabinow, P. (1993) 'Can There be a Science of Existential Structure and Social Meaning', in C. Calhoun, E. Lipuma and M. Postone (eds) *Bourdieu: Critical Perspectives*, Oxford: Blackwell, 35–44.

Dyck, I. (1990) 'Space, Time, and Renegotiating Motherhood: An Exploration of the Domestic Workplace', *Environment and Planning D: Society and Space*, 8(4): 459–83.

Eco, U. (1986) *Travels in Hyperreality, and Other Essays*, translated by W. Weaver, New York: Harcourt Brace Jovanovich.

Edgley, C. and Turner, R. (1975) 'Masks and Social Relations: An Essay on the Sources and Assumptions of Dramaturgical Social Psychology', *Humboldt Journal of Social Relations*, 3: 5–13.

Edwards, D. and Potter, J. (1992) *Discursive Psychology*, London: Sage.

Eissler, K. R. (1963) *Goethe. A Psycho-analytic Study, 1775–1786, 2 Volumes*, Detroit: Wayne State University Press.

Elfenbein, A. (1989) *Women on the Color Line*, Charlottesville: Virginia University Press.

Elias, N. (1956) 'Problems of Involvement and Detachment', *British Journal of Sociology*, 7: 226–52.

Elias, N. (1978) *The Civilizing Process. Volume 1: The History of Manners*, translated by E. Jephcott, New York: Pantheon (first published 1939).

Elias, N. (1982) *The Civilizing Process. Volume 2: State Formation and Civilisation*, translated by E. Jephcott, Oxford: Blackwell (first published 1939).

Elias, N. (1983) *The Court Society*, translated by E. Jephcott, Oxford: Blackwell (first published 1969).

Elias, N. (1987) *Involvement and Detachment*, translated by E. Jephcott, Oxford: Blackwell (first published 1983).

Elias, N. (1991) *The Society of Individuals*, translated by E. Jephcott, Oxford: Blackwell.

Elias, N. (1992) *Time: An Essay*, Oxford: Blackwell (first published 1984).

Elliott, A. (1992) *Social Theory and Psychoanalysis in Transition: Self and Society from Freud to Kristeva*, Oxford: Blackwell.

Elton, O. (1929) *C. E. Montague: A Memoir*, London: Chatto and Windus.

Eribon, D. (1991) *Michel Foucault*, translated by B. Wing, London: Faber and Faber (first published 1989).

Erikson, E. H. (1959) *Identity and the Life Cycle*, New York: International Universities Press.

Estroff, S. E. (1985) *Making It Crazy: An Ethnography of Psychiatric Clients in an American Community*, Berkeley: University of California Press.

Etchell, M. (1868) *Ten Years in a Lunatic Asylum*, London: Simpkin, Marshall and Company, Stationers' Hall Court.

Eyles, J. (1988) 'Mental Health Services, the Restructuring of Care, and the Fiscal Crisis of the State: the United Kingdom Case Study', in C. J. Smith and J. A. Giggs (eds) *Location and Stigma: Contemporary Perspectives on Mental Health and Mental Health Care*, London: Unwin Hyman.

Faderman, L. (1991) *Odd Girls and Twilight Lovers: A History of Lesbian Life in Twentieth-century America*, New York: Penguin.

Fagg, C. C. and Hutchings, G. E. (1930) *An Introduction to Regional Surveying*, Cambridge: Cambridge University Press.

Fanon, F. (1967) *Black Skin, White Masks*, London: Pluto Press (first published 1952).

Featherstone, M. (1990) *Consumer Culture and Postmodernism*, London: Sage.

Felman, S. (1987) *Jacques Lacan and the Adventure of Insight*, Cambridge, Mass.: Harvard University Press.

Fillmore, C. (1976) *Statistical Methods in Linguistics*, Stockholm: Skriptor.

Finch, J. and Summerfield, P. (1991) 'Social Reconstruction and the Emergence of Companionate Marriage, 1945–1959', in D. Clark (ed.) *Marriage, Domestic Life and Social Change*, London: Routledge, 7–32.

Fiske, J. (1986) *Understanding Popular Culture*, London: Routledge.

Fiske, J. (1990) 'Ethnosemiotics: Some Personal and Theoretical Reflections', *Cultural Studies*, 4: 85–99.

Fjellman, S. (1992) *Vinyl Leaves: Walt Disney World and America*, Boulder: West View Press.

Fladmark, J. M. (ed.) (1993) *Heritage: Conservation, Interpretation and Enterprise*, London: Donhead Publishing.

Forrester, J. (1987) 'The Seminars of Jacques Lacan: In Place of an Introduction. Book 1: Freud's Papers on Technique, 1953–1954', *Free Associations*, 10: 63–93.

Foster, H. (1982) 'Subversive Signs', *Art in America*, November: 88–92.

Foucault, M. (1967) *Madness and Civilization: A History of Insanity in the Age of Reason*, translated by Richard Howard, London: Tavistock (first published 1961).

Foucault, M. (1970) *The Order of Things: An Archaeology of the Human Sciences*, London: Tavistock (first published 1966).

Foucault, M. (1972) *The Archaeology of Knowledge*, London: Tavistock (first published 1969).

Foucault, M. (1977) *Discipline and Punish: The Birth of the Prison*, Harmondsworth: Penguin (first published 1975).

Foucault, M. (1979) *The History of Sexuality. Volume 1: An Introduction*, translated by R. Hurley, Harmondsworth: Penguin (first published 1976).

Foucault, M. (1982) 'The Subject and Power', in H. L. Dreyfus and P. Rabinow, *Michel Foucault: Beyond Structuralism and Hermeneutics*, Brighton: Harvester, 208–26.

Foucault, M. (1985) *The History of Sexuality. Volume 2: The Use of Pleasure*, translated by R. Hurley, Pantheon: New York (first published 1984).

Foucault, M. (1988) 'Technologies of the Self', in L. H. Martin, H. Gutman and P. H. Hutton (eds) *Technologies of the Self: A Seminar with Michel Foucault*, London: Tavistock, 16–49.

Foucault, M. (1990) *The History of Sexuality. Volume 3: The Care of the Self*, translated by R. Hurley, Harmondsworth: Penguin (first published 1984).

Fraser, R. (1984) *In Search of a Past*, London: Verso.

Freud, S. (1905) 'Fragment of an Analysis of a Case of Hysteria: "Dora"', in *Case Histories I: 'Dora' and 'Little Hans'*, Harmondsworth: Penguin Freud Library (volume 8, published 1977), 35–164.

Freud, S. (1925) 'Negation', in *On Metapsychology: The Theory of Psychoanalysis*, Harmondsworth: Penguin Freud Library (volume 11, published 1984), 437–42.

Freud, S. (1927a) 'Fetishism', in *On Sexuality: Three Essays on the Theory of Sexuality and Other Works*, Harmondsworth: Penguin Freud Library (volume 7, published 1977), 351–7.

Freud, S. (1927b) 'The Future of an Illusion', in *Civilization, Society and Religion: Group Psychology, Civilization and its Discontents and Other Works*, Harmondsworth: Penguin Freud Library (volume 12, published 1985), 183–241.

Freud, S. (1933) 'Femininity', in *New Introductory Lectures on Psychoanalysis*, Harmondsworth: Penguin Freud Library (volume 2, published 1973), 145–69.

Frosh, S. (1987) *The Politics of Psychoanalysis: An Introduction to Freudian and Post-Freudian Theory*, Basingstoke: Macmillan.

Frosh, S. (1991) *Identity Crisis: Modernity, Psychoanalysis and the Self*, London: Macmillan.

Frye, M. (1983) *The Politics of Reality: Essays in Feminist Theory*, Trumansburg, NY: Crossing Press.

Fuss, D. (1989) *Essentially Speaking: Feminism, Nature and Difference*, New York: Routledge.

Fuss, D. (1991) 'Inside/out', in D. Fuss (ed.) *Inside/Out: Lesbian Theories, Gay Theories*, New York: Routledge, 1–10.

Fuss, D. (1992) 'Fashion and the Homospectatorial Look', *Critical Inquiry*, 18: 713–37.

Gallop, J. (1988) *Thinking Through the Body*, New York: Columbia University Press.

Game, A. (1991) *Undoing the Social: Towards a Deconstructive Sociology*, Milton Keynes: Open University Press.

Gane, M. (1993) *Baudrillard Live: Selected Interviews*, London: Routledge.

Gange, J. and Johnstone, S. (1993) 'Believe Me, Everybody has Something Pierced in California: An Interview with Nayland Blake', *New Formations*, 19: 51–68.

Garber, M. (1992) *Vested Interests: Cross-dressing and Cultural Anxiety*, New York: Routledge.

Garcia, C. (1963) 'Clinical Studies on Human Fertility', in R. Greep (ed.) *Human Fertility and*

Population Problems, Cambridge, Mass.: Schenkman Publishing Co., 43–75.

Garcia, C., Pincus, G. and Rock, J. (1958) 'Effects of Three 19-Nor Steroids on Human Ovulation and Menstruation', *American Journal of Obstetrics and Gynecology*, 75: 82–97.

Geertz, C. (1989) *Works and Lives: The Anthropologist as Author*, Cambridge: Polity Press.

Geltmaker, T. (1992) 'The Queer Nation Acts Up: Health Care, Politics, and Sexual Diversity in the County of Angels', *Environment and Planning D: Society and Space*, 10: 609–50.

Gergen, K. J. (1989) 'Warranting Voice and the Elaboration', in J. Shotter and K. J. Gergen (eds) *Texts of Identity*, London: Sage, 70–81.

Gergen, K. J. (1991) *The Saturated Self: Dilemmas of Identity in Contemporary Life*, New York: Basic Books.

Gibson, W. (1991) 'Academy Leader', in M. Benedikt (ed.) *Cyberspace: First Steps*, Cambridge, Mass.: MIT Press, 27–30.

Giddens, A. (1985) 'Time, Space and Regionalisation', in D. Gregory and J. Urry (eds) *Social Relations and Spatial Structures*, London: Macmillan, 265–95.

Giddens, A. (1991) *Modernity and Self-Identity*, Cambridge: Polity Press.

Giggs, J. A. (1973) 'The Distribution of Schizophrenics in Nottingham', *Transactions of the Institute of British Geographers*, 59: 55–76.

Gilcraft (1942) *Exploring*, London: C. Arthur Pearson (first published 1930).

Gilman, S. (1985) *Difference and Pathology: Stereotypes of Sexuality, Race and Madness*, Ithaca: Cornell University Press.

Gilroy, P. (1991) 'It Ain't Where You're From, It's Where You're At: The Dialectics of Diasporic Identification', in P. Gilroy (1993) *Small Acts: Thoughts on the Politics of Black Cultures*, London: Serpent's Tail, 120–45.

Gilroy, P. (1993) *The Black Atlantic: Modernity and Double Consciousness*, London: Verso.

Ginzburg, C. (1980) *The Cheese and the Worms: The Cosmos of a Sixteenth Century Miller*, London: Routledge and Kegan Paul.

Glass, J. M. (1989) *Private Terror/Public Life: Psychosis and the Politics of Community*, Ithaca: Cornell University Press.

Glassman, B. (1975) *Anti-Semitic Stereotypes without Jews*, Detroit: Wayne State University Press.

Glover, J. (1988) *I: The Philosophy and Psychology of Personal Identity*, Harmondsworth: Pelican.

Goethe, J. W. von (1913) *Wilhelm Meister's Theatrical Mission*, translated by G. A. Page, London: Heinemann.

Goethe, J. W. von (1977) *Wilhelm Meister's Years of Apprenticeship, Six Volumes*, translated by H. M. Waidson, London: Calder (originally published in German 1795–6).

Goffman, E. (1959) *The Presentation of Self in Everyday Life*, New York: Anchor Books.

Goffman, E. (1961) *Asylums: Essays on the Social Situation of Mental Patients*, Harmondsworth: Penguin.

Golding, S. (1993a) 'Quantum Philosophy, Impossible Geographies and a Few Small Points About Life, Liberty and the Pursuit of Sex (All in the Name of Democracy)', in M. Keith and S. Pile (eds) *Place and the Politics of Identity*, London: Routledge, 206–19.

Golding, S. (1993b) 'The Excess: An Added Remark on Sex, Rubber, Ethics and Other Impurities', *New Formations*, 19: 23–8.

Goodey, J. S. (1993) 'Fear of Crime, Children and Gendered Socialization', British Criminology Conference, Cardiff.

Goodey, J. S. (forthcoming) 'Fear of Crime: What Can Children Tell Us?' *International Review of Victimology*, 3(3).

Gordon, C. (1991) 'Governmental Rationality: An Introduction', in G. Burchell, C. Gordon and P. Miller (eds) *The Foucault Effect: Studies in Governmentality*, Hemel Hempstead: Harvester Wheatsheaf.

Graham, D. (1981) 'Signs', *Artforum*, April: 38–43.

Gratton, C. (1992) 'Is there Life after EuroDisney?', *Leisure Management*, April: 24–7.

Green, G. H. (1921) *Psychoanalysis in the Class Room*, London: University of London Press.

Green, G. H. (1939) *The Healthway Books*, London: University of London Press.

Green, M. (1977) *Children of the Sun: A Narrative of 'Decadence' in England after 1918*, London: Constable.

Green, M. (1986) *Mountain of Truth: The Counterculture Begins*, London/Hanover: University Press of New England.

Gregory, D. (1994) *Geographical Imaginations*, Oxford: Blackwell.

Griffiths, R. (1983) *Fellow Travellers of the Right: British Enthusiasts for Nazi Germany 1933–9*, Oxford: Oxford University Press.

Groning, G. and Wolschke-Bulmahn, J. (1987) 'Politics, Planning and the Protection of Nature: Political Abuse of Early Ecological Ideas in Germany, 1933–45', *Planning Perspectives*, 2: 127–48.

Gross, E. (1990) 'The body of signification', in J. Fletcher and A. Benjamin (eds) *Abjection, Melancholia and Love: The Work of Julia Kristeva*, London: Routledge, 80–103.

Grossberg, L. (1988) 'Wondering Audiences, Nomadic Critics', *Cultural Studies*, 2: 377–91.

Grosz, E. (1989) *Sexual Subversions: Three French Feminists*, London: Routledge.

Grosz, E. (1990a) 'Contemporary Theories of Power and Subjectivity', in S. Gunew (ed.) *Feminist Knowledge: Critique and Construct*, London: Routledge, 59–120.

Grosz, E. (1990b) *Jacques Lacan: A Feminist Introduction*, London: Routledge

Grosz, E. (1992) 'Kristeva, Julia', in E. Wright (ed.) *Feminism and Psychoanalysis: A Critical Dictionary*, Oxford: Blackwell.

Grosz, E. (1993) 'Merleau-Ponty and Irigaray in the Flesh', *Thesis Eleven*, 36: 37–59.

Grosz, E. (1994) 'A Million Tiny Sexes: Feminisim and Rhizomatics', in L. V. Boundas and D. Olkowski (eds) *Gilles Deleuze and the Theatre of Philosophy*, New York: Routledge, 187–210.

Hacking, I. (1994) 'Memoro-politics, Trauma and the Soul', *History of the Human Sciences*, 7: 29–52.

Hall, J. (1891) 'January Searle', *Yorkshire County Magazine*, 1: 40–2.

Hall, M. (1989) 'Private Experiences in the Public Domain: Lesbians in Organizations', in J. Hearn, D. Sheppard, P. Tancred-Sheriff and G. Burrell (eds) *The Sexuality of Organization*, London: Sage, 125–38.

Hall, S. (1991a) 'The Local and the Global: Globalisation and Ethnicity', in A. D. King (ed.) *Culture, Globalisation and the World System*, London: Macmillan, 19–40.

Hall, S. (1991b) 'Old and New Identities, Old and New Ethnicities', in A. D. King (ed.) *Culture, Globalisation and the World System*, London: Macmillan, 41–68.

Hall, S. (1995) 'New Cultures for Old', in D. Massey and P. Jess (eds) *A Place in the World: Places, Culture and Globalization*, Oxford: Oxford University Press.

Hall, S. and Jefferson, T. (eds) (1975) *Resistance through Rituals*, London: Hutchinson.

Hall, S., Critcher, C., Jefferson, T., Clarke, J. and Roberts, B. (1978) *Policing the Crisis: Mugging the State and Law and Order*, Basingstoke: Macmillan.

Hannerz, U. (1987) 'The World in Creolisation', *Africa*, 57(4): 546–59.

Hannerz, U. (1992) *Cultural Complexity*, New York: Columbia University Press.

Haraway, D. (1988) 'Situated Knowledges: The Science Question in Feminism and the Privilege of Partial Perspective', in D. Haraway, *Simians, Cyborgs and Women: The Reinvention of Nature*, London: Free Association Books, 183–201.

Haraway, D. (1990) 'A Manifesto for Cyborgs: Science, Technology and Socialist Feminism in the 1980s', in L. Nicholson (ed.) *Feminism/Postmodernism*, London: Routledge, 190–233.

Haraway, D. (1991) *Simians, Cyborgs and Women: The Reinvention of Nature*, London: Free Association Books.

Haraway, D. (1992) 'Ecce Homo, Ain't (Ar'n't) I a Woman, and Inappropriate/d Others: The Human in a Post-Humanist Landscape', in J. Butler and J. Scott (eds) *Feminists Theorize the Political*, New York: Routledge, 86–100.

Hardt, M. (1993) *Gilles Deleuze: An Apprenticeship in Philosophy*, Minneapolis: University of Minnesota Press.

Harland, R. (1987) *Superstructuralism: The Philosophy of Structuralism and Poststructuralism*, London: Methuen.

Harré, R. (1979) *Social Being*, Oxford: Blackwell.

Harré, R. (1991) *Physical Being*, Oxford: Blackwell.

Harré, R. (1993) *Social Being* (second edition), Oxford: Blackwell.

Harré, R. (ed.) (1986) *The Social Construction of Emotions*, Oxford: Blackwell.

Harré, R., Clarke, D. and de Carlo, N. (1985) *Motives and Mechanisms*, London: Methuen.

Hartsock, N. (1990) 'Foucault on Power: A Theory for Women', in L. J. Nicholson (ed.) *Feminism/Postmodernism*, New York: Routledge, 157–75.

Harvey, S. (1983) 'Community Mental Health Services: A Review of Legislative Developments, Literature and Theory', School of Geography, University of Leeds, Working Paper no. 364.

Heidegger, M. (1983) *Being and Time*, translated by J. Macquarrie and E. Robinson, Oxford: Blackwell.

Hennelly, M. M. (1984) '*Jane Eyre*'s Reading Lesson', *English Literary History*, 51(3): 693–717.

Henriques, J., Hollway, W., Urwin, C., Venn, C. and Walkerdine, V. (1984) *Changing the Subject: Psychology, Social Regulation and Subjectivity*, London: Methuen.

Herek, G. and Berrill, K. (eds) (1992) *Hate Crimes: Confronting Violence Against Lesbians and Gay Men*, London: Sage.

Herf, J. (1984) *Reactionary Modernism: Technology, Culture and Politics in Weimar and the Third Reich*, Cambridge: Cambridge University Press.

Heritage, J. (1984) *Garfinkel and Ethnomethodology*, Cambridge: Polity Press.

Higson, A. (1984) 'Space, Place and Spectacle', *Screen*, 25: 2–21.

Hill, H. (1980) *Freedom to Roam: The Struggle for Access to Britain's Moors and Mountains*, Ashbourne: Moorland Publishing.

Hillman, M., Adams, J. and Whitelegg, J. (1990) *One False Move: A Study of Children's Independent Mobility*, London: Policy Studies Institute.

Hoggett, P. (1992) 'A Place for Experience: A Psychoanalytic Perspective on Boundary, Identity and Culture', *Environment and Planning D: Society and Space*, 10: 345–56.

Holzer, J. (1985) in 'Jenny Holzer's Language Games: Interview by Jeanne Siegel', *Arts Magazine*, December: 64–8.

Holzer, J. (1989) *Laments*, New York: Dia Art Foundation.

Holzer, J. (1990) 'Space, Language and Time – an A&D Interview with Jenny Holzer', *Art and Design*, 6: 30–9.

Hopkins, J. (1990) 'Landscapes of Myths and Elsewhereness', *The Canadian Geographer*, 34: 2–17.

Horder of Ashford, Lord (1938) 'Quiet – a Physician Prescribes', in C. Williams-Ellis (ed.) *Britain and the Beast*, London: Dent, 176–82.

Howard of Penrith, Lord (1938) 'Lessons from Other Countries', in C. Williams-Ellis (ed.) *Britain and the Beast*, London: Dent, 279–97.

Irigaray, L. (1977) *This Sex Which is Not One*, Ithaca: Cornell University Press.

Irigaray, L. (1985a) *Speculum of the Other Woman*, Ithaca: Cornell University Press.

Irigaray, L. (1985b) *This Sex Which is Not One*, Ithaca: Cornell University Press.

Irigaray, L. (1989) 'The Gesture in Psychoanalysis', in T. Brennan (ed.) *Between Feminism and Psychoanalysis*, London: Routledge, 127–38.

Iversen, M. (1991) 'Shaped by Discourse, Dispersed by Desire: Masquerade and Mary Kelly's *Interim*', *Camera Obscura*, 27: 134–47.

Jacobs, J. (1994) 'The Battle of Bank Junction: The Contested Iconography of Capital' in S. Corbridge, N. J. Thrift and R. L. Martin (eds) *Money, Power and Space*, Oxford: Blackwell, 356–82.

Jacobus, M. (1986) 'Madonna: Like a Virgin', *Oxford Literary Review*, 8: 35–50.

Jacobus, M. (1990) '"Tea Daddy": Poor Mrs. Klein and the Pencil Shavings', *Women*, 1: 160–79.

James, S. (1990) 'Is There a "Place" for Children in Geography?' *Area*, 22: 378–83.

Jameson, F. (1984) 'Postmodernism, or the Cultural Logic of Late Capitalism', *New Left Review*, 146: 53–92.

Jameson, F. (1991) *Postmodernism, or the Cultural Logic of Late Capitalism*, London: Verso.

Jardine, A. (1985) *Gynesis*, Ithaca: Cornell University Press.

Jay, M. (1993) *Downcast Eyes: The Denigration of Vision in Twentieth Century French Thought*, Berkeley: University of California Press.

Jeans, D. (1990) 'Planning and the Myth of the English Countryside', *Rural History*, 1(2): 249–64.

Jenness, V. (1992) 'Coming Out: Lesbian Identities and the Categorization Problem', in K. Plummer (ed.) *Modern Homosexualities: Fragments of Lesbian and Gay Experience*, London: Routledge, 65–74.

Joad, C. E. M. (1934) *A Charter for Ramblers*, London: Hutchinson.

Joad, C. E. M. (1938) 'The People's Claim', in C. Williams-Ellis (ed.) *Britain and the Beast*, London: Dent, 64–85.

Joad, C. E. M. (1946) *The Untutored Townsman's Invasion of the Country*, London: Faber and Faber.

Joad, C. E. M. (1948) *The English Counties*, London: Odhams.

Johnson, D. (1981) 'Disney World as Structure and Symbol: Re-creation of the American Experience', *Journal of Popular Culture*, 15(1): 157–65.

Johnson, L. (1983) 'Bracketing Lifeworlds: Husserlian Phenomenology as Geographical Method', *Australian Geographical Studies*, 21(1): 101–8.

Jones, A. (1991) 'Modernist Logic in Feminist Histories of Art', *Camera Obscura*, 27: 148–65.

Jones, S. (1987) 'State Intervention in Sport and Leisure in Britain between the Wars', *Journal of Contemporary History*, 22(1): 163–82.

Jukes, A. (1993) *Why Men Hate Women*, London: Free Association Books.

Katz, C. (1992) 'All the World is Staged: Intellectuals and the Projects of Ethnography',

Environment and Planning D: Society and Space, 10: 495–510.

Katz, C. (1993) 'Growing Girls/Closing Circles', in C. Katz and J. Monk (eds) *Full Circles*, London: Routledge, 88–106.

Kearney, R. (1988) *The Wake of Imagination: Ideas of Creativity in Western Culture*, London: Hutchinson.

Kearns, R. A. (1986) 'Convergence of Humanistic and Social Thought in Social Geographic Practice', paper presented at the annual Association of American Geographers meeting, Minneapolis.

Kearns, R. A. (1990) 'Coping and Community Life for People with Chronic Mental Disability in Auckland', Department of Geography, University of Auckland, Occasional Paper no. 26.

Kearns, R. A. and Taylor, S. M. (1989) 'Daily Life Experience of People with Chronic Mental Disabilities in Hamilton, Ontario', *Canada's Mental Health*, 37, 4 December: 1–4.

Keith, M. (1993) *Race, Riots and Policing: Lore and Disorder in a Multi-racist Society*, London: UCL Press.

Keith, M. (1994) 'Street Sensibility? Negotiating the Political by Articulating the Spatial', Paper presented at conference on social justice and fin-de-siècle urbanism, 14 and 15 March 1994, University of Oxford, School of Geography.

Keith, M. and Rogers, A. (1991) 'Hollow Promises: Policy, Theory and Practice in the Inner City', in M. Keith and A. Rogers (eds) *Hollow Promises: Rhetoric and Reality in the Inner City*, London: Mansell.

Kellein, T. (1991) *Cindy Sherman*, Basel: Edition Cantz.

Kelly, L. D. (1982) 'Jane Eyre's Paintings and Bewick's *History of British Birds*', *Notes and Queries*, 227: 230–2.

Kennedy, J., Smith, W. A. and Johnson, A. F. (1926) *Dictionary of Anonymous and Pseudoanonymous Literature, Volume VI*, Edinburgh: Oliver and Boyd.

Keynes, J. M. (1938) 'Art and the State', in C. Williams-Ellis (ed.) *Britain and the Beast*, London: Dent, 1–7.

Kirby, K. (1995) *Indifferent Boundaries: Exploring the Space of the Subject*, New York: Guilford Press.

Kirk, J. F. (1891) *A Supplement to Allibone's Critical Directory of English Literature and British and American Authors, Volumes I and II*, Philadelphia: Lippincott.

Kirkham, P. (1986) *Harry Peach: Dryad and the D. I. A.*, London: Design Council.

Klein, M. (1960) *Our Adult World and Its Roots in Infancy*, London: Tavistock Pamphlet, 2.

Klineberg, O. (1951) 'The Scientific Study of National Stereotypes', *UNESCO Social Science Bulletin*, III(3): 505–11.

Korosec-Serfaty, P. (1984) 'The Home from Attic to Cellar', *Journal of Environmental Psychology*, 4: 172–9.

Kristeva, J. (1979) 'Women's Time', in T. Moi (ed.) (1987) *The Kristeva Reader*, Oxford: Blackwell.

Kristeva, J. (1982) *Powers of Horror*, New York: Columbia University Press.

Kristeva, J. (1983) 'Freud and Love', in T. Moi (ed.) (1987) *The Kristeva Reader*, Oxford: Blackwell.

Kristeva, J. (1987) (trans. 1989) *Black Sun*, New York: Columbia University Press.

Kroker, A. and Cook, D. (1988) *The Postmodern Scene: Excremental Culture and Hyper-Aesthetics* (second edition), London: Macmillan Educational.

Kruger, B. (1991) 'An Interview with Barbara Kruger', *Critical Inquiry*, 17: 434–48.

Kruger, B. and Mariani, P. (1989) 'Introduction', in B. Kruger and P. Mariani (eds) *Remaking Histories*, Seattle: Bay Press, ix–xi.

Kuhn, A. (ed.) (1990) *Alien Zone: Cultural Theory and Contemporary Science Fiction in Cinema*, London: Verso.

Kynaston, D. (1994) *The City of London, Volume 1: A World of its Own 1815–1890*, London: Chatto and Windus.

Lacan, J. (1957) 'The Agency of the Letter in the Unconscious of Reason Since Freud', in J. Lacan (1977) *Ecrits: A Selection*, London: Tavistock, 146–78.

Lacan, J. (1958a) 'The Meaning of the Phallus', in J. Mitchell and J. Rose (eds) *Feminine Sexuality*, London: Macmillan (English translation published 1982), 74–85.

Lacan, J. (1958b) 'The Signification of the Phallus', in J. Lacan, *Ecrits: A Selection*, London: Tavistock (English translation published 1977), 281–91.

Lacan, J. (1972–3) 'God and the *Jouissance* of The Woman', in J. Mitchell and J. Rose (eds) *Feminine Sexuality: Jacques Lacan and the Ecole Freudienne*, London: Macmillan (English translation published 1982), 138–48.

Lacan, J. (1973) *The Four Fundamental Concepts of Psycho-analysis*, Harmondsworth: Peregrine (English translation published 1986).

Laclau, E. (1994) 'Introduction', in E. Laclau (ed.) *The Making of Political Identities*, London: Verso, 1–8.

Laclau, E. and Mouffe, C. (1985) *Hegemony and Socialist Strategy*, London: Verso.

Lalvani, S. (1993) 'Photography, Epistemology and the Body', *Cultural Studies*, 7: 442–65.

Land, N. (1992) *The Thirst for Annihilation: George Bataille and Virulent Nihilism (An Essay in Atheistic Religion)*, London: Routledge.

Lang, R. (1985) 'The Dwelling Door: Towards a Phenomenology of Transition', in D. Seamon and R. Mugerauer (eds) *Dwelling, Place and Environment: Towards a Phenomenology of Person and World*, Dordrecht: Martinus Nijhoff Publishers, 201–14.

Laplanche, J. and Leclaire, S. (1966) (trans. 1972) 'The Unconscious', *Yale French Studies*, 48: 118–75.

Latour, B. (1986) 'The Powers of Association', in J. Law (ed.) *Power, Action and Belief: A New Sociology of Knowledge?*, London: Routledge and Kegan Paul, 264–80.

Latour, B. (1988) 'Opening One Eye While Closing the Other … A Note on Some Religious Paintings', in G. Fyfe and J. Law (eds) *Picturing Power: Visual Depiction and Social Relations*, London: Routledge, 15–38.

Latour, B. (1991) 'Technology is Society Made Durable', in J. Law (ed.) *A Sociology of Monsters. Essays on Power, Technology and Dominations*, London: Routledge, 103–32.

Latour, B. (1993) *We Have Never Been Modern*, Hemel Hempstead: Harvester Wheatsheaf.

Law, J. (1994) *Organising Modernity*, Oxford: Blackwell.

Laws, G. and Dear, M. (1988) 'Coping in the Community: A Review of Factors Influencing the Lives of Deinstitutionalised Ex-psychiatric Patients', in C. J. Smith and J. A. Giggs (eds) *Location and Stigma: Contemporary Perspectives on Mental Health and Mental Health Care*, London: Pion, 83–102.

Lawson, H. (1985) *Reflexivity: The Postmodern Predicament*, London: Hutchinson.

Lea, C. (1978) *Emancipation, Assimilation, and Stereotypes*, Bonn: Grundmann.

Lederer, W. (1986) *The Kiss of the Snow Queen: Hans Christian Andersen and Man's Redemption by Woman*, Berkeley: University of California Press.

Lee, S. (1896) *Dictionary of National Biography, Volume XLV*, London: Smith, Elder and Company.

Leisure Management (1992) 'The Art of Simulation', *Leisure Management*, January: 48–9.

Lester, Lord A. (chair) (1993) *Political Speech and Race Relations in a Liberal Democracy: Report of an Inquiry into the Conduct of the Tower Hamlets Liberal Democrats in Publishing Allegedly Racist Election Literature between 1990 and 1993*, London: Liberal Democratic Party.

Lewis, P. (1985) 'Men On Pedestals', *Ten: 8*, 17: 22–9.

Light, A. (1991) *Forever England: Femininity, Literature and Conservatism Between the Wars*, London: Routledge.

Linker, K. (1990) *Love for Sale: The Words and Pictures of Barbara Kruger*, New York: Harry N. Abrams.

Lippman, W. (1947) *Public Opinion*, New York: Macmillan.

Livingstone, D. (1992) *The Geographical Tradition*, Oxford: Blackwell.

Lonsdale, S. (1993) 'Feminists Feel the Backlash of Opinion', *Observer*, 4 April: 3.

Lorraine, T. (1990) *Gender, Identity and the Production of Meaning*, Boulder, Colorado: Westview Press.

Lowe, G., Foxcroft, D. R. and Sibley, D. (1993) *Adolescent Drinking and Family Life*, Chur, Switzerland: Harwood Academic Publishers.

Lowerson, J. (1980) 'Battles For The Countryside', in F. Gloversmith (ed.) *Class, Culture and Social Change: A New View of the 1930s*, Brighton: Harvester, 258–80.

Luckin, B. (1990) *Questions of Power: Electricity and Environment in Inter-war Britain*, Manchester: Manchester University Press.

Lukes, S. (1985) 'Conclusion', in M. Carrithers, S. Collins and S. Lukes (eds) *The Category of the Person: Anthropology, Philosophy, History*, Cambridge: Cambridge University Press, 282–301.

Macmillan, D. (1958a) 'Mental Health Services of Nottingham', *The International Journal of Social Psychiatry*, 4(1).

Macmillan, D. (1958b) 'Community Treatment of Mental Illness', *The Lancet*, July 26: 201–4.

Mangan, J. A. and Walvin, J. (eds) (1987) *Manliness and Morality: Middle-class Masculinity in Britain and America 1800–1940*, Manchester: Manchester University Press.

Marcus, G. and Fisher, M. (1986) *Anthropology as Cultural Critique: An Experimental Moment in the Human Sciences*, Chicago: University of Chicago Press.

Marcus, S. (1992) 'Fighting Bodies, Fighting Words: A Theory and Politics of Rape Prevention', in J. Butler and J. Scott (eds) *Feminists Theorize the Political*, New York: Routledge, 385–403.

Margolis, J. (1991) *The Truth about Relativism*, Oxford: Blackwell.

Martin, B. (1992) *Matrix and Line: Derrida and the Possibilities of Postmodern Social Theory*, New York: State University of New York Press.

Martin, E. (1990) 'The End of the Body?' *American Ethnologist*, 19: 121–40.

Marx, K. (1852) 'The Eighteenth Brumaire of Louis Bonaparte', in K. Marx, *Surveys from Exile*, Harmondsworth: Penguin (English translation published 1973), 143–249.

Mason, P. (1990) *Deconstructing America: Representations of the Other*, London: Routledge.

Mass, B. (1976) *Population Target: The Political Economy of Population Control in Latin America*, Toronto: Women's Press.

Masson, J. (1989) *Against Therapy*, London: Fontana.

Massumi, B. (1992) *A User's Guide to Capitalism and Schizophrenia: Deviations from Deleuze and Guattari*, London: MIT Press.

Mathews, F. (1991) *The Ecological Self*, London: Routledge.

Mathy, J. P. (1993) *Extrême-Occident: French Intellectuals and America*, Chicago: Chicago University Press.

Matless, D. (1990a) 'Ages of English Design: Preservation, Modernism and Tales of their History', *Journal of Design History*, 3(4): 203–12.

Matless, D. (1990b) 'Definitions of England, 1928–89', *Built Environment*, 16(3): 179–91.

Matless, D. (1990c) 'The English Outlook: A Mapping of Leisure, 1918–1939', in N. Alfrey and S. Daniels (eds) *Mapping the Landscape*, Nottingham: Castle Museum/University Art Gallery, 28–32.

Matless, D. (1991) 'Nature, the Modern and the Mystic: Tales from Early Twentieth Century Geography', *Transactions of the Institute of British Geographers*, 16: 272–86.

Matless, D. (1992) 'Regional Surveys and Local Knowledges: The Geographical Imagination in Britain, 1918–39', *Transactions of the Institute of British Geographers*, 17: 464–90.

Matless, D. (1994) 'Moral Geography in Broadland', *Ecumene*, 1(2): 127–56.

Matthews, J. J. (1990) 'They Had Such a Lot of Fun: The Women's League of Health and Beauty Between the Wars', *History Workshop Journal*, 30: 22–54.

Maury, M. (ed.) (1963) *Birth Rate and Birth Right*, New York: Macfadden Books.

Mauss, M. (1985) 'A Category of the Human Mind: The Notion of Person, the Notion of Self', in M. Carrithers, S. Collins, and S. Lukes (eds) *The Category of the Person: Anthropology, Philosophy, History*, Cambridge: Cambridge University Press, 1–25.

May, E. (1988) *Homeward Bound: American Families in the Cold War Era*, New York: Basic Books.

May, E. (1989) 'Explosive Issues: Sex, Women and the Bomb', in L. May (ed.) *Recasting America: Culture and Politics in the Age of Cold War*, Chicago and London: University of Chicago Press, 154–70.

Mayhew, H. (1850) 'The Green Markets of London', *Morning Chronicle*, 5 December: 5–6.

Mayhew, H. (1861–2) *London Labour and the London Poor, Four Volumes*, London: Griffin Bohn.

McDonald, M. (1993) 'The Construction of Difference: An Anthropological Approach to Stereotypes', in S. Macdonald (ed.) *Inside European Identities*, Oxford: Berg, 219–36.

McDowell, L. and Court, L. (1994) 'Gender divisions of labour in the post-Fordist economy: the maintenance of occupational sex segregation in the financial services sector', *Environment and Planning A*, 26: 1397–418.

McKim-Smith, G., Andersen-Bergdoll, G. and Newton, R. (1988) *Examining Velázquez*, New Haven: Yale University Press.

McLaren, A. (1990) *A History of Contraception*, Oxford: Blackwell.

McLuhan, M. (1962) *The Gutenberg Galaxy*, Toronto: University of Toronto Press.

McNay, L. (1992) *Foucault and Feminism*, Cambridge: Polity Press.

McPherson, C. B. (1962) *The Political Theory of the Possessive Individualism*, London: Oxford University Press.

Mead, G. H. (1934) *Mind, Self and Society*, Chicago: Chicago University Press.

Megill, A. (1985) *Prophets of Extremity: Nietzsche, Heidegger, Foucault, Derrida*, London: University of California Press.

Mennell, S. (1992) *Norbert Elias: An Introduction*, Oxford: Blackwell.

Merleau-Ponty, M. (1962a) *The Phenomenology of Perception*, translated by C. Smith, London: Routledge and Kegan Paul.

Merleau-Ponty, M. (1962b) *The Visible and the Invisible*, Evanston: Northwestern University Press.

Merquior, J. G. (1986) *From Prague to Paris: A Critique of Structuralist and Poststructuralist Thought*, London: Verso.

Miles, A. (1981) *The Mentally Ill in Contemporary Society: A Sociological Introduction*, Oxford: Martin Robertson.

Miles, R. (1993) *Gothic Writing, 1750–1820: A Genealogy*, London: Routledge.

Miller, C. L. (1993) 'The Postidentitarian Predicament in the Footnotes of *A Thousand Plateaus*: Nomadology, Anthropology and Authority', *Diacritics*, 23: 6–35.

Miller, M., Moen, P. and Dempster-McClain, D. (1991) 'Motherhood, Multiple Roles, and Maternal Well-Being: Women in the 1950s', *Gender and Society*, 5(4): 565–82.

Mills, S. (1992) *Discourses of Difference: An Analysis of Women's Travel, Living and Colonialism*, London: Routledge.

Mintz, S. and Kellogg, S. (1988) *Domestic Revolutions: A Social History of American Family Life*, New York: Free Press.

Minuchin, S. (1974) *Families and Family Therapy*, London: Tavistock.

Mitchell, J. (1974) *Psychoanalysis and Feminism: A Radical Reassessment of Freudian Psychoanalysis*, Harmondsworth: Penguin.

Mitchell, J. (1984) 'The Rise of Capitalist Woman', in J. Mitchell (ed.) *Women: the Longest Revolution: Essays on Feminism, Literature and Psychoanalysis*, London: Virago.

Mitchell, J. and Rose, J. (1982) *Feminine Sexuality: Jacques Lacan and the* Ecole Freudienne, Basingstoke: Macmillan.

Moffitt, J. R. (1983) 'Velázquez in the Alcázar Palace in 1656: The Meaning of the Mise-en-scène of Las Meninas', *Art History*, 6: 271–300.

Mohan, J. (1992) *Financial Pressure on the Voluntary Sector in Tower Hamlets*, QMW, University of London: Department of Geography Working Paper.

Moi, T. (1985) *Sexual/Textual Politics*, London: Methuen.

Moi, T. (1989a) 'Men against Patriarchy', in L. Kaufman (ed.) *Gender and Theory. Dialogues on Feminist Criticism*, Oxford: Blackwell, 181–8.

Moi, T. (1989b) 'Patriarchal Thought and the Drive for Knowledge', in T. Brennan (ed.) *Between Feminism and Psychoanalysis*, London: Routledge, 189–205.

Moller Okin, S. (1980) *Women in Western Political Thought*, London: Virago.

Monroe, W. S. (1899) 'The Status of Child Study in Europe', *Pedagogical Seminary*, 6(3): 372–81.

Montague, C. E. (1924) *The Right Place: A Book of Pleasures*, London: Chatto and Windus.

Moon, G. (1988) 'Is There One Around Here? – Investigating Reaction to Small-scale Mental Health Hostel Provision in Portsmouth, England', in C. J. Smith and J. A. Giggs (eds) *Location and Stigma: Contemporary Perspectives on Mental Health and Mental Health Care*, London: Unwin Hyman, 203–23.

Moorcock, M. (1988) *Mother London*, London: Martin, Secker and Warburg.

Moore, A. (1980) 'Walt Disney World: Bounded Ritual Space and Playful Pilgrimage Centre', *Anthropological Quarterly*, 53: 207–18.

Moore, R. (1986) *Childhood's Domain: Play and Place in Child Development*, London: Croom Helm.

Moore, S. (1988) 'Getting a Bit of the Other: The Pimps of Postmodernism' in F. Mort (ed.) *Dangerous Sexualities*, London: Lawrence and Wishart.

Morden, T. (1983) 'The Pastoral and the Pictorial', *Ten: 8*, 12: 18–25.

Morley, M. (1993) *Television, Audiences and Cultural Studies*, London: Routledge.

Morris, M. (1988) 'Banality in Cultural Studies', *Discourse X*: 3–29.

Morss, J. R. (1990) *The Biologising of Childhood. Developmental Psychology and the Darwinian Myth*, Hove: Lawrence Erlbaum.

Mort, F. (1988) 'Boy's Own: Masculinity, Style and Popular Culture', in R. Chapman and J. Rutherford (eds) *Male Order: Unwrapping Masculinity*, London: Lawrence and Wishart.

Morton, H. V. (1944) *In Search of England*, London: Methuen (first published 1927).

Muirhead, J. H. (1900) 'The Founders of Child Study in England', *Paidologist*, 2(2): 114–24.

Muller, J. P. (c. 1930) *My Sun Bathing and Fresh Air System*, London: Athletic Publications.

Mulvey, L. (1975) 'Visual Pleasure and Narrative Cinema', in L. Mulvey (1989) *Visual and Other Pleasures*, London: Macmillan, 14–26.

Mulvey, L. (1989) *Visual and Other Pleasures*, London: Macmillan.

Mulvey, L. (1991) 'A Phantasmagoria of the Female Body: The Work of Cindy Sherman', *New Left Review*, 188: 136–50.

Myers, M. (1989) 'Servants As They Are Now Educated', *Essays in Literature*, 16(1): 51–69.

Nairne, S. (1990) *State of the Art: Ideas and Images in the 1980s*, London: Chatto and Windus in collaboration with Channel 4.

Nash, C. (1993) 'Remapping and Renaming: New Cartographies of Identity, Gender and Landscape in Ireland', *Feminist Review*, 44: 39–57.

National Fitness Council (1939) *The National Fitness Campaign*, London: National Fitness Council.

Neisser, U. (1976) *Cognition and Reality*, San Francisco: Freeman.

Nelson, C. (1991) *Boys Will Be Girls: The Feminine Ethic in British Children's Fiction, 1857–1917*, New Brunswick: Rutgers University Press.

Nicholson, L. (ed.) (1990) *Feminism/Postmodernism*, London: Routledge.

Nicholson, T. R. (1983) *Wheels on the Road: Maps of Britain for the Cyclist and Motorist 1870–1940*, Norwich: Geo Books.

Nietzsche, F. (1969) *Thus Spake Zarathustra*, translated by G. Hollingdrake, Harmondsworth: Penguin.

Nussbaum, F. (1989) *The Autobiographical Subject: Gender and Ideology in Eighteenth Century England*, Baltimore: Johns Hopkins University Press.

O'Donnell, E. (1977) *Northern Ireland Stereotypes*, Dublin: College of Industrial Relations.

O'Rourke, P. (1990) 'Past Imperfect', *Leisure Management*, May: 12–14.

Oerton, S. (1993) 'A Safer Place in Which to Work?': The Disclosure of Lesbianism in the Feminist Research Process', paper presented at British Sociological Association Conference, University of Essex, 5–8 April.

Ogborn, M. J. (1991) 'Can You Figure It Out? Norbert Elias's Theory of the Self', in C. Philo (compiler) *New Words, New Worlds: Reconceptualising Social and Cultural Geography. Conference Proceedings*, Department of Geography, St David's University College, Lampeter, 78–87.

Okin, S. M. (1980) *Women in Western Political Thought*, London: Virago.

Olson, D., Russell, C. and Sprenkle, D. (1983) 'Circumplex Model of Marital and Family Systems: VI. Theoretical Update', *Family Process*, 22: 69–83.

Olsson, G. (1980) *Birds In Egg: Eggs In Bird*, London: Pion.

Olsson, G. (1991) *Lines of Power: Limits of Language*, Minneapolis: University of Minnesota Press.

Orwell, G. (1965) *The Road to Wigan Pier*, London: Heinemann (first published by Left Book Club, 1937).

Osborn, F. J. (1943) 'Introduction', in V. Cornish *The Beauties of Scenery*, London: Frederick Muller, 9–14.

Owens, C. (1984) 'The Medusa Effect or, the Specular Ruse', *Art in America*, January: 97–105.

Palin, M. (1986) 'Review', *The Toronto Globe and Mail*, 13 December: C3.

Palomino, A. (1724) *El Museo Pictórico y Escala Optica: Part III: El Parnaso Español Pintoresco Laureado*, Madrid.

Parfit, D. (1984) *Reasons and Persons*, Oxford: Clarendon Press.

Parker, I. (1989) *The Crisis in Social Psychology – And How to End It*, London: Routledge.

Parker, I. (1992) *Discursive Psychology*, London: Routledge.

Parmelee, M. (1927) *The New Gymnosophy: The Philosophy of Nudity as Applied in Modern Life*, New York: Hitchcock.

Parr, H. (1991) 'The Impact of Mental Health Care Facilities Upon Two Urban Residential Districts In Nottingham', unpublished B.A. dissertation, Department of Geography, St David's University College, Lampeter.

Parr, H. (1994) '"Sane" and "Insane" Spaces: New Geographies of Deinstitutionalisation', paper presented at the annual Association of American Geographers meeting, San Francisco.

Parry-Jones, W. Ll. (1972) *The Trade in Lunacy: A Study of Private Madhouses in England in the Eighteenth and Nineteenth Centuries*, London: Routledge and Kegan Paul.

Pateman C. (1988) *The Sexual Contract*, Stanford, CA: Stanford University Press.

Patton, C. (1990) *Inventing AIDS*, New York: Routledge.

Peabody, D. (1985) *National Characteristics*, Cambridge: Cambridge University Press.

Peach, H. H. (1930) *Let Us Tidy Up*, Leicester: Dryad Press.

Peach, H. H. and Carrington, N. (eds) (1930) *The Face of the Land*, London: George Allen and Unwin.

Perez, R. (1990) *On An(archy) and Schizoanalysis*, New York: Autonomedia.

Perin, C. (1988) *Belonging in America*, Madison: University of Wisconsin Press.

Peterson, D. (ed.) (1982) *A Mad People's History of Madness*, Pittsburgh: University of Pittsburgh Press.

Pettigrew, T. (1981) 'Extending the Stereotype Concept', in D. Hamilton (ed.) *Cognitive*

Processes in Stereotyping and Intergroup Behaviour, Hillsdale: Erlbaum, 303–31.

Pheterson, G. (1993) 'Historical and Material Determinants of Psychodynamic Development', in J. Adleman and G. Enguidanos (eds) *Racism in the Lives of Women*, New York: Haworth Press.

Phillips, C. (1871) *Echoes of Life*, Manchester: privately printed.

Phillips, M. (1993) 'A Lesson in the Value of Relatives', *The Guardian*, 3 April: 24.

Philo, C. (1986) 'The Same and the Other: On Geographies, Madness and Outsiders', Department of Geography, Loughborough University of Technology, Occasional Paper no. 11.

Philo, C. (1987) '"Fit Localities For an Asylum": The Historical Geography of the "Mad-business" in England as Viewed Through the Pages of the *Asylum Journal*', *Journal of Historical Geography*, 13: 398–415.

Philo, C. (1992a) 'Foucault's Geography', *Environment and Planning D: Society and Space*, 10: 137–61.

Philo, C. (1992b) *The Space Reserved for Insanity: Studies in the Historical Geography of the Mad-Business in England and Wales*, unpublished Ph.D. dissertation, Department of Geography, University of Cambridge.

Philo, C. (1995) 'Journey to Asylum: The Contexts of a Medical-geographical Idea', forthcoming in *Journal of Historical Geography*.

Phythian, M. (1926) 'January Searle in Yorkshire', *Journal of Adult Education*, 1: 145–57.

Piercy, M. (1992) *Body of Glass*, New York: W. W. Norton.

Pile, S. (1991) 'Practising Interpretative Geography', *Transactions of the Institute of British Geographers*, 16: 458–69.

Pile, S. (1993) 'Human Agency and Human Geography Revisited: A Critique of "New Models" of the Self', *Transactions of the Institute of British Geographers*, 18(1): 122–39.

Pile, S. (1994) 'Masculinism, the Use of Dualistic Epistemologies, and Third Spaces', *Antipode*, 26: 255–77.

Pincus, G. (1958) 'Fertility Control with Oral Medication, Second Oliver Bird Lecture', *Studies on Fertility*, X: 3–26.

Pincus, G., Rock, J., Chang, M. and Garcia, C. (1959) 'Effects of Certain 19-Nor Steroids on Reproductive Processes and Fertility', *Federation Proceedings*, 18: 1051–6.

Pincus, G., Rock, J., Garcia, C., Rice-Wray, E., Paniagua, M. and Rodriguez, I. (1958) 'Fertility Control with Oral Medication', *American Journal of Obstetrics and Gynecology*, 75: 1333–46.

Plumwood, V. (1993) *Feminism and the Mastery of Nature*, London: Routledge.

Pollock, G. (1988) *Vision and Difference: Feminity, Feminism and the Histories of Art*, London: Routledge.

Poovey, M. (1984) *The Proper Lady and the Woman Writer*, Chicago: University of Chicago Press.

Poster, M. (1990) *The Mode of Information: Post-Structuralism and Social Context*, Cambridge/Oxford: Polity Press/Blackwell.

Potts, A. (1989) '"Constable Country" Between the Wars', in R. Samuel (ed.) *Patriotisms*, London: Routledge, 160–86.

Pratt, G. (1992) 'Spatial Metaphors and Speaking Positions', *Environment and Planning D: Society and Space*, 10: 241–4.

Pratt, G. (1994) 'Phallocentrism', in R. J. Johnston, D. Gregory and D. M. Smith (eds) *The Dictionary of Human Geography*, Oxford: Blackwell, 437–8.

Preyer, W. (1890a) *The Mind of the Child: Part I: The Senses and the Will*, New York: Appleton (originally published in German 1882).

Preyer, W. (1890b) *The Mind of the Child: Part II: The Development of the Intellect*, New York: Appleton (originally published in German 1882).

Price, V. (1992) *Communication Concepts 4: Public Opinion*, Newbury Park: Sage.

Probyn, E. (1990) 'Travels in the Postmodern: Making Sense of the Local', in L. J. Nicholson (ed.) *Feminism/Postmodernism*, New York: Routledge, 176–89.

Probyn, E. (1993) *Sexing the Self: Gendered Positions in Cultural Studies*, London: Routledge.

Prout, A. and James, A. (1990) 'A New Paradigm for the Sociology of Childhood? Provenance, Promise and Problems', in A. James and A. Prout (eds) *Constructing and Reconstructing Childhood*, London: Falmer Press, 7–34.

Puget, J. (1988) 'Social Violence and Psychoanalysis in Argentina: The Unthinkable and the Unthought', *Free Associations*, 13.

Rainwater, L. (1966) 'Fear and the House-as-haven in the Lower Class', *Journal of the American Institute of Planners*, 32(1): 23–30.

Ramirez de Arrallano, A. and Seipp, C. (1983) *Colonialism, Catholicism and Contraception: A History of Birth Control in Puerto Rico*, Chapel Hill and London: University of North Carolina Press.

Rapport, N. J. (1987) *Talking Violence: An Anthropological Interpretation of Conversation in the City*, St John's: ISER Press, Memorial University.

Rapport, N. J. (1990) 'Ritual Conversation in a Canadian Suburb: Anthropology and the Problem of Generalisation', *Human Relations*, 43(9): 849–64.

Rapport, N. J. (1993) *Diverse World-Views in an English Village*, Edinburgh: Edinburgh University Press.

Registers of Admissions: County and Borough Asylums, 1855–8 and 1859–61 (PRO.MH94/16 and PRO.MH94/17, Public Records Office, Kew, London).

Relph, E. (1976) *Place and Placelessness*, London: Pion.

Reynolds, E. E. (1950) *The Scout Movement*, Oxford: Oxford University Press.

Rheingold, H. (1991) *Virtual Reality*, London: Mandarin.

Rich, A. (1986a) 'Notes Towards a Politics of Location', in *Blood, Bread and Poetry*, New York: W. W. Norton.

Rich, A. (1986b) *Blood, Bread and Poetry*, London: Virago, 210–31.

Richards, B. (1992) 'Themed Trends', *Leisure Management*, December: 40–1.

Richards, E. P. (1935) *Report with Proposals on the Recreational Use of Gathering Grounds*, unpublished report for the Council for the Preservation of Rural England.

Richards, J. (1987) ' "Passing the Love of Women": Manly Love and Victorian Society', in J. A. Mangan and J. Walvin (eds) *Manliness and Morality*, Manchester: Manchester University Press, 92–122.

Riesman, D. (1958) 'The Suburban Sadness', in W. Dobriner (ed.) *The Suburban Community*, New York: Putnam, 375–408.

Robbins, B. (1993) *The Servant's Hand: English Fiction from Below*, Durham: Duke University Press (first published 1986).

Robbins, D. (1987) 'Sport, Hegemony and the Middle Class: The Victorian Mountaineers', *Theory, Culture and Society*, 4: 579–601.

Roberts, H. (1942a) *British Rebels and Reformers*, London: Collins.

Roberts, H. (1942b) *The Practical Way To Keep Fit*, London: Odhams.

Robinson, E. (ed.) (1983) *John Clare's Autobiographical Writings*, Oxford: Oxford University Press.

Rodaway, P. (1994) *Sensuous Geographies: Body, Sense and Place*, London: Routledge.

Rojek, C. (1990) 'Baudrillard and Leisure', *Leisure Studies*, 9: 7–20.

Rojek, C. (1993a) *Ways of Escape: Modern Transformations in Leisure and Travel*, London: Macmillan.

Rojek, C. (1993b) 'Disney Culture', *Leisure Studies*, 12: 121–35.

Romanes, G. (1888) *Mental Evolution in Man*, London: Kegan Paul.

Rose, G. (1993a) *Feminism and Geography: The Limits of Geographical Knowledge*, Cambridge: Polity Press.

Rose, G. (1993b) 'Some Notes Towards Thinking about the Spaces of the Future', in J. Bird, B. Curtis, T. Putnam, G. Robertson and L. Tickner (eds) *Mapping the Futures: Local Cultures, Global Change*, London: Routledge, 70–83.

Rose, G. (1994) 'The Cultural Politics of Place: Local Representation and the Oppositional Discourse in Two Films', *Transactions of the Institute of British Geographers*, 19(1): 46–60.

Rose, J. (1984) *The Case of Peter Pan, or, The Impossibility of Children's Fiction*, London: Macmillan.

Rose, J. (1986) *Sexuality in the Field of Vision*, London: Verso.

Rosen, S. (1987) *Hermeneutics as Politics*, Oxford: Oxford University Press.

Rosenberger, N. R. (ed.) (1992) *Japanese Sense of Self*, Cambridge: Cambridge University Press.

Rosenthal, M. (1986) *The Character Factory: Baden-Powell and the Origins of the Boy Scout Movement*, London: Collins.

Roth, M. S. (1987) *Psycho-analysis as History: Negation and Freedom in Freud*, Ithaca: Cornell University Press.

Rothman, B. (1982) *The 1932 Kinder Trespass*, Timperley: Willow Publishing.

Rowe, S. and Wolch, J. (1990) 'Social Networks in Time and Space: Homeless Women in Skid Row, Los Angeles', *Annals of the Association of American Geographers*, 80: 184–204.

Runnymede Trust (1993) *Race Relations in Tower Hamlets*, London: Runnymede Trust.

Russell, B. (1963) 'Population Pressure and War', in S. Mudd (ed.) *The Population Crisis and the Use of World Resources*, The Hague: Dr W. Junk Publishers, 1–5.

Russo, M. (1986) 'Female Grotesques: Carnival and Theory', in T. de Lauretis (ed.) *Feminist Studies/Critical Studies*, London: Macmillan, 213–29.

Said, E. (1988) *Orientalism*, Harmondsworth: Peregrine.

Said, E. (1992) *Culture and Empire*, New York: Knopf.

Samuel, R. (1989) 'Ewan MacColl (obituary)', *The Independent*, 30 October.

Sandow, E. (c.1919) *Life is Movement*, London: The National Health Press.

Sartre, J.-P. (1943) *Being and Nothingness: An Essay on Phenomenological Ontology*, London: Routledge (this translation first published 1958).

Scarman, Lord L. (1981) *The Scarman Report*, Harmondsworth: Penguin.

Schlenker, B. (1985) *The Self and Social Life*, New York: McGraw-Hill.

Schneider, B. (1988) 'Invisible and Independent Lesbians' Experiences in the Workplace', in A. Stromberg and S. Harkess, *Women Working: Theories and Facts in Perspective*, California: Mayfield.

Searle, G. R. (1971) *The Quest for National Efficiency: A Study in British Politics and Political Thought, 1899–1914*, Oxford: Blackwell.

Searle, J. R. (1980) 'Las Meninas and the Paradoxes of Pictorial Representation', *Critical Inquiry*, 6: 477–88.

Sedgwick, E. K. (1991) *Epistemology of the Closet*, Hemel Hempstead: Harvester Wheatsheaf.

Sedgwick, E. K. (1993) 'Queer Performativity: Henry James's *The Art of the Novel*', *GLQ*, 1: 1–16.

Sehlinger, B. (1992) *The Unofficial Guide to Walt Disney World*, New York: Prentice Hall.

Seidler, V. J. (1986a) *Kant, Respect and Injustice: The Limits of Liberal Moral Theory*, London: Routledge and Kegan Paul.

Seidler, V. J. (1986b) *Rediscovering Masculinity: Reason, Language and Sexuality*, London: Routledge and Kegan Paul.

Seidler, V. J. (1994) *Unreasonable Men: Masculinity and Social Theory*, London: Routledge.

Seidler, V. J. (ed.) (1992) *Men, Sex and Relationships: Writings From Achilles Heel*, London: Routledge.

Shand, J. (1984) 'The Reichsautobahnen: Symbol for the Third Reich', *Journal of Contemporary History*, 19: 189–200.

Sheail, J. (1981) *Rural Conservation in Inter-War Britain*, Oxford: Clarendon.

Sherif, M. (1967) *Social Interaction*, Chicago: Aldine.

Shields, R. (1991) *Places on the Margin*, London: Routledge.

Shilling, C. (1993) *The Body and Social Theory*, London: Sage.

Shotter, J. (1984) *Social Accountability and Selfhood*, Oxford: Blackwell.

Shotter, J. (1993a) *Cultural Politics of Everyday Life*, Milton Keynes: Open University Press.

Shotter, J. (1993b) *Conversational Realities: Constructing Life through Language*, London: Sage.

Showalter, E. (1987) 'Critical Cross-Dressing: Male Feminists and the Woman of the Year', in A. Jardine and P. Smith (eds) *Men in Feminism*, New York: Methuen.

Sibley, D. (1988) 'Purification of Space', *Environment and Planning D: Society and Space*, 6: 409–21.

Siebenschuh, W. R. (1976) 'The Image of the Child and the Plot of *Jane Eyre*', *Studies in the Novel*, 8(3): 304–17.

Sims, G. R. and Scott, C. (1885) *Jack in the Box*, British Library, Lord Chamberlain's Collection, Add. Ms. 5334.C8.

Singer, L. (1993) *Erotic Welfare: Sexual Theory and Politics in the Age of Epidemic*, New York: Routledge.

Sixsmith, A. (1988) 'A Humanistic Approach to Medical Geography', in D. C. D. Pocock (ed.) *Humanistic Approaches In Geography*, University of Durham, Department of Geography Occasional Publications, no. 22: 12–31.

Skidelsky, R. (1967) *Politicians and the Slump: The Labour Government 1929–1931*, London: Macmillan.

Smith, C. J. (1975) *The Residential Neighbourhood as Therapeutic Community*, unpublished Ph.D. dissertation, Department of Geography, University of Michigan.

Smith, C. J. (1977) 'Geography and Mental Health', *Association of American Geographers, Commission on College Geography, Resource Paper 76*, 4.

Smith, C. J. (1983) 'Innovation in Mental Health Policy: Community Mental Health in the United States of America, 1965–1980', *Environment and Planning D: Society and Space*, 1: 447–68.

Smith, C. J. and Hanham, R. Q. (1981) 'Any Place But Here! – Mental Health Facilities as Noxious Neighbours', *Professional Geographer*, 33: 326–34.

Smith, N. (1993) 'Homeless/Global: Scaling Places', in J. Bird, B. Curtis, T. Putnam, G. Robertson and L. Tickner (eds) *Mapping the Futures: Local Cultures, Global Change*, London: Routledge, 87–119.

Smith, P. (1988) *Discerning the Subject*, Minneapolis: University of Minnesota Press.

Smyth, C. (1992) *Lesbians Talk Queer Notions*, London: Scarlet Press.

Snyder, J. and Cohen, T. (1980) 'Reflexions on Las Meninas: Paradox Lost', *Critical Inquiry*, 7: 429–47.

Snyder, M. (1987) *Public Appearances, Private Realities: The Psychology of Self-monitoring*, New York: W. H. Freeman.

Soja, E. (1989) *Postmodern Geographies: The Reassertion of Space in Social Critical Theory*, London: Verso.

Soper, K. (1986) *Humanism and Anti-Humanism*, London: Hutchinson.

Spender, S. (1953) *World Within World*, London: Readers Union.

Spivak, G. C. (1987) 'Subaltern Studies: Deconstructing Historiography', in G. C. Spivak *In Other Worlds: Essays in Cultural Politics*, London: Methuen, 197–221.

Spivak, G. C. (1988) 'Can the Subaltern Speak?', in C. Nelson and L. Grossberg (eds) *Marxism and the Interpretation of Culture*, Basingstoke: Macmillan, 271–313.

Spivak, G. C. (1991) 'Time and Timing: Law and History', in J. Bender and D. E. Wellberg (eds) *Chronotypes. The Construction of Time*, Stanford: Stanford University Press, 99–117.

Springer, C. (1991) 'The Pleasures of the Interface', *Screen*, 32: 316–29.

Springhall, J. (1977) *Youth, Empire and Society*, London: Croom Helm.

Squiers, C. (1987) 'Diversionary (Syn)Tactics: Barbara Kruger has her Way with Words', *ARTnews*, February: 77–85.

Squirrell, G. (1989) 'In Passing ... Teachers and Sexual Orientation', in S. Acker (ed.) *Teachers, Gender and Careers*, London: Falmer.

Stainton Rogers, R. and Stainton Rogers, W. (1992) *Stories of Childhood: Shifting Agendas of Child Concern*, London: Harvester Wheatsheaf.

Stallybrass, P. and White, A. (1986) *The Politics and Poetics of Transgression*, London: Methuen.

Stamp, W. (1949) '*Doctor Himself*': An Unorthodox Biography of Harry Roberts, London: Hamish Hamilton.

Stapledon, R. G. (1935) *The Land: Now and Tomorrow*, London: Faber and Faber.

Stapledon, R. G. (1943) *The Way of the Land*, London: Faber and Faber.

Stedman, J. W. (1966–70) 'Charlotte Brontë and Bewick's "British Birds"', *Brontë Society Transactions*, 15(76–80): 36–40.

Stedman Jones, G. (1989) 'The "Cockney" and the Nation, 1780–1988', in D. Feldman and G. Stedman Jones (eds) *Metropolis London: Histories and Representations Since 1800*, 272–324.

Steedman, C. (1982) *The Tidy House*, London: Virago.

Steedman, C. (1986) *Landscape for a Good Woman: A Story of Two Lives*, London: Virago.

Steedman, C. (1990) *Childhood, Culture and Class in Britain. Margaret McMillan, 1860–1931*, London: Virago.

Stephenson, T. (ed.) (1946) *The Countryside Companion*, London: Odhams (first published 1939).

Stevens, K. (1906) 'Child Study in Great Britain', *Pedagogical Seminary*, 13(2): 245–9.

Stevenson, J. (1984) *British Society 1914–45*, Harmondsworth: Penguin.

Stone, S. (1991) 'The Empire Strikes Back: A Posttranssexual Manifesto', in J. Epstein and K. Straub (eds) *Body Guards: The Cultural Politics of Gender Ambiguity*, New York: Routledge, 280–304.

Strathern, M. (1992) *After Nature*, Cambridge: Cambridge University Press.

Sulloway, F. (1979) *Freud: Biologist of the Mind*, New York: Basic Books.

Sully, J. (1896) *Studies of Childhood*, London: Longman.

Sully, J. (1897) *Children's Ways: Being Selections from the Author's 'Studies of Childhood'*, London: Longman.

Szasz, M. D. (1974) *The Myth of Mental Illness*, New York: Harper and Row (first published 1960).

Tagg, J. (1988) *The Burden of Representation: Essays on Photographies and Histories*, London: Macmillian.

Tansley, A. G. (1920) *The New Psychology*, London: George Allen and Unwin.

Taussig, M. (1987) *Shamanism, Colonialism and the Wild Man: A Study in Terror and Healing*, Chicago: University of Chicago Press.

Taussig, M. (1993) *Mimesis and Alterity: A Particular History of the Senses*, London: Routledge.

Taylor, C. (1989) *Sources of the Self: The Making of Modern Identity*, Cambridge: Cambridge University Press.

Taylor, C. (1993) 'To Follow a Rule', in C. Calhoun, E. Lipuma and M. Postone (eds) *Bourdieu: Critical Perspectives*, Cambridge: Polity Press, 45–60.

Taylor, S. M. (1988) 'Community Reactions to Deinstitutionalisation', in C. J. Smith and J. A. Giggs (eds) *Location and Stigma: Contemporary Perspectives on Mental Health and Mental Health Care*, London: Unwin Hyman, 224–45.

Tedeschi, J. (1986) 'Private and Public Experiences of Self', in R. Baumeister (ed.) *Public Self and Private Self*, New York: Springer-Verlag, 1–20.

Terry, L. (1992) 'Malcolm Ross: An Interview with the Vice President of Operations for EuroDisney', *Leisure Management*, April: 22–4.

Thrift, N. J. (1983a) 'On the Determination of Social Action in Space and Time', *Environment and Planning D: Society and Space*, 1(1): 23–57.

Thrift, N. J. (1983b) 'Literature, the Production of Culture and the Politics of Place', *Antipode*, 15: 12–24.

Thrift, N. J. (1986) 'Little Games and Big Stories: The Practices of Political Personality in the 1945 General Election', in K. Hoggart and E. Kofman (eds) *Politics, Geography and Social Stratification*, Beckenham: Croom Helm, 90–155.

Thrift, N. J. (1994a) 'A Hyperactive World', in R. J. Johnston, P. J. Taylor and M. Watts (eds) *Geographies of Global Change*, Oxford: Blackwell.

Thrift, N. J. (1994b) 'Inhuman Geographies: Landscapes of Speed, Light and Power', in P. J. Cloke, M. A. Doel, D. Matless, M. Phillips and N. J. Thrift *Writing the Rural*, London: Paul Chapman.

Thrift, N. J. and Leyshon, A. (1992) 'In the Wake of Money: The City of London and the Accumulation of Value', in L. Budd and S. Whimster (eds) *Global Finance and Urban Living*, London: Routledge, 282–311.

Thrift, N. J. and Leyshon, A. (1994) 'A Phantom State? The Detraditionalisation of Money, the International Financial System and International Financial Centres', *Political Geography*, 13: 299–327.

Todd J. and Butler M. (eds) (1989) *The Works of Mary Wollstonecraft, Seven Volumes*, London: William Pickering, Volume 2, *Elements of Morality for the Use of Children, with an Introductory Address to Parents*, translated from the German of the Rev. C. G. Salzman, two volumes (first published 1790) and *Young Grandson: A Series of Letters from Young Persons to their Friends*, translated from the Dutch of Madame de Cambon, with Alterations and Improvements, in two volumes (first published 1790); Volume 4, *Original Stories from Real Life: with Conversations Calculated to Regulate the Affections and Form the Mind to Truth and Goodness* (first published 1788); Volume 5, *Vindication of the Rights of Woman with Strictures on Political and Moral Subjects* (first published 1792).

Topoi (1988) 'Topos: who comes after the subject?', J.-L. Nancy (ed.) *Topoi*, 7(2): 87–185.

Torgovnick, M. (1990) *Gone Primitive: Savage Intellects, Modern Lives*, Chicago: University of Chicago Press.

Tower Hamlets Homeless Families Campaign (THFC) (1993) *Tower Hamlets Policies on Homelessness*, London: THFC.

Travers, A. (1993) 'An Essay on Camp and Self', *Theory, Culture and Society*, 10: 127–43.

Trevelyan, G. M. (1929) *Must England's Beauty Perish?* London: Faber and Gwyer.

Trevelyan, G. M. (1930) *Clio, A Muse*, London: Longman.

Trevelyan, G. M. (1931) *The Call and Claims of Natural Beauty*, London: University College London.

Trinh, M.-h. (1988) 'Not You/ Like You: Post-colonial Women and the Interlocking Questions of Identity and Difference', *Inscriptions*, 3(4): 71–7.

Trinh, M.-h. (1990) 'Cotton and Iron', in R. Ferguson, M. Gever, M.-h. T. Trinh and C. West (eds) *Out There: Marginalization and Contemporary Cultures*, New York and Cambridge, Mass.: New Museum of Contemporary Art and Massachusetts Institute of Technology, 327–36.

Trinh, M.-h. (1991) *When the Moon Waxes Red: Representation, Gender and Cultural Politics*, London: Routledge.

Tseelon, E. (1992) 'Is the Presented Self Sincere? Goffman, Impression Management and the Postmodern Self', *Theory, Culture and Society*, 9: 115–28.

Turkle, S. (1992) *Psychoanalytic Politics: Jacques Lacan and Freud's French Revolution* (second edition), London: Free Association Books.

Turner, B. (1992) *Regulating Bodies: Essays in Medical Sociology*, London: Routledge.

Tyler, C.-A. (1991) 'Boys Will Be Girls: The Politics of Gay Drag', in D. Fuss (ed.) *Inside/Out:*

Lesbian Theories, Gay Theories, New York: Routledge, 32–70.

Uzzell, D. (ed.) (1989) *The Heritage Experience: Volume 2 – The visitor experience*, Chichester: Belhaven Press.

Vahlne, B. (1982) 'Velázquez's Las Meninas: Remarks on the Staging of a Royal Portrait', *Konsthistorisk Tidskrift*, 51: 21–8.

Vale, V. and Juno, A. (1989) *Modern Primitives: An Investigation of Contemporary Adornment and Ritual*, San Francisco: Re/Search.

Valentine, G. (1990) 'Women's Fear and the Design of Public Space', *Built Environment*, 16(4): 288–303.

Valentine, G. (1993a) '(Hetero)sexing Space: Lesbian Perceptions and Experiences of Everyday Spaces', *Environment and Planning D: Society and Space*, 11: 395–413.

Valentine, G. (1993b) 'Negotiating and Managing Multiple Sexual Identities: Lesbian Time-Space Strategies', *Transactions of the Institute of British Geographers*, 18: 237–48.

Valentine, G. (1993c) 'Desperately Seeking Susan: A Geography of Lesbian Friendships', *Area*, 25: 109–16.

van Horn, S. (1988) *Women, Work and Fertility, 1900-1986*, New York: New York University Press.

van Krieken, R. (1989) 'Violence, Self-discipline and Modernity: Beyond the "Civilizing Process"', *Sociological Review*, 37: 193–218.

van Krieken, R. (1990) 'The Organisation of the Soul: Elias and Foucault on Discipline and the Self', *Archives Européenes de Sociologie*, 31: 353–71.

Varey, J. E. (1984) 'The Audience and the Play at Court Spectacles: The Role of the King', *Bulletin of Hispanic Studies*, 61: 399–406.

Virilio, P. (1984) *War and Cinema*, London: Verso.

Virilio, P. (1991) *The Lost Dimension*, New York: Semiotext(e).

Virilio, P. (1994) *The Vision Machine*, London: British Film Institute.

Vogt, W. (1949) *The Road To Survival*, London: Victor Gollancz.

Wakefield, N. (1990) *Postmodernism: The Twilight of the Real*, London: Pluto Press.

Waldman, D. (1989) *Jenny Holzer*, New York: Solomon R. Guggenheim Museum and Harry N. Abrams.

Walkerdine, V. (1984) 'Dreams From an Ordinary Childhood', in L. Heron (ed.) *Truth, Dare or Promise: Girls Growing Up in the Fifties*, London: Virago.

Walkerdine, V. (1991) *Schoolgirl Fictions*, London: Verso.

Walkerdine, V. and Lucey, H. (1989) *Democracy in the Kitchen: Regulating Mothers and Socialising Daughters*, London: Virago.

Wallace, A. D. (1993) *Walking, Literature and English Culture: The Origins and Uses of the Peripatetic*, Oxford: Oxford University Press.

Wallace, M. (1985) 'Mickey Mouse History: Portraying the Past at Disney World', *Radical History Review*, 32: 33–57.

Waller, R. (1962) *Prophet of the New Age: The Life and Thought of Sir George Stapledon*, London: Faber and Faber.

Ward, C. (1977) *The Child in the City*, Harmondsworth: Penguin.

Ward, R. and Jenkins, R. (1984) *Ethnic Communities in Business: Strategies for Economic Survival*, Cambridge: Cambridge University Press.

Warren, A. (1987) 'Popular Manliness: Baden-Powell, Scouting and the Development of Manly Character', in J. A. Mangan and J. Walvin (eds) *Manliness and Morality*, Manchester: Manchester University Press, 199–219.

Watney, S. (1987) 'The Spectacle of AIDS', *October*, 43: 71–86.

Watney, S. (1990) 'The Subject of AIDS', in T. Boffin and S. Gupta (eds) *Ecstatic Antibodies: Resisting the AIDS Mythology*, London: Rivers Oram Press.

Weeks, J. (1990) 'Post-modern AIDS?', in T. Boffin and S. Gupta (eds) *Ecstatic Antibodies: Resisting the AIDS Mythology*, London: Rivers Oram Press, 133–41.

Weisner, T.S. (1984) 'Ecocultural Niches of Middle Childhood: A Cross-cultural Perspective', in W. A. Collins (ed.) *Development During Middle Childhood*, Washington, D. C.: National Academy Press, 335–69.

Whitford, M. (1989) 'Rereading Irigaray', in T. Brennan (ed.) *Between Feminism and Psychoanalysis*, London: Routledge, 106–26.

Wigley, M. (1992) 'Untitled: The Housing of Gender', in B. Colomina (ed.) *Sexuality and Space*, New York: Princeton Architectural Press, 327–89.

Wilder, D. (1981) 'Perceiving Persons as a Group', in D. Hamilton (ed.) *Cognitive Processes in Stereotyping and Intergroup Behaviour*, Hillsdale: Erlbaum, 213–57.

Williams, P. (1986) 'Social Relations, Residential Segregation and the Home', in K. Hoggart and

E. Kofman (eds) *Politics, Geography and Social Stratification*, London: Croom Helm, 247–73.

Williams, R. (1990) *Notes on the Underground: An Essay on Technology, Society, and the Imagination*, Cambridge, Mass.: MIT Press.

Williams-Ellis, C. (1928) *England and the Octopus*, London: Geoffrey Bles.

Williams-Ellis, C. (1934) 'Our Physical Environment', in C. E. M. Joad (ed.) *Manifesto*, London: George Allen and Unwin, 216–47.

Williams-Ellis, C. (1971) *Architect Errant*, London: Constable.

Williamson, J. (1986a) *Consuming Passions: The Dynamics of Popular Culture*, London: Marion Boyars, 91–114.

Williamson, J. (1986b) 'The Problems of being Popular', *New Socialist*, September.

Wilson, E. (1985) *Adorned in Dreams: Fashion and Modernity*, London: Virago.

Wilson, E. (1993) 'Is Transgression Transgressive?', in J. Bristow and A. Wilson (eds) *Activating Theory: Lesbian, Gay, Bisexual Politics*, London: Lawrence and Wishart, 107–17.

Wilson, L. N. (1898) 'Bibliography of Child Study', *Pedagogical Seminary*, 5(4): 541–89.

Winner, L. (1986) *The Whale and the Reactor: A Search for Limits in the Age of High Technology*, Chicago: Chicago University Press.

Winnicott, D. W. (1957) *The Child and the Family: First Relationships*, London: Tavistock.

Winnicott, D. W. (1974) *Playing and Reality*, Harmondsworth: Penguin.

Winnicott, D. W. (1975) *Through Pediatrics to Psychoanalysis*, London: Hogarth Press.

Wittgenstein, L. (1953) *Philosophical Investigations*, Oxford: Blackwell.

Wittig, M. (1977) *Le Corps lesbien*, Paris: Minuit.

Wolch, J. R. (1979) 'Residential Location and the Provision of Human Services: Some Directions for Geographic Research', *Professional Geographer*, 31: 271–7.

Wolch, J. R. (1980) 'Residential Location of the Service Dependent Poor', *Annals of the Association of American Geographers*, 70: 330–41.

Wolch, J. R. (1990) *The Shadow State: Government and Voluntary Sector in Transition*, New York: The Foundation Center.

Wolpert, E. and Wolpert, J. (1974) 'From Asylum to Ghetto', *Antipode*, 6(3): 63–76.

Wolpert, J. (1976) 'Opening Closed Spaces', *Annals of the Association of American Geographers*, 66: 1–13.

Women In MIND (1986) *Finding Our Own Solutions: Women's Experience of Mental Health Care*, London: MIND.

Wood, D. (1992) *The Power of Maps*, London: Routledge.

Woolf, J. (1993) 'On the Road Again: Metaphors of Travel in Cultural Criticism', *Cultural Studies*, 7: 224–39.

Wordsworth, W. (1951) *A Guide Through the District of the Lakes in the North of England*, London: Rupert Hart-Davis (first published in complete form 1835).

Wright, E. (1989) 'Thoroughly Postmodern Feminist Criticism', in T. Brennan (ed.) *Between Feminism and Psychoanalysis*, London: Routledge, 141–52.

Wright, P. (1992) *A Journey Through Ruins: The Last Days of London*, London: Radius.

Wright Mills, C. (1956) *The Sociological Imagination*, Oxford: Oxford University Press.

Wrightsman, L. (1977) *Social Psychology*, Monterey: Brooks/Cole.

Yingling, T. (1991) 'AIDS in America: Postmodern Governance, Identity, and Experience', in D. Fuss (ed.) *Inside/Out: Lesbian Theories, Gay Theories*, New York: Routledge.

Young, A. (1990) *Femininity in Dissent*, London: Routledge.

Youth Hostels Association (1939) *Handbook For 1939*, Welwyn Garden City: Youth Hostels Association.

Zijderveld, A. (1979) *On Clichés*, London: Routledge and Kegan Paul.

INDEX